Environmental UV Radiation: Impact on Ecosystems and Human Health and Predictive Models

NATO Science Series

A Series presenting the results of scientific meetings supported under the NATO Science Programme.

The Series is published by IOS Press, Amsterdam, and Springer (formerly Kluwer Academic Publishers) in conjunction with the NATO Public Diplomacy Division.

Sub-Series

I. **Life and Behavioural Sciences**	IOS Press
II. **Mathematics, Physics and Chemistry**	Springer (formerly Kluwer Academic Publishers)
III. **Computer and Systems Science**	IOS Press
IV. **Earth and Environmental Sciences**	Springer (formerly Kluwer Academic Publishers)

The NATO Science Series continues the series of books published formerly as the NATO ASI Series.

The NATO Science Programme offers support for collaboration in civil science between scientists of countries of the Euro-Atlantic Partnership Council. The types of scientific meeting generally supported are "Advanced Study Institutes" and "Advanced Research Workshops", and the NATO Science Series collects together the results of these meetings. The meetings are co-organized by scientists from NATO countries and scientists from NATO's Partner countries — countries of the CIS and Central and Eastern Europe.

Advanced Study Institutes are high-level tutorial courses offering in-depth study of latest advances in a field.
Advanced Research Workshops are expert meetings aimed at critical assessment of a field, and identification of directions for future action.

As a consequence of the restructuring of the NATO Science Programme in 1999, the NATO Science Series was re-organized to the four sub-series noted above. Please consult the following web sites for information on previous volumes published in the Series.

http://www.nato.int/science
http://www.springer.com
http://www.iospress.nl

Series IV: Earth and Environmental Sciences – Vol. 57

Environmental UV Radiation: Impact on Ecosystems and Human Health and Predictive Models

edited by

Francesco Ghetti
CNR Istituto di Biofisica, Pisa, Italy

Giovanni Checcucci
CNR Istituto di Biofisica, Pisa, Italy

and

Janet F. Bornman
Danish Institute of Agricultural Sciences,
Research Centre Flakkebjerg,
Slagelse, Denmark

Published in cooperation with NATO Public Diplomacy Division

Proceedings of the NATO Advanced Study Institute on
Environmental UV Radiation: Impact on Ecosystems and Human Health
and Predictive Models
Pisa, Italy
June 2001

A C.I.P. Catalogue record for this book is available from the Library of Congress.

ISBN-10 1-4020-3696-5 (PB)
ISBN-13 978-1-4020-3696-5 (PB)
ISBN-10 1-4020-3695-7 (HB)
ISBN-13 978-1-4020-3695-8 (HB)
ISBN-10 1-4020-3697-3 (e-book)
ISBN-13 978-1-4020-3697-3 (e-book)

Published by Springer,
P.O. Box 17, 3300 AA Dordrecht, The Netherlands.

www.springer.com

Printed on acid-free paper

Printed in the Netherlands.

CONTENTS

PREFACE

This volume originates from the NATO Advanced Study Institute *Environmental UV Radiation: Impact on Ecosystems and Human Health and Predictive Models*, held in Pisa, Italy in June 2001. The Institute was sponsored and mainly funded by the NATO Scientific Affairs Division, whose constant contribution in favour of the cooperation among scientists from different countries must be acknowledged. Other Institutions substantially contributed to the success of the ASI and our thanks and appreciation go to the Italian National Research Council (Consiglio Nazionale delle Ricerche), the Italian Space Agency (Agenzia Spaziale Italiana), the European Society for Photobiology and the bank Banca Toscana.

In the last two decades of the past century, concern has been growing for the possible effects on the biosphere of the stratospheric ozone depletion, due to anthropogenic emissions of ozone-destroying chemicals. The ozone loss causes an increase in the biologically important part of the solar ultraviolet radiation (UV) reaching the Earth's surface, which constitutes a threat to the biosphere, because of UV damaging effects on humans, animals and plants.

The international agreements have reduced the production of ozone-destroying compounds, which, however, are still present in high concentrations in the stratosphere, mainly because of their longevity, and thus ozone depletion will likely continue for several decades.

It is, therefore, critical to adequately predict how changes in UV radiation and other environmental variables will affect ecosystems and the biological processes needed to sustain life on Earth and to provide useful hints for future actions of governmental and international agencies, as well as non-governmental organizations.

This book offers not only basic information on the action mechanisms of UV radiation on ecosystems and various biological systems, but also a picture of the possible scenarios of the long-term global increase of environmental UV radiation and emphasises the research aspects aimed at the proper quantitative assessment of risk factors and the formulation of reliable predictive models.

The book is structured in four sections: the first one is devoted to a general overview of the consequences of ozone depletion and to the basic concepts of radiation measurements and monitoring; the other three sections are devoted to the effects on plants, aquatic ecosystems and human health. At the end a few abstracts of contributions from students who attended the school are included.

For the latest information on the ongoing research about the environmental effects of ozone depletion and its interactions with climate change, the reader is referred to the last full assessment by the UNEP EEAP (United Nations Environment Programme, Environmental Effects Assessment Panel), published in *Photochemical & Photobiologica. Sciences* 2: 1-72 (2003), and to its following updates, published in *Photochem. Photobiol. Sci.* 3: 1-5 (2004) and *Photochem. Photobiol. Sci.* 4: 177-184 (2005).

HISTORICAL OVERVIEW OF OZONE TRENDS AND FUTURE SCENARIOS

JANET F. BORNMAN
Department of Genetics and Biotechnology, Research Centre Flakkebjerg, Danish Institute of Agricultural Sciences (DIAS), Slagelse, Denmark

1. Ozone distribution

Stratospheric ozone (O_3) is created over low latitudes by the action of ultraviolet radiation of wavelengths shorter than ca 240 nm. An oxygen molecule (O_2) reacts with the high energy radiation and two oxygen atoms are formed in the reaction. A third molecule (M), e.g. another oxygen or nitrogen, is required to remove the excess kinetic energy in the following way:

$$O_2 + UV\ (< 240\ nm) \rightarrow O + O$$

$$O_2 + O + M \rightarrow O_3 + M$$

The destruction of ozone results in its breakdown to molecular oxygen and atomic oxygen. In equilibrium, these two events of synthesis and degradation have in the past resulted in an average ozone content of ca 300 DU (Dobson unit, DU = 1 mm of ozone at STP). However, with the loading of the atmosphere with halogen compounds containing Cl and Br from industrial activities, the balance is no longer in place, since Br, BrO, Cl and ClO take part in catalytic breakdown cycles involving ozone.

Most of the stratospheric ozone occurs between 10 and 30 km above the surface of the earth, providing an effective filter against harmful ultraviolet radiation. This high energy radiation can cause erythema (sunburn), skin cancers, cataracts, and changes in immune response, etc. UV also modifies terrestrial and aquatic life forms, as well as detrimentally affecting synthetic and natural materials. The filtering layer typically removes 70-90% of the UV radiation. Furthermore, since ozone absorbs solar energy, the ozone layer is an important controlling factor of upper stratospheric temperature. Between 1979 and 1997, the annual global average temperature decreased by 0.6 Kelvin per decade in the lower stratosphere, and by 3 K per decade in the upper stratosphere[1].

2. Progression of changes in stratospheric ozone

Reductions in ozone have been recorded in the Antarctic, Arctic, and mid-latitudes in both hemispheres. This thinning of the ozone is also not confined only to the polar spring, which is the period of minimum ozone concentration. The ozone measurements in the Antarctic started in the mid-1950s. Up to the 1970s, the apparently

F. Ghetti et al. (eds.), Environmental UV Radiation: Impact on Ecosystems and Human Health and Predictive Models, 1–3.

normal ozone cycles in the Antarctic gave values of ca 300 DU during winter and until late spring in October. Thereafter a rise to ca 400 DU by the beginning of December was typical, with a gradual falling off again towards March. Since the 1980s this cycle of events has been replaced by a superimposed thinning of the ozone with values below 100 DU during the Antarctic spring. This spring-time decrease was first reported by Farman and co-workers[2] of the British Antarctic Survey team in 1985. They had observed a mean loss during October 1984 from ca 300 to 180 DU. Subsequent recordings have shown values of less than 100 DU. A press release by NASA in September 2000 announced that an ozone depletion area three times larger than the land mass of the United States, had occurred. This depleted area spanned ca 28 million km^2, and was considerably larger than that noted two years previously.

3. Contributing factors for the ozone loss

There is general consensus that the ozone layer is being destroyed mainly as a consequence of the release of chlorofluorocarbons (CFCs) into the atmosphere by industry[3]. The chlorine monoxide molecule (ClO) and other important chemical species such as bromine monoxide (BrO) and nitrogen oxides are involved. Reservoirs of bromine, nitrogen oxide and chlorine are transported to the upper atmosphere, although bromine reservoirs are more unstable than chlorine, and occur mainly as Br and BrO. The chlorine reservoirs may consist of hydrochloric acid and chlorine nitrate, while dinitrogen pentoxide and nitric acid are the nitrogen oxide reservoirs. Catalytic breakdown reactions of ozone involve heterogeneous interactions with bromine, chlorine and nitrogen compounds. Of importance are also sulphate aerosol particles and sulphur dioxide, the latter of which has been found in high amounts following volcanic eruptions. During polar spring conditions of ca –80°C, the sulphate aerosols take up water and nitric acid and form the so-called polar stratospheric clouds (PSC) which become solidified as temperatures drop further. These clouds serve as catalytic surfaces where ozone-degrading substances are released and become concentrated, and react with ozone when the polar regions warm up.

Another important climatic factor is the increasing trend of CO_2 levels, which also negatively affect the ozone chemistry by preventing re-emission of the radiation from the surface of the earth. This greenhouse phenomenon cools the upper atmosphere, which in turn results in favourable conditions for the formation of the polar stratospheric clouds over the polar regions, especially over the Antarctic.

4. Antarctic versus Arctic climates

The geographical features of the Arctic region apparently result in more irregular polar vortex winds and higher temperatures compared to the Antarctic, which is surrounded by ocean rather than by mountainous continents. Consequently, polar stratospheric clouds are not as frequent in the Arctic. Despite this, recent recordings of ozone losses exceeding 60% have coincided with increased sightings of PSCs and a more stable polar vortex. The lower temperatures found in the Arctic in recent years show a positive correlation with stratospheric ozone loss[4]. These findings suggest that the recovery of the ozone layer may be delayed longer than predicted.

5. Actions against ozone depletion

The Convention for the Protection of the Ozone Layer, agreed upon in Vienna in 1985, was the start of further efforts among countries to analyse global environmental problems. Following the publication in Nature by Farman and co-workers[2] the Montreal Protocol on Substances that Deplete the Ozone Layer was signed in 1987, and has been adjusted and amended by subsequent meetings including those held in London (1990), Copenhagen (1992), Vienna (1995) and Beijing (1999). The broad aim of The Parties to the Montreal Protocol is the protection of the environment and human health from anthropogenic activities which may modify the ozone layer. The Montreal Protocol includes provision for developing countries to delay their compliance for up to 10 years based on economics. The main goal of the protocol is to phase out the use of CFCs in the next few decades.

6. Conclusions

While the international agreements have reduced production of ozone-destroying compounds, high concentrations are still present in the stratosphere mainly due to their longevity, and thus ozone depletion will likely continue for several decades[3]. Monitoring networks are continuously being set up to study biological effects of ozone depletion and to determine correlations between ozone reduction and UV increases and to couple these to the biological impact.

References

1. SORG (1999) *Stratospheric Ozone,* United Kingdom Stratospheric Ozone Review Group. HMSO, London, UK.
2. Farman, J.C., Gardiner, B.G. & Shanklin, J.D. 1985. Large losses of total ozone in Antarctica reveal seasonal ClO_x/NO_x interaction. Nature 315, 207 – 210.
3. WMO (1998) *Scientific Assessment of Ozone Depletion. Report 44.* Geneva, Switzerland.
4. Rex, M., Dethloff, K., Handorf, D., Herber, A., Lehmann, R., Neuber, R., Notholt, J., Rinke, A., von der Gathen, P., Weisheimer, A. and Gernandt, H. (2000) Arctic and Antarctic ozone layer observations - chemical and dynamical aspects of variability and long-term changes in the polar stratosphere. *Polar Research,* 19: 193-204.
5. Schulz, A., Rex, M., Steger, J., Harris, N., Braathen, G.O., Reimer, E., Alfier, R., Beck, A., Alpers, M., Cisneros, J., Claude, H., De Backer, H., Dier, H., Dorokhov, V., Fast, H., Godin, S., Hansen, G., Kondo, Y., Kosmidis, E., Kyro, E., Molyneux, M.J., Murphy, G., Nakane, H., Parrondo, C., Ravagnani, F., Varostos,C., Vialle, C., Yushkov, V., Zerefos, C. and von der Gathen, P. (2000) Match observation in the Arctic winter 1996/97: High stratospheric ozone loss rates correlate with low temperatures deep inside the polar vortex. *Geophys. Res. Lett.*, 27: 205-208.

Further reading

Madronich, S., McKenzie, R.L., Björn, L.O. & Caldwell, M.M. 1998. Changes in biologically active ultraviolet radiation reaching the Earth's surface. *J. Photochem. Photobiol. B:Biol.*, 46: 5-19.

BASIC CONCEPTS OF RADIATION

DAVID H. SLINEY, ERIN CHANEY
US Army Center for Health Promotion and Preventive Medicine
Aberdeen Proving Ground, MD, USA

1. Introduction

Light is an electromagnetic wave that propagates through space at the phenomenal velocity of 300 km/s, but it represents only a tiny fraction of what we term the electromagnetic spectrum as shown in Figure 1. To describe any wave, there are three basic parameters: frequency, wavelength and velocity of propagation. Since all electromagnetic waves travel at the velocity of light c, the fundamental relationship is: $c = \lambda \nu$, where λ is the wavelength in meters (m), and ν is the frequency in Hertz (Hz), i.e., in wave cycles per second. The wavelength is measured between any two corresponding points along the wave, as shown in Figure 2. Figure 2 also shows the orthogonal (right-angle) relationship of the electric vector and magnetic vector of this electromagnetic wave.

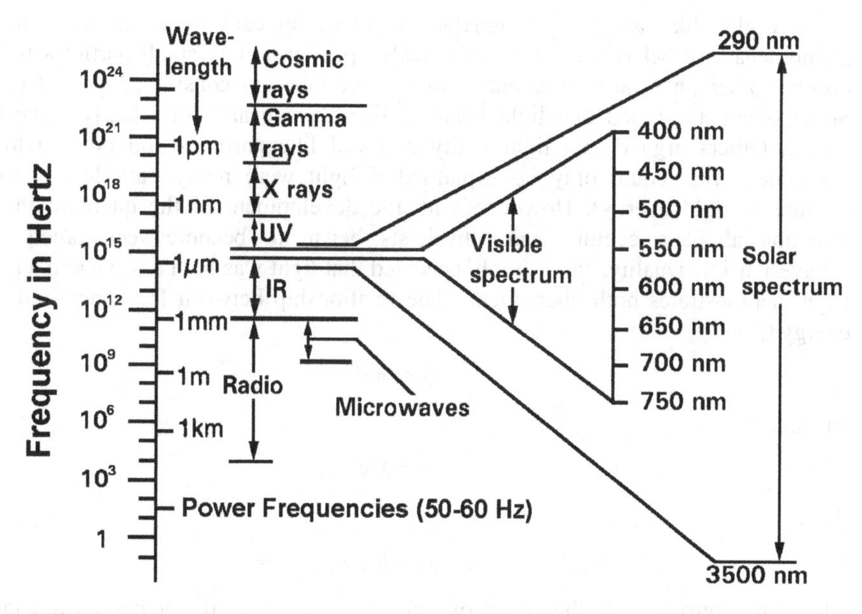

Figure 1. The electromagnetic spectrum.

F. Ghetti et al. (eds.), Environmental UV Radiation: Impact on Ecosystems and Human Health and Predictive Models, 5–23.

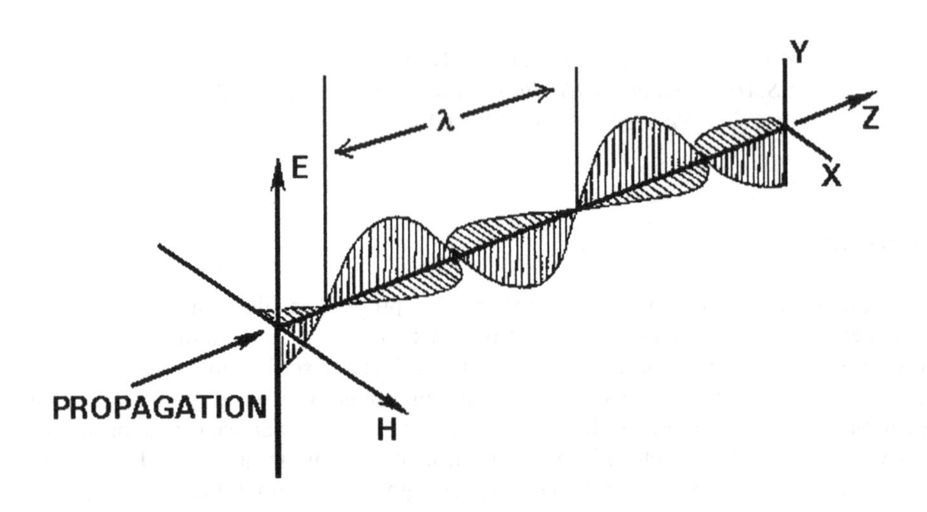

Figure 2. The wavelength is the distance between two corresponding points at the same location from cycle to cycle.

Light, like all electromagnetic radiation, appears to be a wave in some phenomena, but behaves as if it were made up of infinitely small particles when we observe other physical phenomena. At one time this was considered a great paradox. Some scientists argued that light behaved like a wave and was clearly wave-like in nature. Others argued that light really behaved like particles and pointed to some phenomena that could only be explained if light were really particles, or photons (quanta of light energy). However, with the development of the quantum theory of radiation about a century ago, physicists began to become accustomed to this "Wave-Particle Duality," and simply accepted that light was both a wave and a particle! Light demonstrates both phenomena. The relationship between frequency and photon energy Q is:

$$Q_v = h \cdot v \tag{1}$$

and since:

$$c = \lambda \cdot v \tag{2}$$

then:

$$Qv = h \cdot c / \lambda \tag{3}$$

where the constants are the speed of light = c = 3.00×10^8 meters per second; and Planck's Constant = h = 6.63×10^{-34} joule per second.

When one enters the world of radiation biophysics and the effects of electromagnetic radiation upon biological systems, it is frequently useful to separate the entire EM spectrum into ionizing and non-ionizing radiation. Ionizing radiation has enough photon energy to create ion pairs; that is, to make an atom gain or lose electrons. Examples of ionizing radiation include: X rays and gamma rays. Non-ionizing radiation includes optical radiation (ultraviolet, visible and infrared radiation) and radio-frequency radiation. The optical spectrum is frequently broken into smaller spectral bands. The band definitions depend upon whether one is interested in optical engineering, meteorological optics, photobiology or some other technical area. The division of ultraviolet spectrum by optical engineers could be divided by the transmission characteristics of soft glass materials (330-400 nm), quartz optics (middle ultraviolet: 180-320 nm) and the absence of transmission in quartz and air (far ultraviolet or vacuum ultraviolet at wavelengths below 180 nm). From the standpoint of meteorological optics, the bands are defined relative to the transmission of water bands in the infrared and ozone in the ultraviolet. However, from the standpoint of the photobiologist, the absorbance and transmission bands of proteins and water are critical. Thus, it is important always to define what is meant by terms such as "near ultraviolet" or far "ultraviolet. When considering photobiological effects, it is useful to employ the convention of the International Commission on Illumination (CIE) for spectral bands. The CIE has designated 315 to 400 nm as UV-A, 280 to 315 nm as UV-B, and 100-280 nm as UV-C (CIE, 1987). Light (visible radiation) overlaps the UV-A and IR-A extending from 380 nm to at least 780 nm.[1] [IR-A extends from 770 nm to 1400 nm; IR-B, from 1400 to 3000 nm; and IR-C from 3000 nm to 1 mm.] The CIE photobiological bands are summarized in Table 1.

Table 1. CIE Photobiological Spectral Bands

CIE Spectral Band Designation	Wavelength Interval	Characteristics
UV-C	100 nm - 280 nm	Superficial absorption in tissue; significant protein absorption, particularly at 250-280 nm
UV-B	280 nm - 315 nm	Penetrates, still actinic, most photocarcinogenic
UV-A	315 nm - 400 nm	Deeper penetration, less absorption; single photons generally do not interact except in photodynamic role
Light (visible)	380 nm - 780 nm	Photopic (day) and Scotopic (night) vision
IR-A	780 nm - 1400 nm	Deeply penetrating, water transmits
IR-B	1400 nm - 3000 nm (1.4 µm - 3.0 µm)	Water strongly absorbs, very slight penetration—generally to less than ~ 1 mm
IR-C	3 µm - 1,000 µm (1 mm or 300 GHz)	Very superficial absorption, generally less than 0.1 mm

From a biological point of view, wavelengths below 180 nm (vacuum UV) are of little practical significance since they are readily absorbed in air; hence, the spectral region below ~ 180 nm is also referred to as "the vacuum ultraviolet." UV-C wavelengths are more photochemically active, because these wavelengths correspond to the most energetic photons, are strongly absorbed in certain amino acids and therefore by most proteins; whereas, UV-B wavelengths are somewhat less photochemically active, but more penetrating in most tissues.[2-6] UV-A wavelengths are far less photobiologically active, but are still more penetrating than UV-B wavelengths and often play an interactive role when exposure occurs following UV-B exposure.[2-6]

Although useful, it is very important to keep in mind that these photobiological spectral bands are merely "short-hand" notations, and they can be used to make general (but not absolute) statements about the relative spectral effectiveness of different parts of the UV spectrum in producing effects. The dividing lines, while not arbitrary, are certainly not fine dividing lines between wavelengths that may or may not elicit a given biological effect. One should always provide a wavelength band or spectral emission curve for the UV source being used and not rely totally on these spectral terms. There are also many authors who use 320 nm rather than the CIE defined dividing line of 315 nm to divide UV-A from UV-B. Some authors also may divide the UV-A band into two regions: UV-A1 and UV-A2, with a division made at about 340 nm. The exposure limits often overlap bands or do not even make use of them. For this reason, providing action spectra and lamp spectra in research reports are critical.[7]

2. Optical phenomena

Polarization

An electromagnetic wave is a transverse wave. The fluctuating electric and magnetic fields are perpendicular to each other and to the direction of propagation (orthogonal). When all of the waves in a beam of light have the electric vector in only one plane, e.g., the wave has only Y displacements, we say the wave is linearly polarized.

Reflection

When light arrives at an interface, a certain fraction of the incident radiant energy is reflected. This reflection may be very orderly as from a mirror or a sheet of glass (specular reflection), or the light rays may be reflected in random directions and produce a diffuse reflection.

Absorption

When energy is not transmitted or reflected from a medium, it must be absorbed. In simple media where generally only one scattering of a photon takes place, the absorption follows Beer's Law, or "the exponential law of absorption." Beer's Law is expressed as:

$$\Phi/\Phi_0 = e^{-\alpha x} \qquad (4)$$

where Φ is the radiant power (radiant flux) exiting the medium, Φ_0 is the initial radiant power, α is the absorption coefficient, and x is the thickness of the medium. The absorption coefficient is expressed as inverse units of length, e.g., cm^{-1} and the reciprocal of α is termed the penetration depth, where about 63% of the incident energy is absorbed, and at twice that depth, only 13.5% of the incident flux remains, etc.

3. Radiometric quantities and units

Radiant energy

The radiant energy (CIE/ISO Symbol: Q) is the energy emitted, transferred or received in the form of radiation. The SI unit: is the joule (J). The symbol Q is used to identify radiant energy. If there is an e subscript (Q_e) we know that we are using radiometric units in joules, but if the Q has a subscript v (Q_v), then we know that this is luminous energy in lumens, and the energy is dependent upon the CIE action spectrum for daylight vision (the eye's photopic response). The CIE visual response functions will be discussed in a later section.

The fundamental quantity of energy is actually defined in physics in terms of mechanical work; work is the product of a displacement d and the constant force F in the direction of d, such that Work = Force • Displacement. The work done by the constant force F on an object is equal to the change in energy of that object. As force is expressed in the units of $kg \cdot m/s^2$, the fundamental unit of the joule is basically defined as: 1 Joule = $(1\ kg \cdot m/s^2) \cdot 1\ m$. In quantum mechanics, the energy of single photons can also be described as noted in Equation [1]. The annex describes radiometric quantities.

Radiant power

Radiant power (also termed in some circles, "radiant flux") is a measurement of the radiant energy passing a point in a time interval (seconds). The CIE/ISO standard symbol may be either P (for "power") or the Greek letter Φ for "flux." A watt is a unit of power, such that: 1 watt = 1 joule/second (J/s). Remembering that energy is defined in terms of work, it is easy to see that power is work/time. This energy can also be converted to heat, and calorimeters are used to measure mechanical heat, or they can be used to measure the heat energy produced by the absorption of radiant energy, thus the fundamental quantities are related in optics, mechanics and thermal physics.

Radiant exposure

Radiant exposure is the radiant energy incident on a surface divided by the area (projected to the normal) of the surface. The SI unit is $Joules/m^2$. The CIE/ISO symbol is H. In photobiology radiant exposure is also called the "exposure dose."

Irradiance

The irradiance on a surface is the radiant power incident on a surface divided by the area of the surface. Irradiance multiplied by the exposure duration, the time t in seconds, is the radiant exposure. The CIE/ISO symbol is E. Irradiance is known as the "exposure dose rate" in photobiology.

Fluence and fluence rate

In theory, fluence and fluence rate are the most important and useful radiometric concepts in photochemistry and photobiology, since they actually best describe the dose or dose-rate, respectively. However, these terms are frequently misused, and fluence is frequently mistaken for radiant exposure H because the same units of J/m^2. However, the fundamental definition of fluence differs from that of radiant exposure. Fluence is the radiant energy passing through a small unit area, but it does not include the cosine response and fluence includes backscatter. This unit was virtually designed for use in radiation biology and biochemistry. It is particularly useful in photochemistry and photobiology to describe the microscopic exposure of a molecule in a scattering medium. Likewise, fluence rate is also mistaken for irradiance E, since it has the same units of W/m^2, but it includes backscatter and ignores a cosine response. Thus, it is also useful in photochemistry and photobiology in a scattering medium.

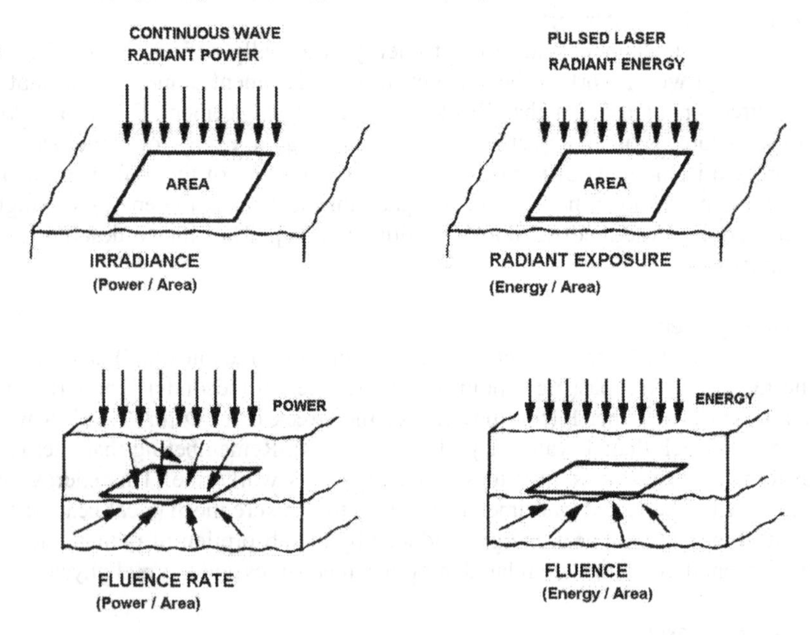

Figure 3. Schematic comparison of irradiance and fluence rate, both expressed as power per unit area; and radiant exposure and fluence, both expressed as energy per unit area.

4. Action spectra

The CIE visual response functions: scotopic & photopic

Examples of action spectra are the CIE visual response functions. The scotopic function describes the eye's response to low light levels experienced at night and this action spectrum peaks around 505 nm. The photopic luminous efficacy function describes the relative spectral response of the CIE "standard observer" (an average of a number of individuals) to daylight; this response peaks around 555 nm.[1,2,7] The entire science of photometry revolves around these CIE luminous efficacy functions as shown in Figure 4. The development of photometers, which are radiometers with a spectral response function that closely mimics the CIE luminous efficacy function V(λ), requires sophisticated filters for accurate spectral matching. Uncertainty ratings for photometers are measured by how well the spectral response fits the desired function; however, for UV photobiology, a single, simple figure of merit number is not sufficient, since errors in the long-wavelength response generally will result in a greater error than if the same error exists in the short-wavelength region. It is generally better to calibrate a UV radiometer against several types of reference sources (e.g., tungsten-halogen lamp, xenon-arc lamp, solar simulator, etc.) and use this calibration factor with similar types of source.

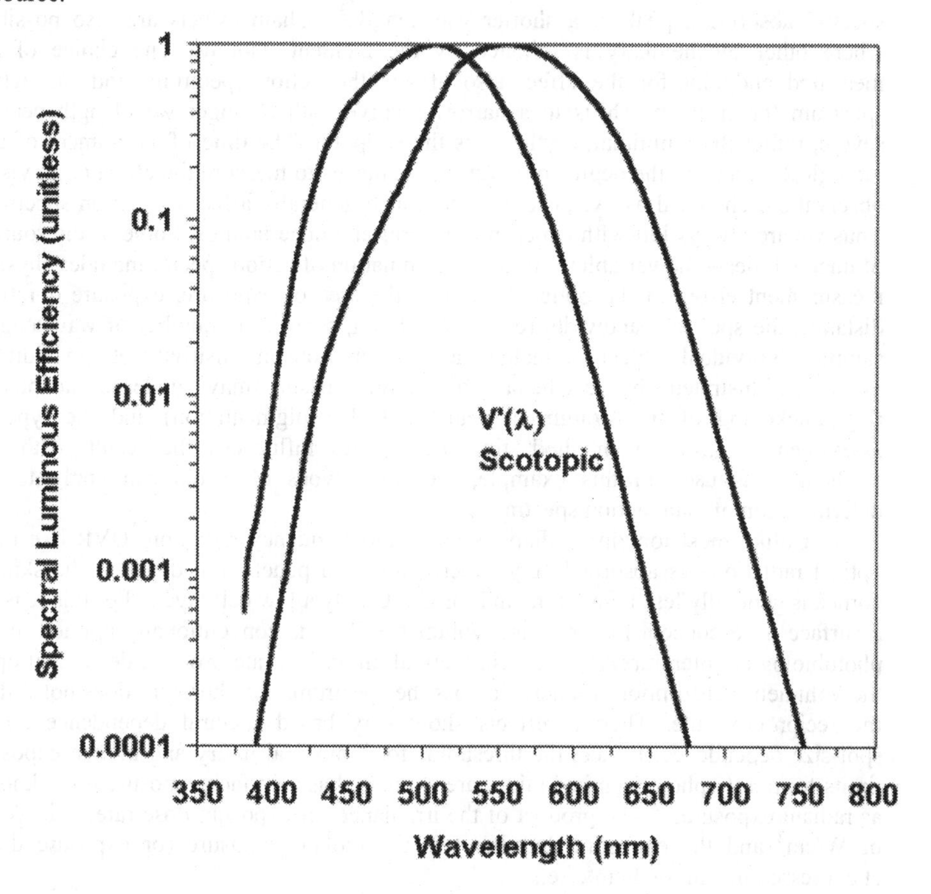

Figure 4. The CIE Photopic and Scotopic visual response functions.

Action spectra: general aspects

An action spectrum is defined as the relative spectral effectiveness of monochromatic radiation in eliciting a defined response relative to a wavelength of maximal effectiveness. Provided that multiple action spectra do not overlap (i.e., one chromophore is present), weighting a broad spectrum source by an action spectrum can predict the effectiveness of the broad spectrum source. The shape of the action spectrum is determined by a number of factors. Most important is the target molecule itself, the chromophore. The action spectrum of a pure solution of the chromophore will provide the fundamental action spectrum; however, other biological factors can alter this fundamental action spectrum. Optical reflection, absorption or scattering prior to absorption of photons by the chromophores will frequently shift the peak of the action spectrum, as in the case of erythema (skin reddening, or "sunburn").[3-5] The proteins of the stratum corneum (outermost, horny surface layer) of the skin spectrally filters the UV photons incident upon the skin and tend to block the transmission of wavelengths much less than 295 nm, even though the most important chromophore - DNA - has a spectral absorption peak at a shorter wavelength.[3,6] Chain effects are also possible, where other biochemistry is initiated by the incident photons. The choice of the measured endpoint for the effect also affects the action spectrum, and the action spectrum for erythema shifts to a narrower curve with a longer wavelength peak if severe, rather than minimal, erythema is the endpoint. The time of assessment of this biological effect and the degree of severity, the means to measure the effect (e.g., visual observation, chemical assay, histology, etc.) also generally affect the action spectrum. Thus we are always left with experimental error and uncertainties. Some of the sources of uncertainties—the variables—in the determination of action spectra include: physical measurement errors of the optical radiation, the area of exposure, exposure duration, distance, the spectral bandwidth (e.g., laser, 1 nm, 5 nm), the number of wavelengths sampled, individual subject (or anatomical site) variations in sensitivity, etc. In addition, as well illustrated by erythema, the target tissue may undergo adaptation (e.g., thickening of the stratum corneum and skin pigmentation) and the type of assessment (e.g., color, method, time delay, etc) influences the result. Although erythema was used in this example, the same types of errors can apply to the determination of plant action spectra.

Unlike most ionizing radiations and radio frequency radiation, UVR-like most optical radiation—is absorbed very superficially and penetration depth in the skin or cornea is generally less than 1 mm and for UV-C only a few cell layers. For this reasons a surface dose concept rather than a volumetric dose is conventionally applied in the photobiological literature. By contrast, thermal injury is a rate process, dependent upon the volumetric absorption of energy across the spectrum, and therefore does not follow the reciprocity rule. Thermal effects show very broad spectral dependence and a spot-size dependence. Hence, the thresholds for biological injury and human exposure limits for purely photochemical injury are expressed as a surface exposure dose, known as radiant exposure.[1-2] The product of the irradiance (or exposure dose rate) E in W/m^2 or W/cm^2 and the exposure duration t is the radiant exposure (or exposure dose) H expressed in J/m^2 or J/cm^2, i.e.,

$$H = E \cdot t \tag{5}$$

This product always must result in the same radiant exposure or exposure dose over the total exposure duration to produce a threshold injury. This is termed the Rule of Reciprocity, or the Bunsen-Roscoe Law.[5] Chemical recombination over long periods (normally hours) will lead to reciprocity failure, and in biological tissue, photochemical damage may be repaired by enzymatic and other repair mechanisms and cellular apoptosis.[7]

Luminous exposure (photometric) and radiant exposure H are both quantities used to describe a total exposure dose from a flashlamp or from a more lengthy exposure. While both E and H may be defined over the entire optical spectrum, the luminous exposure is only defined over the visible spectrum and is therefore not of value in UV photobiology. Where radiant energy is more penetrating, as in the visible and IR-A spectral bands, it is sometimes useful to apply the radiometric concepts of fluence and fluence rate as shown in Figure 3.[7] For all photobiology, it is necessary to employ an action spectrum for photochemical effects.

The relative action spectra for both UV hazards to the eye (acute cataract and photokeratitis) are shown in Figure 5.[2,7-9] These effects will be discussed in greater detail later. The UV safety function S(λ) is also an action spectrum, which is an envelope curve for protection of both eye and skin, is shown in Figure 6.[2,7]

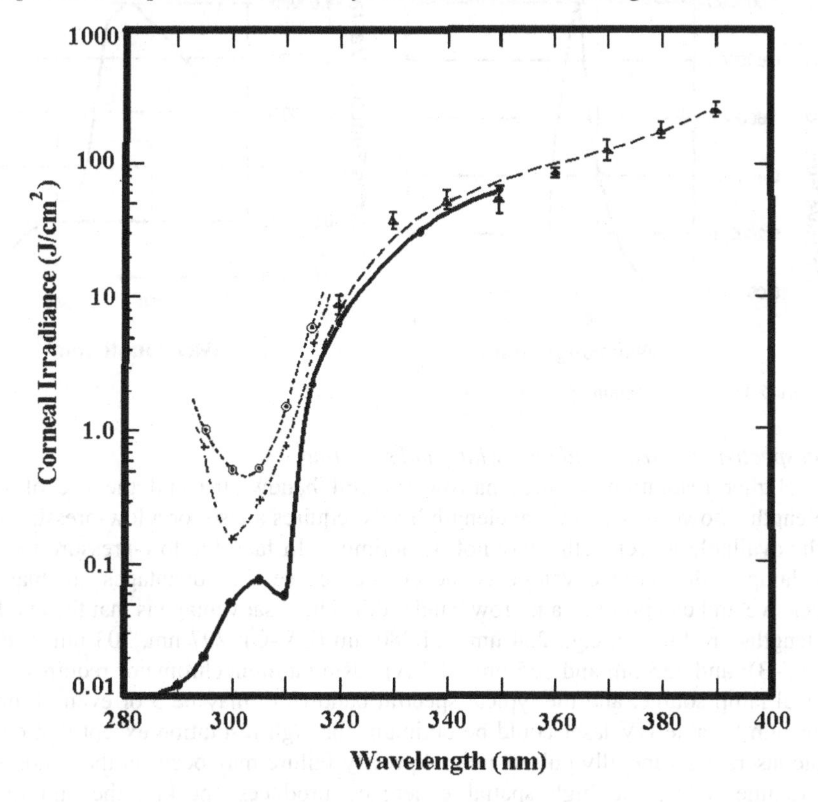

Figure 5. Action spectra for injury to the lens and cornea by near UV radiation. Lower solid curve is the data of Pitts, Cullen and Hacker (1977), for photokeratitis; upper dashed line is the data for permanent lenticular opacities; intermediate dot-dash line with open squares is the threshold for temporary lenticular opacities. The open circles with dashed lines represent threshold for corneal injury by Zuclich and Kurtin (1977).

The $S(\lambda)$ curve of Figure 6 is an action spectra which is used to spectrally weight the incident UVR to determine an effective irradiance for comparison with the threshold value or exposure limit.[13] With modern computer spread-sheet programs, one can readily develop a method for spectrally weighting a lamp's spectrum by a variety of photochemical action spectra. The computation may be tedious, but straightforward:

$$E_{eff} = \sum E_{\lambda} \cdot S(\lambda) \cdot \Delta\lambda \qquad (6)$$

The exposure limit is then expressed as a permissible effective irradiance E_{eff} or an effective radiant exposure. One then can compare different sources to determine relative effectiveness of the same irradiance from several lamps for a given action spectrum.

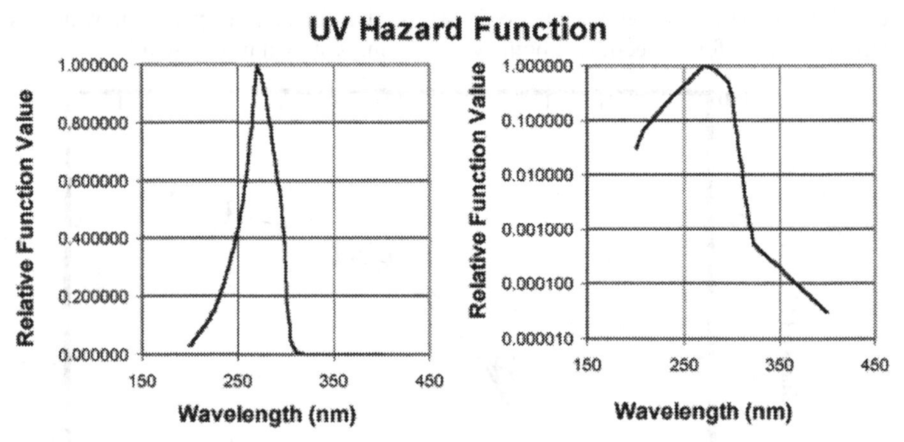

Figure 6. UV Hazard Action Spectrum, $S_{UV}(\lambda)$

Action spectra: spectral band bandwidth and sampling

Higher resolution requires narrow spectral bandwidths and the use of many wavelengths, however narrow wavelength bands requires a laser or a low-pressure lamp and the available wavelengths may not be optimum. In fact, the low-pressure mercury vapor lamp with quartz envelope is the classic source. Its advantages are that it is inexpensive and can produce a narrow bandwidth. The disadvantage is that the available wavelengths are limited, e.g., 254 nm and 280 nm (UV-C); 297 nm, 303 nm, and 313 nm (UV-B); and 335 nm and 365 nm (UV-A). Using a monochromator requires a very powerful lamp source, and the typical spectral bandwidth may be 5 or even 10 nm. A wavelength tunable UV laser could be optimum for high resolution except that current tunable lasers are generally pulsed and reciprocity failure may occur in the nanosecond time regime. Also, the high spatial coherence produces speckle, the shimmering granular appearance of diffusely reflected laser light, and may provide uncertainties. In addition, radiant power may not allow large-spot area irradiation. Nevertheless, careful laser studies have resulted in refined action spectra for erythema (i.e., "sunburn").[10]

A high-intensity monochromator is frequently employed today to obtain action spectra. The advantages are that they are continuously tunable from wavelength to wavelength and they can have a variable bandwidth. However, disadvantages also exist;

the choice of a narrow bandwidth leads to low "throughput" of spectral radiant energy and therefore long exposure sessions. Also, "stray light" outside the spectral band-pass - particularly from gratings - lead to errors from out-of-band radiation.[11,12]

Distorted action spectra can be produced when one uses high-intensity monochromators using arc-lamp sources. Overly broad bandwidth or stray light will lead to a distorted, erroneous action spectrum. Plotting threshold data at the monochromator wavelength setting is reasonable only if the action spectrum is not changing rapidly. Changing action spectrum requires narrow spectral bandwidths for resolution.[2,7]

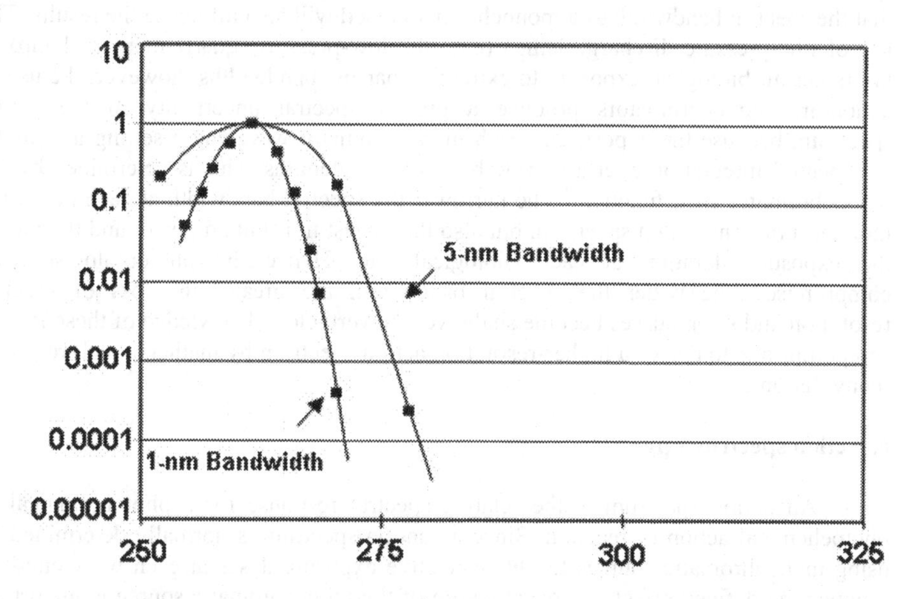

Figure 7. Example of a distorted action spectrum. The "true" action spectrum is the 1 nm bandwidth inner curve. A 5-nm bandwidth produces a wider, erroneous curve.

The accuracy, and reliability of any hazard evaluation or risk assessment of an optical radiation hazard depends strongly upon the accuracy and precision of the relevant action spectra employed. The occupational and environmental health committees of experts which derive human exposure limits used in safety standards and risk assessments rely heavily upon the accuracy of the action spectra obtained by photobiological research. It is therefore important to recognize the factors that influence the quality of the original photobiological research, the possible sources of error and the levels of uncertainty in applying laboratory action spectra to human health risk assessment. One need only study the variation in the reported action spectra published by different laboratories for the same biological effect to recognize that the derivation of sound action spectra is surely fraught with some problems. Both biological and physical factors influence the variation in published action spectra. Biological factors which may affect the actual biological effect of exposure to optical radiation include: variation in response among species, variation in response due to individual adaptation, influence of foods and pharmaceuticals, the biological endpoint applied in each study, and the means and time of assessment of the endpoint in the animal or human subjects. The physical factors which influence any reported action spectrum relate to the accuracy of radiometric and spectroradiometric measurements, the type of light source and the geometry and spectral bandwidth of each exposure used in the biological experiment.

5. Resolution

While it is obvious that the resolution of the final action spectrum depends upon the total number of wavelength intervals used during the experiment, it is less obvious that the spectral bandwidth of a monochromator used will also influence the results. The use of low-pressure discharge lamps (e.g., the low-pressure, quartz-mercury lamp) or lasers permit biological exposure to extremely narrow bandwidths; however, the use of xenon-arc monochromators produce a greater spectral uncertainty in the action spectrum, because the exposure at each monochromator wavelength setting is actually the spectral integration over a narrow band of wavelengths. This is determined by the monochromator's slit function.[2] The narrower the spectral bandwidth at each point, the more accurate the action spectrum, but also the lowest transmitted power and the longer the exposure duration at each biological site. Hence the photobiologist must compromise. The wider the spectral bandwidth, the greater the loss of spectral resolution and steep curves become shallower. Nevertheless, knowledge of these factors can permit one to derive a higher-resolution action spectrum by mathematical treatment (convolution).

6. Action spectroscopy

An action spectrum is the relative spectral response for a photobiological or photochemical action or reaction. Since an action spectrum is normally determined by using monochromatic sources to obtain relative exposure doses at each wavelength to produce the defined effect, the exact nature of the monochromatic source is important. A radiant exposure at the target surface is measured and the underlying assumption is that reciprocity of irradiance and time exists (i.e., the Bunsen-Roscoe Law holds). The action spectrum will differ *in situ* from that measured on an exterior surface if intervening molecules do not have a neutral absorption spectrum. For example erythema, cataract, and retinal effects produced by ultraviolet radiation are mediated by intervening tissues which absorb some of the energy with shorter wavelengths generally more attenuated, thereby reshaping the action spectrum.

The spectral bandwidth of the source can affect the resulting action spectrum as will be shown later. Two types of monochromatic sources are frequently chosen: either a low-pressure lamp, such as the mercury quartz lamp where very narrow wavelength lines can be selected by the use of filters or a monochromator, as was traditional in photobiology of the 1930's, and the use of a xenon-arc monochromator. The advantage of the low-pressure line source is that the spectral bandwidth of each line is less than one nanometer. However, the individual emission lines of a lamp are not equally distributed and may not be near peaks and minima of action spectra. Hence, it is highly desirable to have a tunable monochromatic source. In recent decades, the high-pressure xenon-arc has been sufficiently intense that when used with a grating monochromator, can produce relatively narrow bandwidth monochromatic emission wavelengths.

The problems inherent in the use of a monochromator are not always clearly evident to the research photobiologist and it is worthwhile to explain that here. The problems can appear both when the monochromator is used in spectroradiometry of the source or as part of a monochromatic illumination source. Figure 8 illustrates this point. The spectral bandwidth of the emitted radiation has a triangular shape when the monochromator dial is set at a given wavelength. The spectral bandwidth transmitted by the monochromator is defined at 50% points by what is known as the full-width half-maximum (FWHM). When a source which is varying in spectral output, such as the sun or a tungsten-halogen lamp, is used, the transmitted radiation from the monochromator, when multiplied by the action spectrum provides the relative spectral effectiveness and the spectral effectiveness can be shifted as shown in the figure. The lower panels illustrate the use of a xenon arc monochromator where the xenon arc spectrum is relatively flat in the ultraviolet spectrum, but when the dial is set at 305 nm with the full-width half-bandwidth of 5 nm, the effective peak wavelength for producing erythema is actually near 301 nm, although the instrument is set at 305 nm. Hence the investigator writes down that the relative effectiveness of this narrow band of radiation is a given value at 305 nm when in fact, that effectiveness value is more characteristic of a shorter wavelength. The effect of this error is to broaden action spectra as was shown in Figure 7. This problem has long been recognized, and in deriving occupational exposure limits for ultraviolet radiation or to correct erythemal or photokeratitis action spectrum, it was possible to make an adjustment to determine the shifting of the action spectrum to longer wavelengths. Obviously, a tunable laser would be a desirable light source except that most current tunable UV lasers are very expensive and are pulsed; and investigators worry about the potential loss of reciprocity from short-wavelength pulses. Nevertheless, the laser erythemal action spectrum determined with a laser by Anders and her associates (1995) clearly demonstrate the value of using monochromatic sources with a resolution of 1 nm or less.[10]

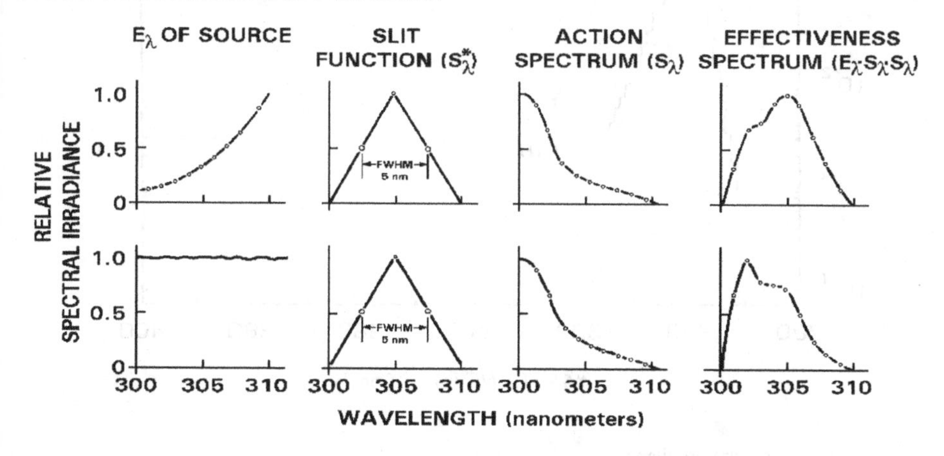

Figure 8. The effect of spectral bandwidth on the effective wavelength of emitted radiation.

The problem of distorted action spectra or distortion of spectral measurements with monochromators is not unique to photobiology. The problem of measuring the solar spectral irradiance (Figure 9) illustrates very clearly the problem of attempting to perform spectral radiometry on a rapidly changing source spectrum. Therefore, when one attempts to use a monochromator in both the monochromatic optical exposure source, for performing the photobiological experiment, and then a monochromator in the spectroradiometer to measure a light source rapidly ascending in spectral irradiance in the same spectral region as shown in Figure 9, the error can be enormous. The error is magnified when the monochromator and the spectral radiometer tends to shift the effective spectral irradiance slightly towards shorter wavelengths because of the identical problem. When spectroradiometry is viewed with this in mind, the potential problems of broadband radiometry do not appear quite so challenging. The experimental errors become comparable.

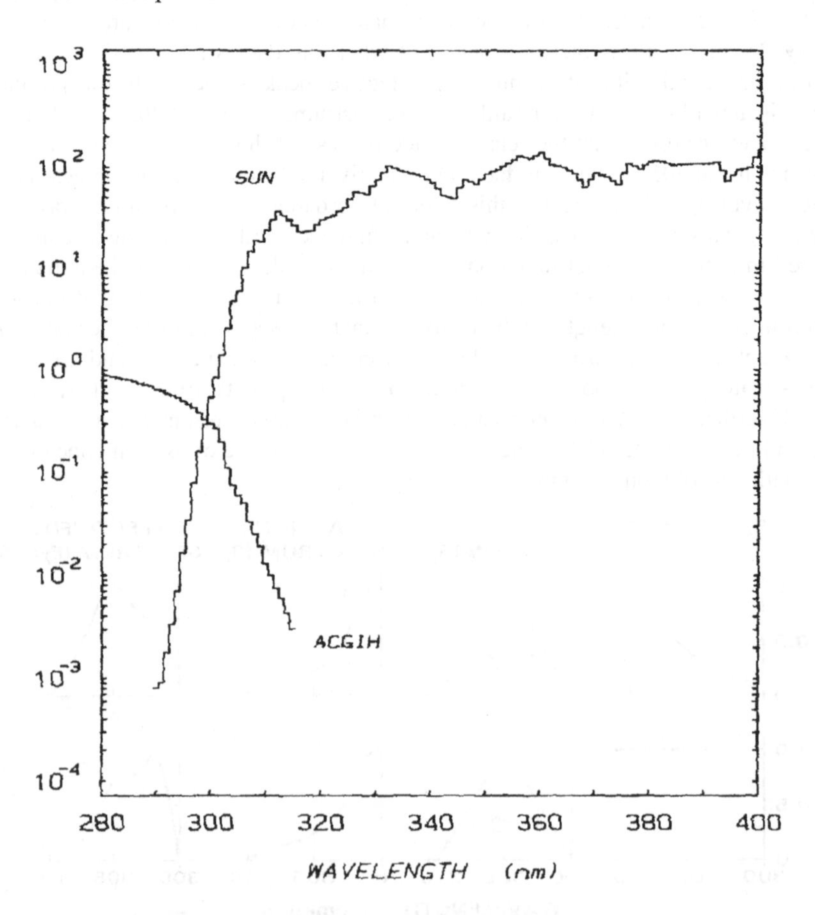

Figure 9. Solar spectral irradiance

Out-of-band radiation is another problem encountered in performing action spectroscopy and in measuring light sources. The term used for out-of-band radiation from a monochromator is stray light, and is illustrated in Figure 10. Stray light poses a serious pitfall when attempting to measure a weak spectral signal near a strong spectral signal, for example in attempting to measure a tungsten lamp at 300 nm or sunlight at 300 nm.[11-12] The monochromator slit function defines the degree of stray light in a fixed wavelength. This is hopefully less than 0.1%. One normally concludes that the best action spectroscopy and the best spectroradiometric measurements with a monochromator should be performed with a narrow bandwidth. This may be slow but should offer more detail. This is clearly more important when a source or action spectrum is rapidly changing. While it is true that the choice of wider bandwidths will allow more energy through the monochromator and allow for a faster measurement or a faster exposure, it offers less detail and can only be used with a source that has a very slowly varying spectrum.

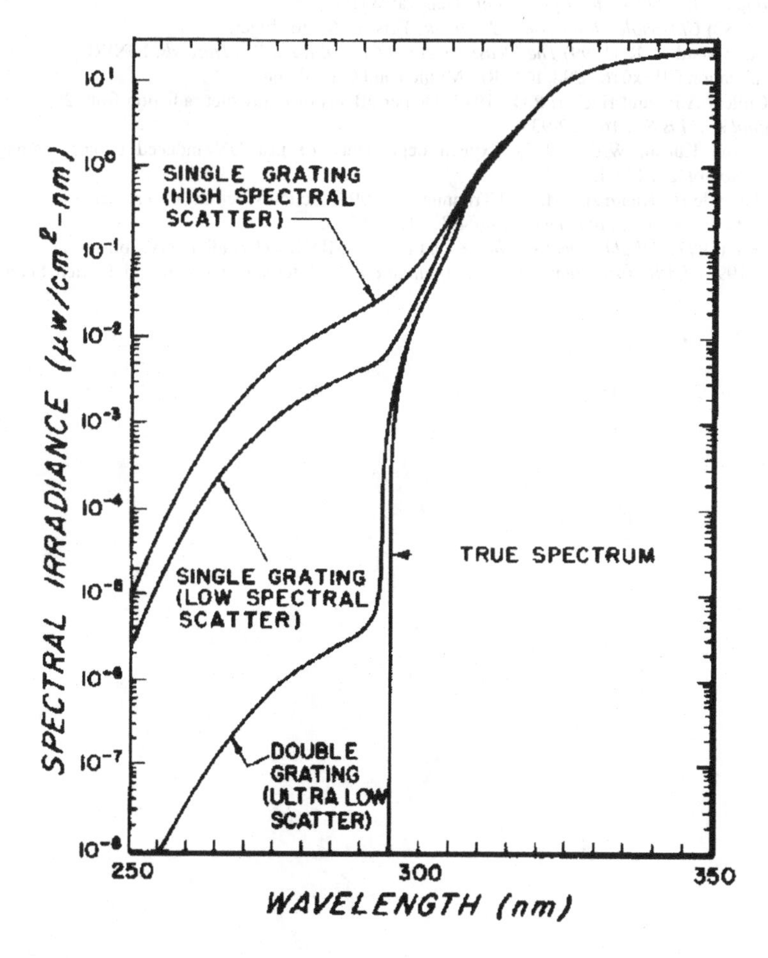

Figure 10. The effects of stray light.

Acknowledgements. This project was supported in part by an appointment to the Internship/Research Participation Program for the U.S. Army Center for Health Promotion and Preventive Medicine administered by the Oak Ridge Institute for Science and Education through an agreement between the U.S. Department of Energy and the USACHPPM.

References

1. Commission International de l'Eclairage (International Commission on Illumination) (1987) *International Lighting Vocabulary, 4th ed.* Pub. CIE No. 17 (E-1.1) Vienna: CIE.
2. Sliney, D.H., and Wolbarsht, M.L. (1980) *Safety with Lasers and Other Optical Sources*, New York: Plenum Publishing Company.
3. Grossweiner, L.I. (1988) Photochemistry of proteins: a review, *Curr Eye Res*, 3: 137-144, 1984.
4. Smith, K.C., *The Science of Photobiology*, New York: Plenum Press.
5. World Health Organization (WHO) (1994) *Environmental Health Criteria No. 160, Ultraviolet Radiation, joint publication of the United Nations Environmental Program, the International Radiation Protection Association and the World Health Organization*, Geneva: WHO.
6. Diffey, B.L. (1982) *Ultraviolet Radiation in Medicine*, Bristol: Adam Hilger.
7. Sliney, D.H., and Matthes, R. (1999) *The Measurement of Optical Radiation Hazards,* ICNIRP Publication 6/98; CIE Publication CIE-x016-1998, ICNIRP: Munich and CIE: Vienna.
8. Pitts, D.G., Cullen, A.P., and Hacker, P.D. (1977) Ocular effects of ultraviolet radiation from 295 to 365 nm, *Inv Ophthal And Vis Sci,* 16: 932-939.
9. Zuclich, J.A. and Kurtin, W.E. (1977) Oxygen dependence of near-UV induced corneal damage, *Photochem Photobiol,* 25: 133-135.
10. Anders, A., Altheide H., Knalmann M., and Tronnier H. (1995) Action spectrum for erythema in humans investigated with dye lasers, *Photochem. Photobiol.*, 61: 200-205.
11. Koskowski, H.J. (1997) *Reliable Spectroradiometry*, La Plata, MD: Spectroradiometry Consulting.
12. Webb, A.R. (1998) *UVB Instrumentation and Applications*, Amsterdam: Gordon and Breach Science Publishers.

ANNEX - Summary of Radiometric Quantities

The following radiometric quantities may be used in photobiology, photochemistry, photodermatology and illuminating science, and are briefly summarized here:

Irradiance (surface dose rate) and ***radiant exposure*** (surface dose) are units specifying power or energy incident upon a plane. These quantities are the most fundamental dose quantities used in all of photobiology. The units most commonly used are W/cm^2 and J/cm^2, respectively. 1 W = 1 J/s.

Fluence rate and ***fluence*** are used in some very sophisticated studies, where the internal surface dose with backscatter is included. These quantities are used correctly most often in theoretical studies of dose distribution and where photochemistry at the molecular level in tissue is enhanced as a result of multiple scattering events in tissue. Unfortunately, these terms are frequently misused to mean irradiance and radiant exposure because the units of W/cm^2 and J/cm^2 are the same.

Radiance (irradiance per solid angle) is an important quantity used by physicists in specifying a source. This quantity limits the ability of lenses and reflective optics in concentrating a light source. For, example, a xenon-arc lamp has a very high radiance and its energy can be focused to produce a very high irradiance on a target tissue. By contrast, a fluorescent lamp tube has a much lower radiance, and its energy cannot be focused to a high concentration. The units are $W/(cm^2 \cdot sr)$.

Radiant Intensity (power per solid angle) is used to indicate how collimated a light source really is. Although useful for specifying searchlights, it normally has very limited use in photobiology. The units are W/sr.

Spectral quantities (units per wavelength) are used for specifying the energy, power or irradiance per wavelength interval. When calculating a *photobiologically effective dose* the spectral quantity must be multiplied by the action spectrum. Examples: spectral radiant power, spectral irradiance, spectral radiant exposure, etc. The units for each quantity are modified by adding "per nanometer," e.g., W/cm^2 becomes $W/(cm^2 \cdot nm)$.

Photon (Quantum) quantities (units of photons) are used primarily in theoretical studies, and in photochemistry. In this case the radiant exposure is specified in $photons/cm^2$ and irradiance is specified in $photon/(cm^2 \cdot s)$.

Table A1. Useful Radiometric Units[1,2]

Term	Symbol	Definition	Unit and abbreviation
Radiant Energy	Q	Energy emitted, transferred, or received in the form of radiation	joule (J)
Radiant Power	Φ	Radiant Energy per unit time	watt (W) defined as $J \cdot s^{-1}$
Radiant Exposure (Dose in Photobiology)	H	Energy per unit area incident upon a given surface	joules per square centimeter ($J \cdot cm^{-2}$)
Irradiance or Radiant Flux Density (Dose Rate in Photobiology)	E	Power per unit area incident upon a given surface	watts per square centimeter ($W \cdot cm^{-2}$)
Integrated Radiant Intensity	IP	Radiant Energy emitted by a source per unit solid angle	joules per steradian ($J \cdot sr^{-1}$)
Radiant Intensity	I	Radiant Power emitted by a source per unit solid angle	watts per steradian ($W \cdot sr^{-1}$)
Integrated Radiance	LP	Radiant Energy emitted by a source per unit solid angle per source area	joules per steradian per square centimeter ($J \cdot sr^{-1} \cdot cm^{-2}$)
Radiance[3]	L	Radiant Power emitted by a source per unit solid angle per source area	watts per steradian per square centimeter ($W \cdot sr^{-1} \cdot cm^{-2}$)
Optical Density	OD	A logarithmic expression for the attenuation produced by a medium $OD = -\log_{10}(\Phi_O/\Phi_L)$ Φ_O is the incident power; Φ_L is the transmitted power	unitless

[1] The units may be altered to refer to narrow spectral bands in which the term is preceded by the word *spectral* and the unit is then per wavelength interval and the symbol has a subscript λ. For example, spectral irradiance E_λ has units of $W \cdot m^{-2} \cdot m^{-1}$ or more often, $W \cdot cm^{-2} \cdot nm^{-1}$.

[2] While the meter is the preferred unit of length, the centimeter is still the most commonly used unit of length for many of the terms below and the nm or µm are most commonly used to express wavelength.

[3] At the source $L = dI/(dA \cdot \cos\theta)$ and at a receptor $L = dE/(d\Omega \cdot \cos\theta)$.

Table A2. UV sources used in photobiological research.

Optical Source	Wavelength Function	Spectral Radiance* ($W/cm^{-2}/sr^{-1}/nm^{-1}$)	Spectral Bandwidth	Advantages
Tungsten Halogen Lamp	monochromator	3×10^{-4}	5-nm	inexpensive
High Pressure Xenon Arc	monochromator	1×10^{-2}	5-nm	relatively flat SPD
Low Pressure Mercury Quartz Lamp	monochromator	4×10^{-7} @ 297 nm	< 1-nm	inexpensive; good spectral purity
Laser	laser transition or non-linear optical tuning	$> 10^{-7}$	<< 1-nm	excellent spectral purity

* The radiance ("source brightness") determines the radiant energy that can be collimated and directed through a monochromator or focused on target tissue.

SOLAR RADIATION AND ITS MEASUREMENT

HARALD K. SEIDLITZ AND ANDREAS KRINS
GSF-Forschungszentrum für Umwelt und Gesundheit
85764 Neuherberg, Germany

1. Introduction

Solar radiation reaching the Earth's surface is a key factor for the development of life. It is characterised by both the intensity and the spectral composition. Its spectrum reaches from approximately 290 nm to 3000 nm. A considerable amount of the energy is contained in the ultraviolet band of which especially the short wave part below 315 nm is considered to be harmful for men, animals and plants. The range between 280 and 315 nm is designated as UV-B radiation, the range between 315 and 400 nm as UV-A radiation (Fig. 1). The exact definitions of radiation quantities is presented in the another contribution in this volume[1].

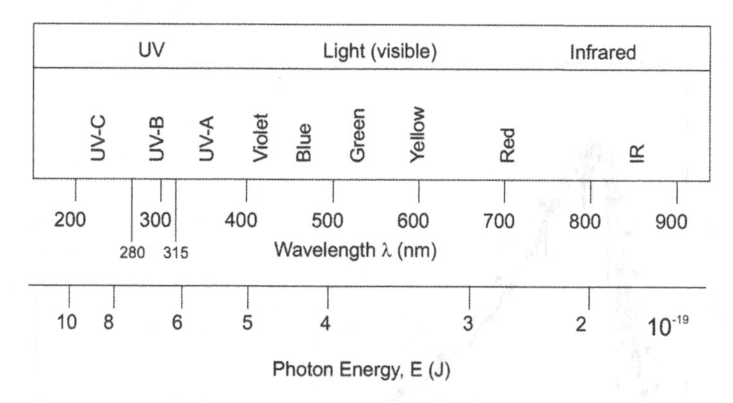

Figure 1. Schematic presentation of the electromagnetic spectrum.

The quantification of the amount and the quality of UV radiation and light reaching the Earth's surface or a given biological system under consideration is of major importance in basic biological research, especially for the assessment of the impact of UV on human health and ecosystems. Especially the study of the complex – coupled - pattern of plants' responses to light and UV radiation, the understanding of the regulatory mechanisms by which plants can adapt to changing environmental conditions call for reliable light and UV measurements.

F. Ghetti et al. (eds.), Environmental UV Radiation: Impact on Ecosystems and Human Health and Predictive Models, 25–38.

The sensitivity of many biological systems typically increases by several orders of magnitude towards shorter wavelengths while the solar spectrum strongly decreases in the same spectral range. Therefore, the requirements for accurate UV and light measurements are very demanding[2].

2. Solar radiation

Extraterrestrial solar radiation

The Sun is the next star in the sky, producing energy by the fusion of hydrogen into helium. Solar radiation originates mainly from the Sun's photosphere which has a temperature of approximately 5800 K. The spectrum can be approximated by that of a black body radiator having the same temperature (Fig. 2). The spectrum is structured by many lines (Fraunhofer lines) which originate from absorption and emission processes in the photosphere, the chromosphere and the corona. The natural variability of the solar flux in the 280 – 400 nm range is less than 1 % over the well known 11-year and 27-day cycle in solar activity. In contrast, the variability of the Sun – Earth distance of about 3.4 % leads to a variation of the irradiance of almost 7 % due to the inverse square law of the irradiance of a point source[3].

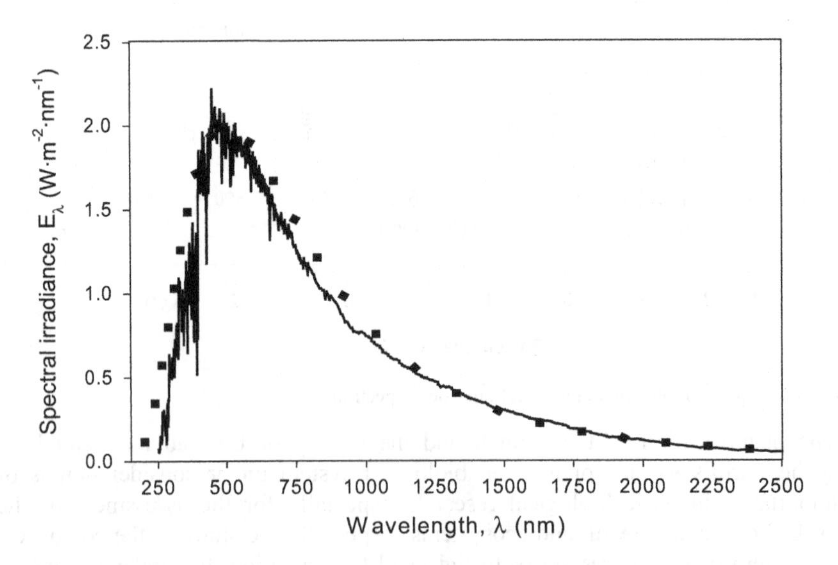

Figure 2. Extraterrestrial solar spectrum (solid line) compared with the black body radiator spectrum at 5780 K (dotted line).

At the top of the Earth's surface, the total irradiance is 1367 $W \cdot m^2$ (solar constant) of which 111 $W \cdot m^2$ or 8 % contribute to the 200 – 400 nm range. The extraterrestrial spectrum was measured by the 'Solar Ultraviolet Spectral Irradiance Monitor' (SUSIM) aboard the Spacelab 2 mission and can be downloaded from the internet site http://www.solar.nrl.navy.mil/susim_atlas_data.html.

Scattering and absorption in the atmosphere

Solar radiation entering the Earth's atmosphere is subject to scattering and absorption processes which greatly influence the spectral composition and spatial distribution of the radiation field on the Earth's surface[4,5]. Scattering is a result of the interaction of small particles with the radiation. It is determined by the wavelength of the radiation and the size of the particle. In Rayleigh scattering the particles are much smaller than the wavelength, e.g. by air or trace gas molecules. The spatial distribution of scattered radiation is the same in the forward and backward direction. The probability for Rayleigh scattering exhibits a $1/\lambda^4$ dependence which results in the blue sky over the day and the red sky at sunrise or sunset. In contrast, Mie scattering, where the particles' size is in the order of the wavelength, e.g. for aerosols, the scattered radiation is strongly peaked to the forward direction.

The main absorbers influencing the UV range of the spectrum at the Earth's surface are oxygen and nitrogen (both atomic and diatomic) and ozone. Between 200 and 300 nm stratospheric ozone is the main absorber of UV radiation, the absorption spectrum of ozone leads to the steep edge of the terrestrial solar spectrum around 290 nm. The amount of UV radiation in the UV-B waveband is strongly dependent on the total ozone column thickness.

As a result of scattering of radiation in the atmosphere, a diffuse radiation field is formed. Besides scattering, absorption is responsible for the attenuation of the Sun's beam which is called direct component of the radiation field. The sum of direct and diffuse radiation defines the global radiation (compare Fig. 3).

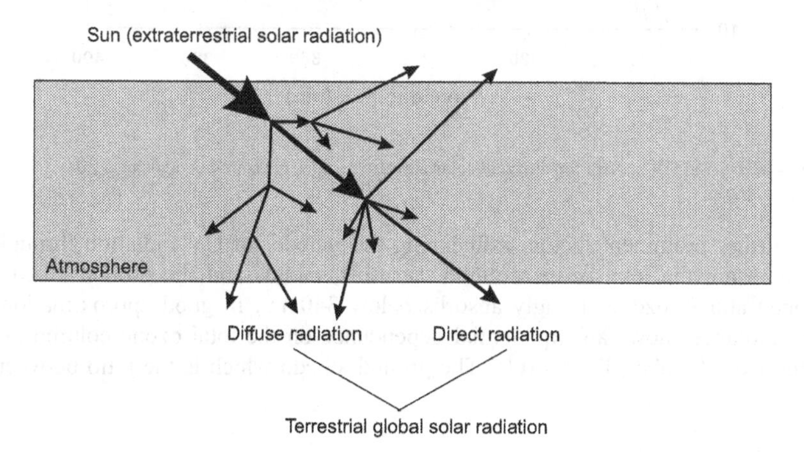

Figure 3. Schematic presentation of atmospheric multiple scattering events and absorption processes.

Terrestrial solar spectra

As a result of the diurnal rotation of the Earth the solar elevation angle and thus the length of the path through the atmosphere changes in the course of the day. This gives rise to a variation of the spectral composition and the intensity of terrestrial solar radiation. Fig. 4 shows the spectral irradiance in the UV range of the solar spectrum on ground at different times of the day respectively at different solar elevation angles, measured at Neuherberg near Munich (11.6° E, 48.22° N, 495 m asl.), Germany, on 17 April 1996. The short wave edge (the 'ozone edge') is shifted towards shorter wavelengths with increasing solar elevation angle. At low solar elevation angles, UV-B radiation is more attenuated as compared with longer wavebands because of the more pronounced scattering and absorption with decreasing wavelength. From Fig. 4 we conclude that the measurement of solar spectral UV irradiances has to cover at least 6 orders of magnitude within a small wavelength range of 30 nm. This implies very stringent demands on the technical equipment used for such measurements as will be discussed in section 3.2.

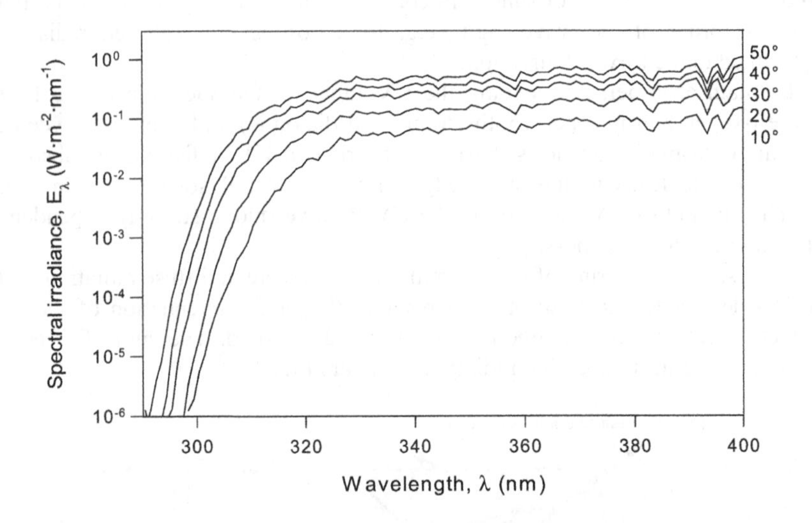

Figure 4. Typical UV spectra for different solar elevation angles measured on a clear sky day[2].

Further prominent factors influencing the transfer of UV radiation through the atmosphere are the total ozone column, aerosols, clouds and the ground albedo. As mentioned above, ozone strongly absorbs below 340 nm. In good approximation, the UV-B irradiance shows an exponential dependence on the total ozone column as it is expected from Lambert-Beer's rule. The ground albedo which is the ratio between the downward and the upward irradiance above ground (resulting from diffuse reflection) strongly depends on the type of surface and the wavelength of the radiation. Typical albedo values are compiled e.g. in Feister and Grewe[6]. Ground albedo values can range from a few percent (e.g. over grassland) to approximately 90 percent above freshly fallen snow. The effect of aerosols and clouds is discussed for example in Seidlitz et al.[2] and Mayer et al.[7].

3. Radiation measurement

General measurement principle

Terrestrial solar radiation is a highly variable quantity. The risk assessment and biological UV research therefore require accurate measurements of the instantaneous radiation environment. Even if the ultraviolet spectral band (280- 400 nm) is of primary interest in this context the visible part should not be neglected, as many organisms especially plants regulate metabolic functions by activating their light receptors which may influence their UV response in turn. Nevertheless the measurement of solar UV-B radiation requires special efforts. The main reason is the very steep decay of the solar spectrum towards wavelength shorter than 300 nm due to the strong filtering effect of stratospheric ozone. Spectral irradiance values drop by 6 orders of magnitude within a bandwidth of 20 nm! The instruments must therefore, exhibit high sensitivity combined with a very large dynamical range. Depending on the spectral resolution we distinguish between spectroradiometry, the method for measuring radiation in narrow wavelength intervals, and broadband radiometry where the irradiance is integrated over a more or less wide waveband. Like any physical measurement radiometry is finally a comparison. In this context the unknown radiation field, e.g. global irradiance is compared with the emission of a well characterised, calibrated standard lamp. Fig. 5 shows the main steps in the assessment of optical radiation (the term designates ultraviolet, visible and infrared radiation).

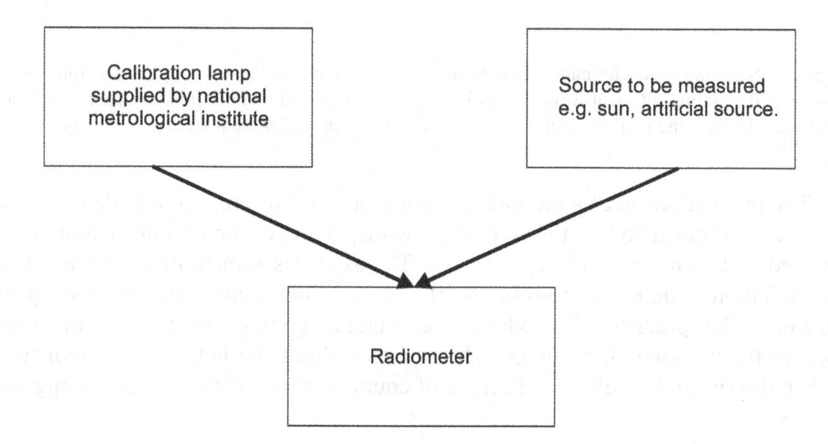

Figure 5. Comparison of detector readings from a known and an unknown source.

Spectroradiometric measurements

The most demanding technique of measuring UV radiation and light is spectroradiometry. It delivers - apart from the polarization state - the full information on the radiation field. A monochromator system consists of different components which are discussed below.

Monochromator. The key component of a spectroradiometer is the monochromator, which disperses radiation incident on the entrance slit. Most instruments use a diffraction grating as dispersing element and mirrors as image-transfer optics. A schematic presentation of a scanning monochromator is shown in Fig. 6. The desired spectral range is scanned by turning the grating over a certain angle. The wavelength projected onto the exit slit depends on the angular position of the grating.

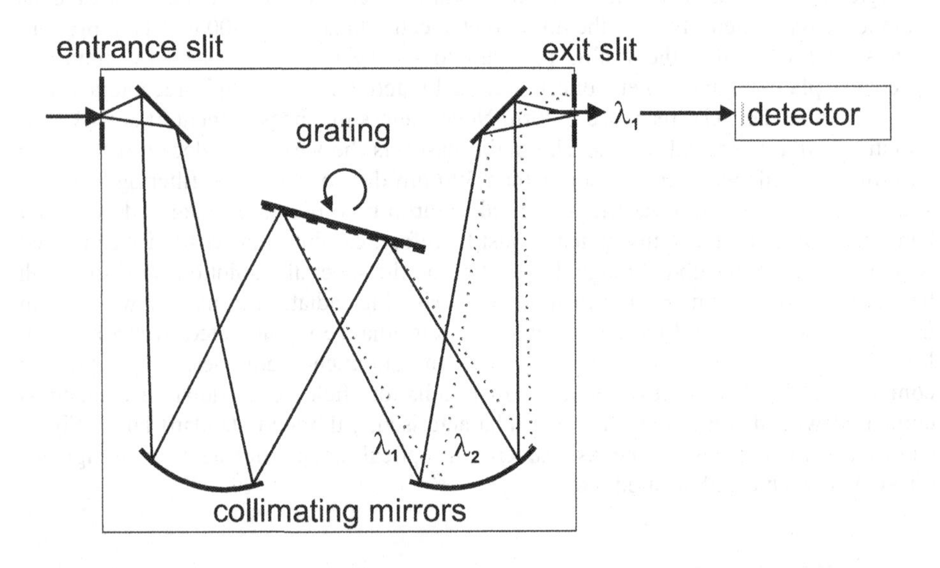

Figure 6. Scanning monochromator. Due to the interference of the 'beams' originating from each groove of the grating different wavelengths λ_1 and λ_2 are projected into different directions. Only a small wavelength band (λ_1) passes the exit slit. In a real system baffles are used to reduce stray light.

The main criterion of a monochromator is spectral resolution. It is determined by the widths of entrance and exit slits and by the dispersion of the grating (which itself is determined by the number of lines per mm). The exact assessment of short wave UV-B global radiation requires a resolution of 0.1 to 1 nm depending on the specific application. The precision by which the angular grating position is maintained determines the wavelength accuracy. The accuracy should be better by a factor two or three than the spectral resolution. There is of course a trade-off between scanning speed and precision.

A great problem in monochromators is stray light. Even tiny fractions of radiation scattered in the wrong direction by imperfections on the grating or by dust particles give rise to false readings especially in the UV-B due to the enormous difference in intensity between wavelengths below and above 300 nm. A tandem arrangement of two monochromators (double monochromator) reduces stray light considerably. A concise review on the main aspects for choosing a monochromator can be found in the article by Domanchi and Gilchrist[8]. The monography by Kostkowski[9] gives a very comprehensive presentation of the subject.

Input optics. A very important component of a radiometer is the input optics. In order to measure irradiance it is necessary that the angular response of the input optic follows the cosine law as close as possible because at oblique angles of incidence the projected sensor area is reduced by the cosine of the angle of incidence (Fig. 7). In order to maintain cosine response usually quartz or Teflon® diffusers or sometimes integrating spheres are employed. Unfortunately all practical devices show more or less large deviations from the ideal cosine response. This can lead to large measurement errors especially at low solar elevations, if the data are not corrected for the cosine mismatch adequately[10]. The direct and the diffuse irradiance components produce different cosine errors. Therefore, their relative portions have to be determined independently by ancillary instruments (e.g. sun tracker, shadow band radiometer, etc.) or model calculations and the correction algorithm has to treat them differently. Azimuth and polarisation dependency of the response may also cause measurement errors. For easy handling the input optics is coupled to the entrance slit by means of a quartz fibre bundle.

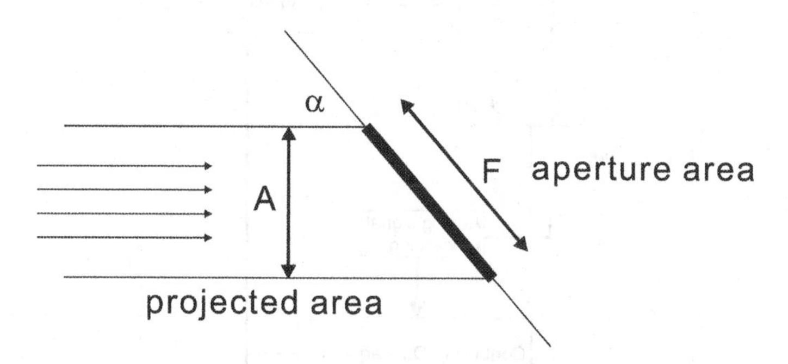

Figure 7. Cosine response of the measurement aperture, $A = F \cdot \cos(\alpha)$.

Detector. After passing input optics, fibre bundle, entrance slit relay optics and the dispersive grating a narrow waveband leaves the monochromator through the exit slit and impinges on to the photodetector. At about 300 nm the absolute values of terrestrial solar irradiance in a 1 nm spectral band are very small ($\approx 10^{-5}$ W^{-2} m^2 nm^{-1}). Therefore, photomultiplier tubes are usually employed because of their excellent noise characteristics. The selection of the photocathode type mainly depends on the waveband of interest. Bi-alkali cathodes are first choice if only UV and blue regions are of primary interest, a trialkali (S20) type covers the UV, visible and near infrared band but it has a somewhat poorer performance than the bialkali cathode. The photomultiplier signal is finally fed into a low-noise dc or chopper amplifier chain and is digitally processed for further use. Photomultipliers require a very constant high voltage. They have a long warm up time up to 48h; therefore, the high voltage should be buffered with a uninterruptible power supply.

System Configuration. The complete set-up of a double monochromator scanning spectroradiometer is depicted in Fig. 8. Particularly double monochromators systems are subject to changes in responsivity due to mechanical strain and temperature changes. They should be mounted on a rigid base. For outdoor use a thermally stabilised cabinet is mandatory.

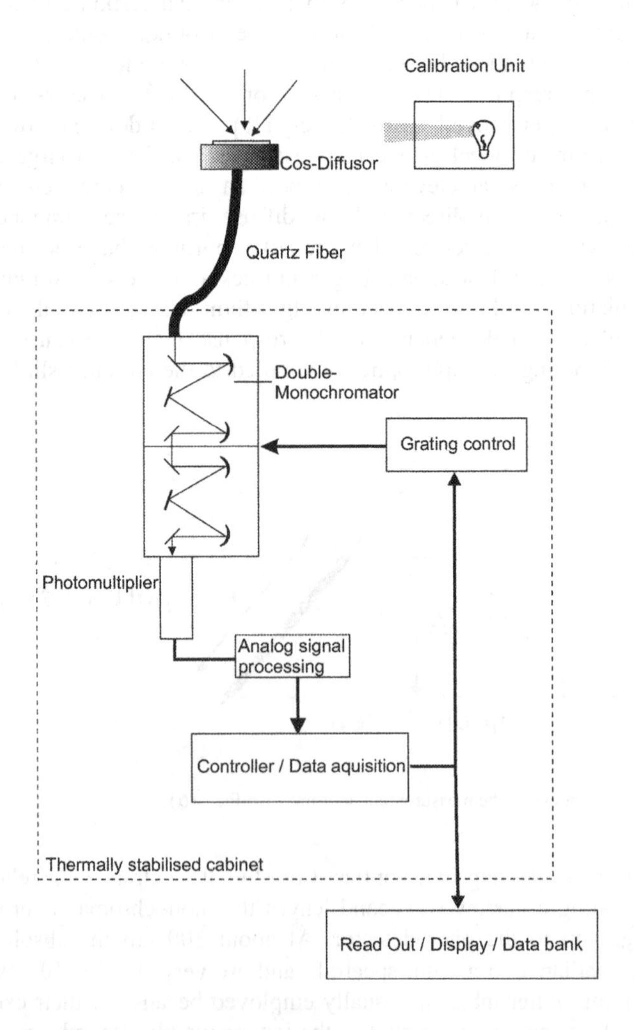

Figure 8. Schematic configuration of a spectroradiometer system.

Calibration. Radiometric calibration requires the spectral characterisation of the system's responsivity. This is facilitated by means of a standard source whose output is well defined[11]. The primary or reference standard is a blackbody. Its spectroradiometric quantities can be calculated from Plank's radiation law. Only national laboratories (like National Institute of Standards and Technology, United States; National Physics Laboratory, United Kingdom; Physikalisch-Technische Bundesanstalt, Germany etc.) maintain blackbody standards. They derive from the blackbody standard secondary standards, usually 1000 W quartz halogen lamps (FEL lamps). Tertiary standards (also FEL lamps) are often transferred to certified suppliers by which the working standards (e.g. FEL lamp or 100 W quartz halogen lamps) are calibrated. Thus the laboratory/working calibration lamp is a forth generation standard (Fig. 9).

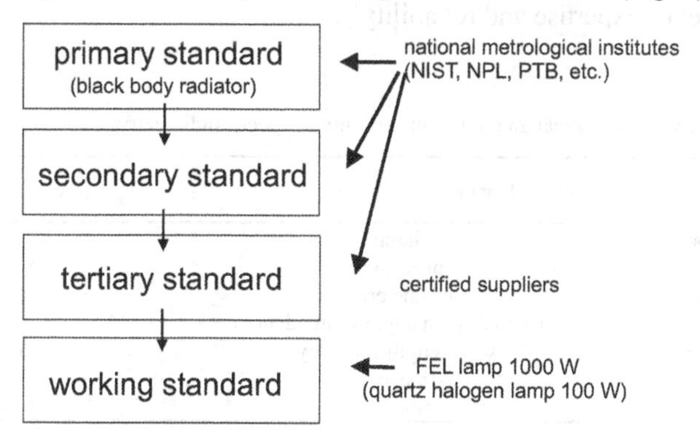

Figure 9. Hierarchy of standards (see text for abbreviations).

The spectral irradiance $E_\lambda(\lambda)$ which is specified by the supplier is defined at a distinct current rating and at a distinct distance from the lamp, these operating parameters have to be maintained as exact as possible. As the lamps will age specifications will hold only for a certain burning time, e.g. 20-50h. After a warm up time of at least 15 minutes the input optics is placed at the given distance, a spectral scan is started and the photosignal $I(\lambda)$ is recorded. The spectral responsivity $\varepsilon(\lambda)$ is then calculated as

$$\varepsilon(\lambda) = I(\lambda) / E_\lambda(\lambda)d\lambda. \tag{1}$$

The spectral irradiance of an unknown source is thus obtained by dividing the measured photocurrent by the spectral responsivitiy.

Small changes in the moving parts of a scanning monochromator may affect the position of the spectrum on the exit slit. Therefore, regular checking of the wavelength setting is required. Wavelength calibration is usually performed with a low pressure discharge lamp (e.g. mercury lamp), which emits narrow spectral lines whose wavelength positions are well defined. An alternative method is the evaluation of the positions of Fraunhofer absorption lines in a solar spectrum. Especially the position of the cadmium line pair (393.37 nm and 396.85 nm) is a good indicator for the UV wavelength performance of the instrument. Narrow spectral lines or laser lines are used to determine shape and width of the monochromator's slit function, which characterises the spectral bandwidth.

Uncertainty Estimation. Spectroradiometric measurements are unfortunately among the least accurate of all physical measurements, because many rather different components contribute considerable measurement uncertainties[12]. The main sources of uncertainty and their approximate contributions are listed in Table1. A total uncertainty of approximately 15% is a realistic estimation. Absolute spectroradiometric measurements are tedious and they require strictly applying the rules of good laboratory practice (GLP)[9,13], i.e. well maintained equipment and the complete documentation of the measurement procedures and calibration sources as well as an appropriate error estimation are obligatory. International instrument intercomparisons are necessary to keep a high level of expertise and reliability[14].

Table 1. Main sources and typical amount of uncertainty in spectroradiometry.

Source of error		Typical relative uncertainty
calibration	lamp	3 – 4 %
	procedure	5 %
input optics	cosine error (depending on angle of incidence)	5 %
radiometer	wavelength accuracy	3 – 10 %
other sources	linearity	1 %
	temperature	1 %
	Total	15 %

Broadband measurements

Biological weighting functions. The amount of energy ('radiant exposure') or the number of photons at a certain wavelength which is necessary to induce a given biological effect usually depends on the wavelength λ of the radiation. This wavelength dependency is described by the action spectrum s(λ). In general, the action spectrum and thus the sensitivity of a biological system increases with decreasing wavelength in the UV-B range. Action spectra exist for a large variety of biological systems and effects, see e.g. Ghetti in this volume[15].

In the case of *monochromatic* irradiation at wavelength λ, the effectiveness of the irradiation for the induction of a certain biological effect is described by the product of the spectral irradiance (or the spectral radiant exposure) and the value of the action spectrum at λ. This multiplication is often called weighting, the product is the weighted spectral irradiance or the weighted spectral radiant exposure. If the sample is irradiated with a broadband (polychromatic) source weighting is performed by multiplication of the action spectrum with the polychromatic source spectrum and integration over the desired wavelength range. Hereby it is tacitly assumed that photons at different wavelength contribute independently to the effect under consideration and the action mechanism is similar for all wavelengths. In order to avoid confusion, it should always be stated exactly which action spectrum was used. Typical examples of biological action spectra are erythemal weighting[16] and the generalised plant action spectrum[17]. The so called integrated UV-B and UV-A irradiances are examples of mathematical weighting, i.e. the weighting function is a step function being equal one between 280 and 315nm for UV-B (315-400 nm for UV-A, respectively) and zero elsewhere.

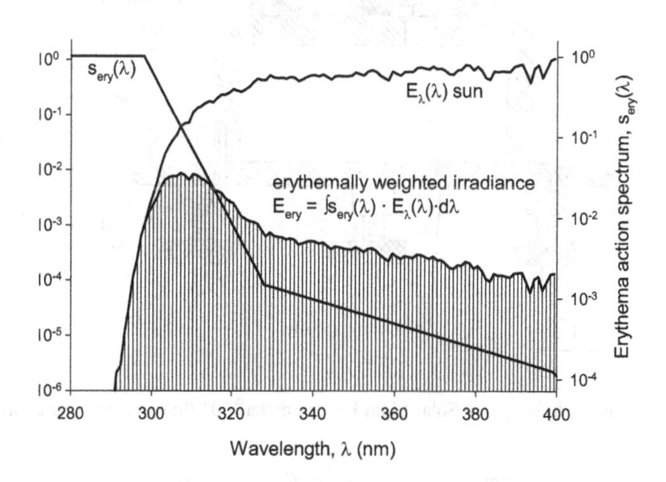

Figure 10. Weighting of an outdoor spectral irradiance with the erythema action spectrum and integrating gives the erythema-effective irradiance (here $E_{ery} = 0.154\ W \cdot m^{-2}$).

Figure 10 shows the example of an erythemally weighted terrestrial solar spectrum.

If only weighted quantities are preferred broadband instruments are a good choice. The requirements are

- a close match of the spectral response of the detector with the desired action spectrum (see below)
- the capability to integrate over the appropriate wavelength range
- a linear response over the dynamical range which has to be considered.

Example: Robertson-Berger meter. An instrument which is frequently used in solar UV measurements is the Robertson-Berger meter (R-B meter)[18]. The basic idea is to convert incoming UV radiation into green light by a phosphor. The green light is then detected by a photodiode. Ambient visible radiation has to be blocked from the photodiode. Therefore, an UG 11 UV filter is located in front of the phosphor (see Figure 11). Unfortunately, this filter has a small transmission for red light. Therefore, a green filter behind the phosphor is used to block the remaining red light. The system is temperature controlled and the entrance optic with cosine response is protected by a quartz dome. The wavelength dependency of the efficiency of the R-B meter quite closely matches the erythema action spectrum. More details on the typical spectral and angular responses of R-B meters can be found in[19,20].

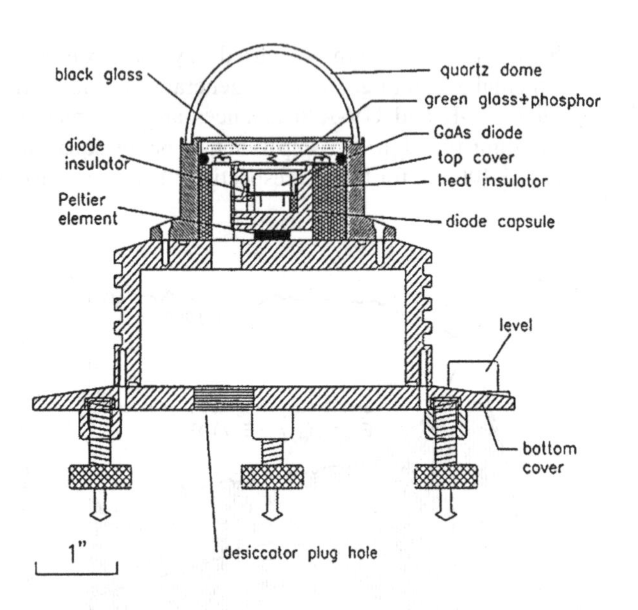

Figure 11. Schematic drawing of a Solar Light UV-Biometer® 501 (http://www.solar.com).

Spectral mismatch. The agreement of the spectral response of the detector and the corresponding action spectrum is of major importance in broadband radiometry. An example illustrates how large errors build up if broad band instruments are employed inadequately. Even if the agreement between the CIE erythemal action spectrum and the actual spectral response of a R-B meter (Figs. 12a and 12b) is quite satisfactory in the UV-B range there is a considerable spectral mismatch at wavelengths below 280 nm. Therefore, the R-B meter reading and the calculated erythema effective irradiance are virtually the same if terrestrial solar radiation is registered (Fig.12a). Measuring the emission of a sun ray lamp with strong lines in the UV-C, however causes detector reading and true erythema effective irradiance to differ by a factor of almost 4. The instrument is certainly not designed for this application. The example however, should draw the attention on the spectral mismatch as a source of error.

Due to their wavelength integrating characteristics broadband meters are not very sensitive to small changes in the spectral composition as those caused by small changes in the stratospheric ozone concentration. Broadband radiometers are however, much easier to handle and less expensive than spectroradiometer systems. Therefore, they are very useful in the long-term monitoring of a variable radiation environment.

Besides photoelectric radiometers UV-dosimeters based on biological or chemical reactions are sometimes employed mainly for personal dosimetry. These devices are reviewed in detail in the article by Horneck in this book[21].

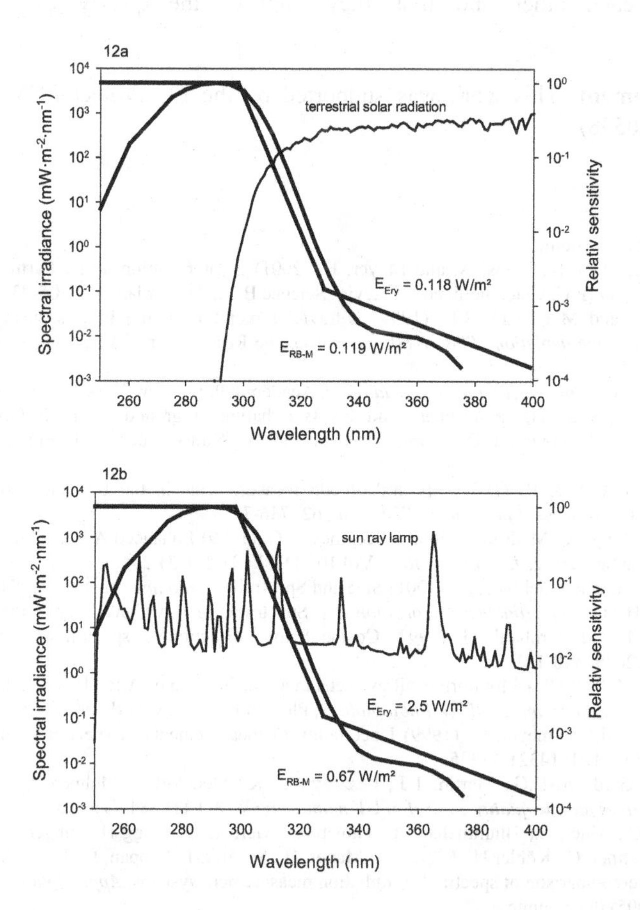

Figure 12. Comparison of the erythema action spectrum according to CIE[16] and the efficiency of a Robertson-Berger meter [manual UV Biometer 501] as an example of a broadband radiometer.
In the case of terrestrial solar radiation (Fig. 12a) the detector reading agrees very well with the true erythema effective irradiance. But for sources with emission in the wavelength range, where the detector efficiency and the action spectrum differ from each other, large deviations between the reading and the true value occur.

The measurement of solar UV radiation is a very exacting task. Due to the steepness of the spectral cut-off at about 300 nm high sensitivity and spectral resolution are required. On the other hand rapidly changing cloud conditions require fast responding instruments. Furthermore, field experiments call for light weight, battery operated devices which are easy to handle. There is until now no "general purpose" instrument which covers all types of applications. Spectroradiometers are necessary to assess the exact position of the UV cut-off and for the determination of biological action spectra. Broadband instruments can be applied much easier in field experiments and in long term monitoring. Ideally both types are employed simultaneously as they complement each other and thus they improve the quality of UV radiation measurements.

Acknowledgement. This work was supported by the EC Project EUVAS (Contract ENV4-CT97-0538)

References

1. Sliney (2005) this volume.
2. Seidlitz, H.K., Thiel, S., Krins, A. and Mayer, H. (2001) Solar radiation at the Earth's surface.In *Sun Protection in Man* (P.U. Giacomoni, ed.), Elsevier Science B.V., Amsterdam, pp. 705-738.
3. Herrman, J.R. and McKenzie, R.L. (1999) Ultraviolet radiation at the Earth's surface. In *Scientific Assessment of ozone depletion:1998*, WMO Global Ozone Research and Monitoring Project- Report 44, Geneva, 9.1-9.56.
4. Iqbal, M. (1983) *An introduction to solar radiation*, Academic Press, New York.
5. Madronich, S. (1993) The atmosphere and UV-B radiation at ground level. In *Environmental UV Photobiology* (A.R. Young, L.O. Björn, J. Moan and W. Nultsch, eds.), Plenum Press New York, pp. 1-39.
6. Feister, U. and Grewe, R. (1995) Spectral albedo measurements in the UV and visible region over different types of surfaces, *Photochem. Photobiol.*, 62: 736-744.
7. Mayer, B., Kylling, A., Madronich, S. and Seckmeyer, G. (1999) Enhanced Absorption of UV Irradiance due to Multiple Scattering, *J. Geophys. Res.*, Vol.103(D23) : 31.241-31.254.
8. Domanchin, J.-L. and Gilchrist, J.R. (2001) Size and Spectrum. *Photonics Spectra*, July 2001: 112-118.
9. Kostkowski, H.J.(1997) *Reliable spectroradiometry,* Spectroradiometry Consulting La Plata, USA 1997
10. Seckmeyer, G. and Bernhard, G. (1993) Cosine Error Correction of spectral UV irradiances. *SPIE Proceedings*, 2049: 140-151.
11. Josephson, W.A.P. (1993) Monitoring ultraviolet radiation, in Young, A.R., Björn, L.O., Moan, J. and Nultsch W. (eds.), *Environmental UV Photobiology*, Plenum Press New York, pp. 73-88
12. Bernhard, G. and Seckmeyer, G. (1999) Uncertainty of measurements of spectral solar UV irradiance, *J. Geophys. Res.*, 104: 14321-14345.
13. Webb, A.R., Gardiner, B.G., Martin, T.J., Leszczynski, K., Metzdorf, J. Mohnen, V.A and Forgan, B. (1998) *Guidelines for site quality control of UV monitoring* WMO/TD 884, Geneva
14. Seckmeyer, G., Thiel, S., Blumthaler, M., Fabian, P., Gerber, S., Gugg-Helminger, A., Häder, D.-P., Huber, M., Kettner, C., Köhler U., Köpke, P., Mayer, H., Schäfer, J., Suppan, P., Tamm, E. and Thomalla, E., (1994) Intercomparison of spectral UV radiation measurement systems. *Appl. Optics*, 33: 7805-7812
15. Ghetti, F. (2005) this volume
16. McKinley, A.F. and. Diffey, B.L. (1978) A reference spectrum for ultraviolet induced erythema in human skin *CIE J.*, 6: 17-22
17. Caldwell, M.M. (1971) Solar ultraviolet radiation and the growth and development of higher plants. In *Photophysiology Vol. 6* (A.C. Giese, ed.), New York, Academic Press, pp. 131-177
18. Berger, D.S (1976) The sunburning ultraviolet meter: Design and performance, *Photochem. Photobiol.*, 24: 587-593
19. Leszczynski, K., Jokela, K., Ylianttila, L., Visuri, R. and Blumthaler, M. (1998) Erythemally weighted radiometers in solar UV monitoring : Results from the WMO/STUK intercomparison, *Photochem. Photobiol.*, 67: 212-221
20. DeLuisi, J., Wendell, J. and Kreiner, F. (1992) An examination of the spectral response characteristics of seven Robertson-Berger meters after long-term field use, *Photochem. Photobiol.*, 56: 115-122.
21. Horneck, G. (2005) this volume

MEDICAL AND ENVIRONMENTAL EFFECTS OF UV RADIATION

B. M. SUTHERLAND
Brookhaven National Laboratory
Upton, NY, 11973, USA

1. Introduction

Organisms living on the earth are exposed to solar radiation, including its ultraviolet (UV) components (for general reviews, the reader is referred to Smith [1] and Young et al. [2]). UV wavelength regions present in sunlight are frequently designated as UVB (290-320 nm) and UVA (320-400 nm). In today's solar spectrum, UVA is the principal UV component, with UVB present at much lower levels. Ozone depletion will increase the levels of UVB reaching the biosphere, but the levels of UVA will not be changed significantly [3]. Because of the high efficiency of UVB in producing damage in biological organisms in the laboratory experiments, it has sometimes been assumed that UVA has little or no adverse biological effects. However, accumulating data [4, 5], including action spectra (efficiency of biological damage as a function of wavelength of radiation; see Section 5) for DNA damage in alfalfa seedlings [6], in human skin [7], and for a variety of plant damages (Caldwell, this volume) indicate that UVA can induce damage in DNA in higher organisms. Thus, understanding the differential effects of UVA and UVB wavebands is essential for estimating the biological consequences of stratospheric ozone depletion.

2. Principles of Light Absorption

Underlying all the biological effects of UV radiation is the absorption of the UV

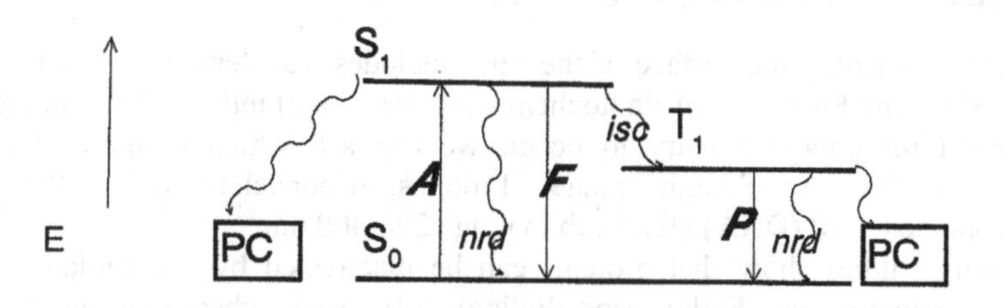

Figure 1. Energy level diagram showing light absorption, radiative and non-radiative emissions and photochemical reactions.

F. Ghetti et al. (eds.), Environmental UV Radiation: Impact on Ecosystems and Human Health and Predictive Models, 39–50.

photon. Figure 1 shows an energy level diagram, with the ground singlet state (S_0), lowest excited singlet state (S_1) and lowest excited triplet state (T_1). In this diagram, the vertical axis corresponds to increasing energy.

In absorption, a photon of at least the same energy spacing as the S_0–S_1 energy gap is absorbed (A) by the biomolecule, raising the molecule to the lowest excited singlet. The energy may be transferred non-radiatively (shown by curved lines; by rotation, translation, etc, without photon emission) to the T_1; this is termed intersystem crossing (isc). Both the S_1 and T_1 may relax by non-radiative emission (nre) to the ground state. The energy may also be emitted promptly (within picoseconds to nanoseconds) from the S_1 by emission of a photon, i.e., producing fluorescence (F). If the molecule has undergone intersystem crossing to the lowest excited triplet, it may emit energy via photon emission, termed phosphorescence (P). Since the latter process involves a change in spin state, it is a "forbidden" transition, and thus may be significantly delayed—from microseconds to as much as seconds, minutes, or even hours—with respect to the original absorption event.

The excited molecule may dissipate the absorbed energy by undergoing photochemistry (PC). Photochemical reactions may occur from either the S_1 or the T_1. In some molecules different reactions may result from photochemistry from specific excited states. The occurrence of photochemistry in an excited molecule competes with but does not preclude energy dissipation by light emission and by non-radiative decay.

This brief summary of photon absorption and energy dissipation points out two important points: first, only a photon with the same energy as the energy difference that matches or exceeds the gap between S_0 and S_1 will be absorbed. (For this simplified discussion, vibrational and rotational states of molecules are ignored.) For a more complete discussion, the reader is referred to classic photochemistry texts such as Calvert and Pitts [8]). This is known as the First Law of Photochemistry, which is sometimes stated as "No photochemical reaction without light absorption." Second, light absorption occurs by quantum events, i.e., by absorption of discreet photons, not of an energy continuum. This is critical in analysis of action spectra, in which we often want to compare an action spectrum (See Section V) with the absorption spectra of candidate target molecules.

3. Potentially Important Target Biomolecules

UV radiation reaching the surface of the earth includes wavelengths in the range 290-400 nm. Since the First Law of Photochemistry (see above) tells us that a photon must be absorbed for photochemistry to occur, we can ask which biomolecules absorb radiation in this wavelength range. Proteins, ribonucleic acid (RNA), and deoxyribonucleic acid (DNA) all absorb in range 290-400 nm.

Many studies show that proteins can be inactivated by UV radiation (for an excellent summary, see Setlow and Pollard [9]). Since absorption must precede photochemistry, amino acids with substantial molar extinction coefficients (ε) for UV absorption would be good candidates for absorbing photons in the ultraviolet range that could lead to protein inactivation by UV radiation. The aromatic amino acids, tryptophan and tyrosine, indeed have εs in the range of 1000-10,000 at ~ 280 nm. However, they have rather low quantum yields (~ 0.005) for photochemical reactions. Nonetheless, proteins with aromatic amino acids (but little cystine, see below) are inactivated by UV and their inactivation action spectrum resembles their absorption spectrum. Cystine is an important target for protein inactivation. Although its molar extinction coefficient is only ~ 100 at 280 nm, its high quantum yield (0.05) and critical role in maintaining protein three-dimensional structure make it the most sensitive residue in producing protein inactivation. Relative to nucleic acids, most proteins are resistant to inactivation by UV. A major source of this resistance is the much lower absorption cross section of proteins, about one-twentieth of that of nucleic acids at ~ 2700 nm. Although some data suggest that Photosystem II of the photosynthetic apparatus is photolabile [10, 11], current data suggest that the effect may not be biologically significant (see Nogues, this volume). RNA is also susceptible to UV damage, forming pyrimidine hydrates as well as pyrimidine photodimers.

Photoproducts are also induced in DNA by UV radiation. Figure 2 shows the induction of a cyclobutane pyrimidine dimers in DNA. UV also induces [6-4] pyrimidine-pyrimidone photoproducts, plus other minority products, principally photoproducts of pyrimidine bases.

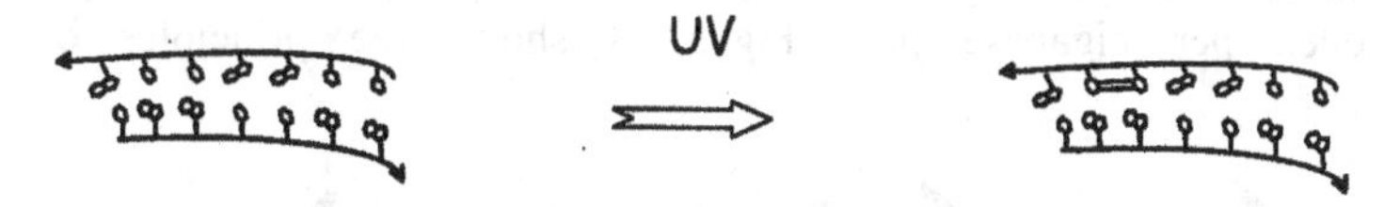

Figure 2. Induction by UV of a cyclobutyl pyrimidine dimer in DNA.

The cyclobutyl pyrimidine dimer (CPD) is the numerically predominant class of UV photolesion in DNA. The 6-4 photoadduct is induced at about 10-30% of the level of CPD, and other photoproducts are formed much lower levels.

Other factors that must be considered are the cellular numerology, and importance of function of the specific molecule. Numerology involves the number of the specific molecules present in a cell. If a thousand of copies of a protein or RNA are present in a cell, and 10% of them are inactivated by UV, the remaining 900 may be quite adequate for cell function. In addition, protein and RNA turnover are normal in cells, and there are well-developed and coordinated paths for metabolism—including salvage paths—of damaged proteins and of RNA. Further, proteins and RNA can be replaced by de novo synthesis, as long as the DNA and cellular machinery for protein synthesis are intact.

4. Measuring DNA Damage

Many approaches have been developed for measuring DNA damage, including those based on chromatography, immunology, and various properties of undamaged and damaged DNA. In the latter category is gel electrophoresis/electronic imaging/number average length analysis. It is a powerful approach of DNA lesion quantitation that stems from physical chemical methods for characterization of polymers. It provides absolute measurements of lesion frequencies (lesions per kilobase or per megabase or even per gigabase). With appropriate experimental modifications, it can be used to quantify lesion frequencies over some six orders of magnitude. It does not require that the lesions have any specific distribution on the DNA, and, in specific, does not require a random distribution of lesions.

Number average length analysis can be used to measure lesions affecting one DNA strand, in which case the DNA is dispersed according to single strand molecular length using denaturing conditions. For assessment of lesions affecting one strand, in which DNAs are dispersed on denaturing gels, it is essential to use fully single-stranded molecules as molecular length standards to establish DNA dispersion curves. Partially denatured molecules can form branched structures, whose electrophoretic mobility may not be a direct function of the DNA molecular length. In the case of lesion determination on denaturing gels, the lesion frequencies are computed in units of damage are lesions per base, or per kilobase (kb, 10^3 bases), or per megabase (Mb, 10^6 bases), or per gigabase (Gb, 10^9 bases), etc. For determinations on non-denaturing gels, the damage frequencies are computed in terms of kilobase pairs (kbp), megabase pairs (Mbp) or gigabase pairs (Gbp).

This method can also measure damages affecting both DNA strands; in this case, the double strand molecular length standard DNAs are electrophoresed on the same gel as the experimental samples. Again, no specific distribution of damages is required. In this case, the damage frequency is given as damages per base pair, per kilobase pair, or even per gigabase pair. Figure 3 shows the principles of damage

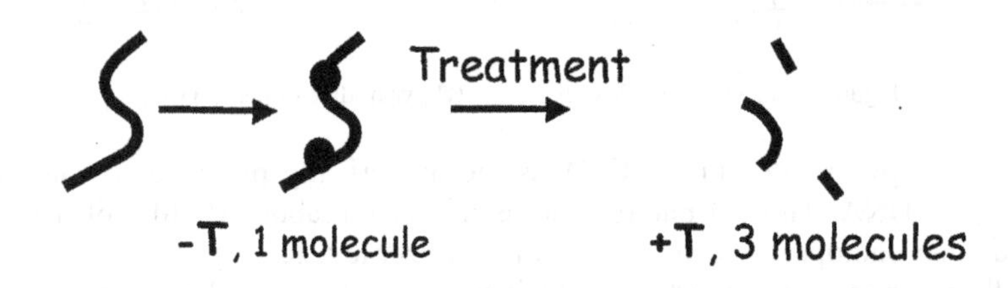

Figure 3. Induction of lesions in a DNA molecule and treatment with a lesion-specific agent.

measurement by these methods. A DNA molecule is damaged by an agent to produce two lesions shown as solid circles. The population of molecules is divided into two, and one portion is treated with an agent that makes a DNA strand break at the site of each lesion. Some examples of such lesion-specific agents are the T4 Endonuclease V, which—among UV-induced lesions—specifically recognizes cyclobutyl pyrimidine dimers [12]. (More recently it has been shown that T4 Endo V also recognizes FapyAdenines, oxidized bases induced by ionizing radiation and at low frequencies by UV [13].) The other portion is incubated without the lesion-specific treatment. We note that the treated population now consists of three molecules, while there is only one molecule in the untreated population. Simple arithmetic shows that the number of lesions (L, two) can be calculated by subtracting the number of molecules in the untreated population (N_u, one) from the number in the treated population (N_t, three).

$$L = N_t - N_u \tag{1}$$

Generally, however, we want to know the Frequency (F) of damages, i.e., lesions per base or per thousand bases or per million bases. This is readily computed by dividing each term by the number of bases (b) in the treated and untreated samples, i.e.

$$F = L/b = N_t/b_t - N_u/b_u \quad (2)$$

Since each term is divided by the quantity of DNA in that sample, the result does not depend on the mass of DNA in each sample, i.e., the procedure is "self-normalizing." Thus the problem of measuring damage levels becomes one of counting the number of molecules in each DNA sample before and after treatment with the lesion-specific agent. One method for "counting" DNA molecules is to separate them according to molecular size by agarose gel electrophoresis, and then to measure the quantity of DNA molecules of each size, computed from a DNA dispersion curve established by molecular length standard DNAs electrophoresed on the same agarose gel [14]. An electronic image is obtained of an electrophoretic gel containing sample DNAs stained with ethidium bromide. A DNA dispersion function is derived from the molecular length standards in the gel. The size distribution of molecules (or number average molecular length, which is the ordinary average of the size of each molecule multiplied by the number of molecules in each size class) in each sample is calculated from the relation

$$L_n = \frac{\int f(x)dx}{\int \frac{f(x)}{L(x)}dx}, \quad (3)$$

where $L(x)$ is the length of the DNA molecules that migrated to position x, and $f(x)dx$ is the intensity of ethidium fluorescence from DNA molecules at that position.

From the L_ns of the treated and untreated populations, the frequency of lesions, ϕ_D, is computed from

$$\phi_D = 1/L_n\ (+\ treatment) - 1/L_n\ (-\ treatment), \quad (4)$$

where $1/L_n$ *(+ treatment)*, and $1/L_n$ *(- treatment)*, are the reciprocals of the L_ns of samples that were treated or not treated, respectively.

Since this method measures the lesion frequency per molecule, the sensitivity of lesion frequency measurement can be increased substantially by increasing the size of the molecules examined. In practical terms this means that improved methods for isolating larger and larger DNA molecules are required. For damages affecting both strands, this means that the level of double strand breaks induced in DNA isolation must be minimized. However, for lesions affecting only one DNA strand, this requires that the level of single strand breaks induced during isolation must be minimized, a much more rigorous requirement. Isolation of DNA by methods that avoid chemical extraction, including enzymatic digestion while the DNA is stabilized mechanically within an agarose "plug" or "button," provides a successful approach to meeting such requirements [15-17].

With high sensitivity methods of detection (quantitative detection by charge coupled device imaging [18-20] of fluorescence from a DNA-bound molecule such as ethidium bromide, whose fluorescence is directly proportional to the mass of DNA over several orders of magnitude, very small quantities of DNA are required for a measurement, and lesions can be measured at very low frequencies. For a homogeneous DNA, i.e., a molecular species of one size and conformation, less than one ng DNA can be used. For damage determination in higher organisms, which contain heterodisperse DNAs, about 50 ng of DNA are used for routine assays [20, 21]. The sensitivity of number average length determination depends on two factors: the size of the DNA molecules and the minimum fraction of molecules that must be broken to be detected in the specific combination of biochemical, electrophoretic, imaging system and computer analysis that is used. For double-strand DNA populations in the size range of ~1 Mbp, damages can be measured down to ~5/Gpb [22].

A wide variety of damages can be quantitated by this approach. Lesions affecting one DNA strand are measured by treatment with a lesion-specific cleaving agent and dispersion under conditions that separate the DNA strands ("denaturing" conditions), and the resulting frequencies are computed as lesions/base. Damages affecting both strands are measured by treatment with the agent as above, but with dispersion under conditions that do not separate the DNA strands, i.e., "non-denaturing" conditions. Strand breaks, whether single strand or double strand breaks, are measured directly, since they in themselves comprise breaks of the phosphodiester backbone.

In addition, lesion–specific or lesion class–specific enzymes that cleave the phosphodiester backbone at each lesion site can be used to convert each lesion to a strand break. Examples of DNA lesions and the cleavage agents that recognize them include: cyclobutyl pyrimidine dimers, T4 endonuclease V [12] or *Micrococcus luteus* UV endonuclease [23, 24]; oxidized purines *Escherichia coli* Fpg protein [25]; oxidized pyrimidines, *Escherichia coli* Endonuclease III [26]. Many of these enzymes have multiple substrates, which must be taken into account. For example, the glycosylase activity of T4 endonuclease V removes the dimer from DNA, and its lyase activity attacks the resulting abasic site(s) and cleaves the DNA [27]. By its lyase activity, this enzyme cleaves DNA at 'regular' abasic sites, i.e., a base-less sugar moiety resulting from cleavage of the N-glycosyl bond to the base [27]. It also recognizes FapyAdenine residues induced by ionizing radiation and in low yields by UV [13].

5. Action Spectroscopy: Powerful Approach For Understanding UV Effects

Action spectra relate the effectiveness of specific wavelengths in producing an effect.

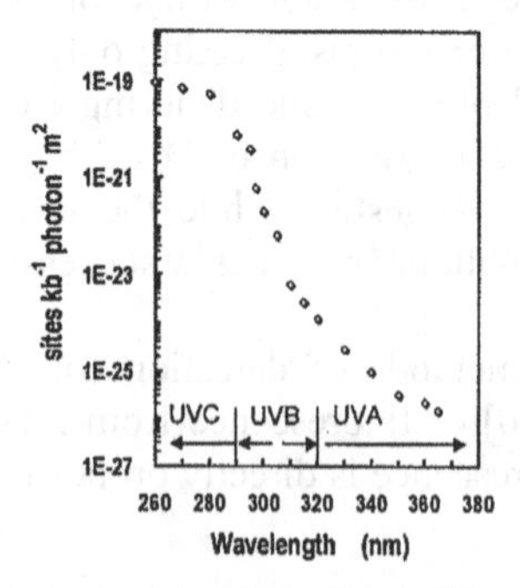

Figure 4. Action spectrum for induction of cyclobutyl pyrimidine dimers in T7 DNA in solution.

For analytical action spectra designed to elucidate the absorbing moiety (chromophore) or mechanism of reaction, Jagger has an excellent description of the criteria that must be met for obtaining valid data [28]. These are:

- The quantum yield must be the same at all wavelengths.
- The absorption spectrum of the chromophore must be the same in vivo as in vitro.
- Intracellular screening, such as absorption or scattering within a cell, must be either negligible or constant at all wavelengths.
- The incident radiation is not entirely absorbed by the sample at any wavelength of interest.
- Reciprocity of time and UV exposure holds under the conditions of the determinations.

For DNA in dilute solution and short path length, it is easy to fulfill these criteria. The data in Figure 4 show an action spectrum for induction of cyclobutyl pyrimidine dimers in purified DNA in dilute solution [29]. The action spectrum in the UVC (220-290 nm) and UVB (290-320 nm) resembles the absorption spectrum of DNA. The high sensitivity of the gel electrophoresis/number average molecular length analysis method, which can range over some 5 or 6 orders of magnitude, allows the quantitation of very low frequencies of dimers induced by UVA (320-400 nm).

Measurement of analytical action spectra in biological systems, especially in higher organisms, is fraught with difficulties. Consideration of the criteria of Jagger [28] outlined above indicates that many are difficult to fulfill in complex systems. Some of these apparent "complications" can be useful in providing critical information

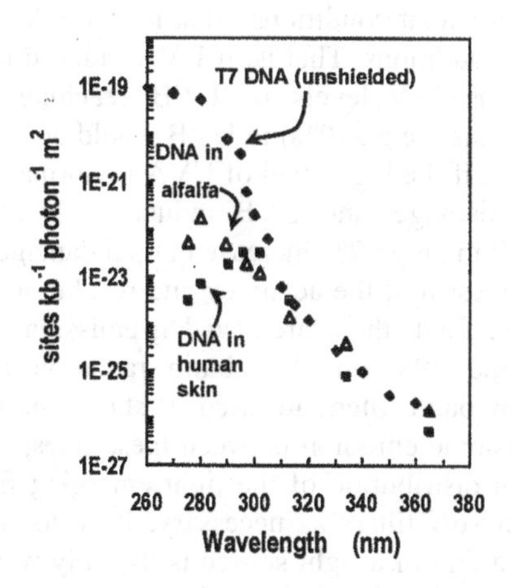

Figure 5. Action spectra for pyrimidine dimer induction in T7 dna in solution, in dna in alfalfa seedlings, and in human skin in situ.

on the biological system. For example, the action spectrum in a complex organism for a DNA damage that results from direct photon absorption in DNA provides information on the internal absorption/shielding/scattering that protects DNA from damage by incident photons. Figure 5 shows action spectra for formation of cyclobutyl pyrimidine dimers in DNA in alfalfa seedlings [6] and in human skin, respectively [7], exposed to narrow band UV radiation. These data are compared with those for T7 DNA in solution (diamonds) [29]. These spectra clearly show that in the shorter UV wavelength region, much lower damage levels are formed in both alfalfa and in human skin than in isolated DNA. Again, the high sensitivity of the number average length analysis method allows the quantitation of dimer induction at long wavelengths. At longer wavelengths, the spectra are rather similar to that of isolated DNA, indicating a measurable but very low level of damage induced by UVA radiation.

The new action spectra for UV- induced damage to plants obtained by Caldwell and his associates (this volume) resemble the alfalfa action spectrum of Quaite et al. [6] (it should be noted that Caldwell's previous 'generalized plant action spectrum' [30-32] differed strikingly from the alfalfa action spectrum, principally because no biological damage data from wavelengths longer than 313 nm was considered in construction of the `generalized plant action spectrum.').

6. Biological Effects of Long Wavelength Environmental UVA

The action spectra for DNA damage induction discussed in Section 5 clearly show that UVA radiation can induce photoproducts in DNA. Most studies evaluating the effect of ozone depletion have centered on UVB, since conditions of ozone depletion increase the number of photons in this wavelength region reaching the surface of the earth. The predominant source of UV in solar radiation is UVA, and the levels of UVA will not be altered significantly by ozone depletion. However, evaluating the biological effect of increased UV requires that we know the damage induced by UVA under normal (in absence of ozone depletion) solar conditions. That is, if UVA under normal (in absence of ozone depletion) solar conditions. That is, if UVA induced no damage, and all DNA damage resulted from the low levels of UVB reaching the biosphere, then a biologically-weighted increase (e.g., 10%) in UVB would be expected to increase DNA damage by 10%. However, if the high level of UVA in normal solar conditions induces 80% of the total DNA damage, and UVB induces only 20%, an additional 10% UVB damage would result in only ~2% increase in total damage.

Current UVA sources limit the accuracy and resolution of action spectra in this critical wavelength range. First, there are few Hg emission lines in this wavelength range: 334, 365, 391 and 398 nm. To obtain radiation from a specific line, a monochromator or narrow band filters are used. It should be noted that Hg-Xe lamps and even Hg lamps have some emission between these lines, and thus it is essential to determine the wavelength distribution of the light emerging from the monochromator and to use appropriate cut-off filters as necessary. It is not valid to assume that the wavelength specified on a dial of a light source is the only wavelength (or wavelength range) emitted by the source, that the wavelength calibration is accurate or that there are no contaminating wavelengths in the beam.

Some UVA wavelengths are available from lasers, but one must be careful that the high peak power does not produce two-photon effects that can result in different photochemical reactions from the one-photon mediated effects from lower intensity sources. Further, some lasers have high brightness (intensity per unit wavelength per unit timer per unit source area per unit solid angle), but do not provide sufficient total photon flux to irradiate a sufficient area for most biological samples. The paucity of lines (or wavelengths from lasers) in this region thus severely limits the resolution of the spectra that can be obtained.

One possible solution is to use a Hg-Xe lamp, which provides a continuum of radiation between the Hg lines. However, in attempting to obtain higher intensity, the slits of the monochromator may be opened wider, thus increasing the range of wavelengths in the incident beam. The pitfalls in using wider bandwidths to measure action spectra in which the slope of the spectrum is changing rapidly in the wavelength region of interest are discussed by Sliney in this volume. It is critical that scattered light from other wavelength ranges be rigorously excluded. If they are not, the shorter wavelength UV may dominate the observed biological effects, which would then be erroneously perceived as resulting from the UVA.

An additional problem is the lack of intensity in the UVA region available from most sources. Since the cross-section for most biological effects in this wavelength range is small, this results in long exposures (many minutes or even hours) for production of observable damage in many biological systems. In isolated molecules, e.g. DNA in solution, as long as nucleases that could degrade the DNA are rigorously excluded by the use of sterile buffers, the presence of EDTA, etc., long exposures are an inconvenience, but do not affect the scientific result.

However, results from biological systems capable of repair may be significantly affected by extended irradiation: simultaneous repair may reduce damages induced by the UVA radiation so that the actual damage level is underestimated. Such repair could be carried out by light-independent repair processes (nucleotide excision repair could be carried out by light-independent repair processes (nucleotide excision repair or base excision repair). Moreover, cells capable of photoreactivation present even more complex responses. Photoreactivation is a one-enzyme repair path, in which a single enzyme, photolyase, binds to a cyclobutyl pyrimidine dimer in DNA; upon absorption by the photolyase-dimer complex of a photon in the visible or UVA range, the dimer is monomerized and the photolyase is liberated to seek another dimer.

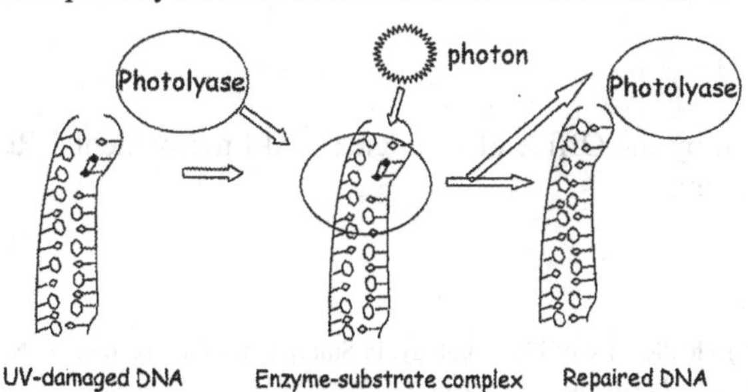

Figure 6. Photolyase repairs cyclobutyl pyrimidine dimers in DNA.

The first step—association of enzyme and substrate—is temperature and concentration dependent [33]. However, the second step—dimer photolysis—is a photochemical reaction, and virtually independent of temperature [34]. As a light-driven, enzymatic reaction, the rate of photorepair depends not only on the number of photolyases per cell but also on the wavelength of the light. Further, this enzyme has two substrates: the dimer and the photolyzing photon. Simple biochemical considerations indicate that the rate of photorepair depends on the substrate concentration, with the maximal rate at a "saturating" concentration of the substrate, i.e., the concentration such that further increases do not increase the reaction rate. Since photolyase has two substrates, the dimer and the photon, the rate of photorepair depends not only on the dimer concentration (frequency) in the cell but also on the wavelengths and intensity of the photoreactivating light.

Thus we can summarize the factors affecting the net effect of UVA as: Net effect of UVA = UVA-induced damage – UVA-mediated repair - Light independent repair. The UVA-induced damage depends on the wavelengths and intensities at each wavelength and the cross-section of the biological system for damage at each wavelength. The UVA-mediated repair term depends on the number of photolyases in a cell, the temperature, the photorepair action spectrum for that photolyase, the wavelength and intensity of the light, as well as the level of dimers per cell. The light-independent repair is dependent on the genotype of the cell, the dimer frequency, the presence of other damages that are also substrate for the repair system, the level of these enzymes in the cell and on the substrates required for that repair, e.g. ATP, used as an energy source.

Thus, obtaining robust data on the effects of UVA is challenging. Even with careful attention to biological, physical and biochemical factors, current light sources are far from ideal, being limited in intensity and in wavelength. An ideal source for UVA radiation would provide a continuous selection of wavelength, high intensity over a substantial area, and the time structure of the radiation should avoid high peak power/low inter-peak modes that produce multi-photon photochemistry. Free electron lasers in the UV range are being developed at several institutions; at least two of these should provide high time-average power in the UVA range [35]. They may offer powerful new next-generation sources for UVA photochemistry and photobiology. The elucidation of the actual biological role of UVA awaits the development of capability of irradiation of biological systems at such facilities.

Acknowledgments

Research supported by the Office of Biological and Environmental Research of the US Department of Energy.

References

1. Smith, K.C., (1989) The Science of Photobiology, in Smith, K.C. (ed.), second ed, New York, Plenum.
2. Young, A.R., Bjorn, L.O., Moan, J., and Nultsch, W. (1993) *Environmental UV Photobiology*, Plenum, New York.
3. Green, A.E.S., Cross, K.R., and Smith, L.A. (1980) Improved analytical characterization of ultraviolet skylight, *Photochem. Photobiol.* **31,** 59-65.
4. Tyrrell, R.M. (1973) Induction of pyrimidine dimers in bacterial DNA by 365 nm radiation, *Photochem Photobiol.* **17,** 69-73.
5. Webb, R.B., and Lorenz, J.R. (1970) Oxygen dependence and repair of lethal effects of near ultraviolet and visible light, *Photochem Photobiol.* **12,** 283-9.
6. Quaite, F.E., Sutherland, B.M., and Sutherland, J.C. (1992) Action spectrum for DNA damage in alfalfa lowers predicted impact of ozone depletion, *Nature* **358**, 576-578.

7. Freeman, S.E., Hacham, H., Gange, R.W., Maytum, D., Sutherland, J.C., and Sutherland, B.M. (1989) Wavelength dependence of pyrimidine dimer formation in DNA of human skin irradiated in situ, *Proc. Natl. Acad. Sci. U.S.A.* **86,** 5605-5609.
8. Calvert, J.G., and Pitts Jr., J.N. (1967) *Photochemistry*, Wiley, New York.
9. Setlow, R.B., and Pollard, E.C. (1962) *Molecular Biophysics*, Addison-Wesley, Reading, MA.
10. Greenberg, B.M., Gaba, V., Canaani, O., Malkin, S., A.K., M., and Edelman, M. (1989) Separate photosensitizers mediate degradation of the 32-kDaphotosystem II reaction center protein in the visible and UV spectral regions, *Proc. Natl. Acad. Sci. U.S.A.* **86,** 6617-20.
11. Friso, G., Barbato, R., Giacometti, G.M., and Barber, J. (1994) Degradation of D2 protein due to UV-B irradiation of the reaction centre of photosystem II, *FEBS Lett.* **339,** 217-21.
12. Higgins, K.M., and Lloyd, R.S. (1987) Purification of the T4 Endonuclease V, *Mutat Res* **183,** 117-21.
13. Dizdaroglu, M., Zastawny, T.H., Carmical, J.R., and Lloyd, R.S. (1996) A novel DNA N-glycosylase activity of E. coli T4 endonuclease V that excises 4,6-diamino-5-formamidopyrimidine from DNA, a UV-radiation and hydroxyl radical-induced product of adenine, *Mutat. Res.* **362,** 1-8.
14. Freeman, S.E., Blackett, A.D., Monteleone, D.C., Setlow, R.B., Sutherland, B.M., and Sutherland, J.C. (1986) Quantitation of radiation-, chemical-, or enzyme-induced single strand breaks innonradioactive DNA by alkaline gel electrophoresis: application to pyrimidine dimers, *Anal. Biochem.* **158,** 119-129.
15. Smith, C.L., and Cantor, C.R. (1987) Purification, Specific Fragmentation, and Separation of Large DNA Molecules, in R. Wu (ed.), *Methods in Enzymology*, Academic, New York, pp. 449-467.
16. Quaite, F.E., Sutherland, J.C., and Sutherland, B.M. (1994) Isolation of high-molecular-weight plant DNA for DNA damage quantitation: relative effects of solar 297 nm UVB and 365 nm radiation, *Plant Molecular Biology* **24,** 475-483.
17. Bennett, P.V., and Sutherland, B.M. (1996) Isolation of High-Molecular-Length DNA from Human Skin, *BioTechniques* **21,** 458-463.
18. Sutherland, J.C., Lin, B., Monteleone, D.C., Mugavero, J., Sutherland, B.M., and Trunk, J. (1987) Electronic imaging system for direct and rapid quantitation of fluorescence from electrophoretic gels: application to ethidium bromide-stained DNA, *Anal. Biochem.* **163,** 446-457.
19. Sutherland, J.C., Sutherland, B.M., Emrick, A., Monteleone, D.C., Ribeiro, E.A., Trunk, J., Son, M., Serwer, P., Poddar, S.K., and Maniloff, J. (1991) Quantitative electronic imaging of gel fluorescence with charged coupled device cameras: applications in molecular biology, *BioTechniques* **10,** 492-497.
20. Sutherland, J.C., Monteleone, D.C., Trunk, J.G., Bennett, P.V., and Sutherland, B.M. (2001) Quantifying DNA Damage by Gel Electrophoresis, Electronic Imaging and Number Average Length Analysis, *Electrophoresis* **22,** 843-854.
21. Sutherland, B.M., Bennett, P.V., and Sutherland, J.C. (1999) Quantitation of DNA Lesions by Alkaline Agarose Gel Electrophoresis, in D. Henderson (ed.), *Methods in Molecular Biology, DNA Repair Protocols*, Humana Press, Totowa NJ, pp. 183-202.
22. Sutherland, B.M., Bennett, P.V., and Sutherland, J.C. (1996) Double Strand Breaks Induced by Low Doses of Gamma Rays or Heavy Ions: Quantitation in Non-radioactive Human DNA, *Anal. Biochem.* **239,** 53-60.
23. Carrier, W.L., and Setlow, R.B. (1970) Endonuclease from *Micrococcus luteus* which has activity toward ultraviolet-irradiated deoxyribonucleic acid: purification and properties, *Journal of Bacteriology* **102,** 178-186.
24. Shiota, S., and Nakayama, H. (1997) UV endonuclease of Micrococcus luteus, a cyclobutane pyrimidine dimer-DNA glycosylase/abasic lyase: cloning and characterization of the gene, *Proc. Natl. Acad. Sci. U.S.A.* **21,** 593-8.
25. Boiteux, S., O'Conner, T.R., Lederer, F., Gouvette, A., and Laval, J. (1990) Homogeneous *Escherichia coli* FPG Protein. A DNA glycosylase which excises imidazole ring-opened purines and nicks DNA at apurinic/apyrimidinic sites, *J. Biol. Chem.* **265,** 3916-3922.
26. Asahara, H., Wistort, P.M., Bank, J.F., Bakerian, R.H., and Cunningham, R.P. (1989) Purification and Characterization of *Escherichia coli* Endonuclease III from the cloned *nth* gene, *Biochemistry* **28,** 4444-4449.
27. Haseltine, W.A., Gordon, L.K., Lindan, C.P., Grafstrom, R.H., Shaper, N.L., and Grossman, L. (1980) Cleavage of pyrimidine dimers in specific DNA sequences by a pyrimidine dimer DNA-glycosylase of *M. luteus, Nature* **285,** 634-641.
28. Jagger, J. (1967) *Introduction to Research in Ultraviolet Photobiology*, Edited by A. Hollaender Biological Techniques Series, Prentice-Hall, Englewood Cliffs, NJ.
29. Emrick, A., and Sutherland, J.C. (1989) Action spectrum for pyrimidine dimer formation in T7 bacteriophage DNA, *Photochem. Photobiol.* **49,** 35S.

30. Caldwell, M.M. (1971) Solar UV irradiation and the growth and development of higher plants, in A. C. Giese (eds.), *Photophysiology*, Academic Press, New York, pp. 131-177.
31. Caldwell, M.M. (1968) Solar Ultraviolet Radiation as an Ecological Factor for Alpine Plants, *Ecol. Monogr.* **38,** 243-268.
32. Caldwell, M.M., L.B., C., Warner, C.W., and Flint, S.D. (1986) Action Spectra and Their Key Role in Assessing Biological Consequences of Solar UV-B Radiation Change, in R. C. Worrest and M. M. Caldwell (eds.), *Stratospheric Ozone Reduction, Solar Ultraviolet Radiation and Plant Life*, Springer-Verlag, Berlin, pp. 87-111.
33. Rupert, C.S., Harm, W., and Harm, H. (1972) Photoenzymatic repair of DNA. II. Physical-chemical characterization of the process, *Johns Hopkins Med J Suppl.* **1,** 64-78.
34. Harm, H. (1969) Analysis of photoenzymatic repair of UV lesions in DNA by single light flashes. 3. Comparison of the repair effects at various temperatures between plus 37 degrees and minus 196 degrees, *Mutat. Res.* **7,** 261-71.
35. Sutherland, J.C. (2000) Can Free Electron Lasers Answer Critical Questions in UltravioletPhotobiology?, *Proceedings of the Society of Photo-Instrumentation Engineers* **3925,** 50-59.

QUANTIFICATION OF BIOLOGICAL EFFECTIVENESS OF UV RADIATION

G. HORNECK, P. RETTBERG, R. FACIUS AND K. SCHERER
German Aerospace Center, Institute of Aerospace Medicine, Radiation Biology, D-51170 Cologne, Germany

1. Introduction

The problem of a progressive depletion of stratospheric ozone, particular in both polar regions, and now identified globally outside the tropics during all seasons of the year[1,2], points at a potentially threat to the biosphere of an increase in solar ultraviolet-B-radiation (UV-B: 280-315 nm) reaching the Earth's surface. There is serious concern on the consequences of an increased UV-B radiation on human health[3], on ecosystem balance[4] and on crop productivity[5,6]. To determine the implications of increased levels of solar UV-B radiation for critical processes of our biosphere in quantitative terms, an instrumentation for UV-measurement is required that weights the spectral irradiance according to the biological responses under consideration.

This chapter gives a brief description of the different approaches available to quantify the biological effectiveness of solar radiation followed by a discussion on the biological dosimeters available and the criteria for their applicability. Then, examples are given for the use of biological UV dosimeters in field and personal monitoring.

2. Approaches to quantify the biological effectiveness of environmental UV

Any electromagnetic radiation can be quantified using radiometric units. These units describe the energy or power that is either emitted from a source or that arrives at a surface. The recommended SI units concerning target incidence units are W/m^2 for irradiance, i.e. flux per unit area incident on a small plane surface, and J/m^2 for the exposure or dose, i.e. energy per unit area incident on one side of a small plane surface. To determine the impact of environmental UV-radiation on critical processes of our biosphere or on human health requires weighting of the spectral irradiance according to the biological spectral responses under consideration, such as erythema, skin cancer, suppression of immune functions, virus induction, ocular damage, reduced productivity of agricultural crops, effects on ecosystem balance and/or biodiversity[7].

Biological effectiveness spectra of environmental UV radiation

The spectrum of extraterrestrial solar UV radiation has been measured during several space missions[8,9] (Figure 1). It contributes to the whole solar electromagnetic

F. Ghetti et al. (eds.), Environmental UV Radiation: Impact on Ecosystems and Human Health and Predictive Models, 51–69.

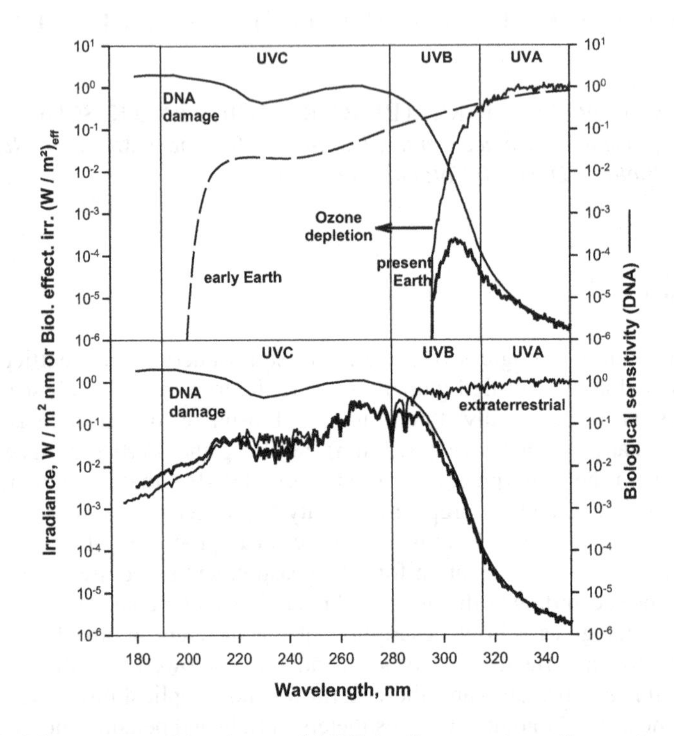

Figure 1. Solar extraterrestrial and terrestrial UV irradiance spectra, action spectrum for DNA damage and biological effectiveness spectra for extraterrestrial and terrestrial conditions.

spectrum by 8.3 %, with 0.5% UV-C (200-280 nm), 1.5% UV-B, and 6.3% UV-A (315-400 nm), respectively[10]. Although the UV-C and UV-B regions make up only 2% of the entire solar irradiance prior to attenuation by the atmosphere, they are mainly responsible for the high lethality of extraterrestrial solar radiation to living organisms. This was demonstrated in spores of *Bacillus subtilis*, that were killed effectively within a few seconds by extraterrestrial sunlight (>190 nm)[11], whereas at the Earth's surface about thousand times longer exposure times were required to reach the same effect (Figure 2). The reason for this high lethality of extraterrestrial solar UV radiation - compared to conditions on Earth - lies in the absorption characteristics of the DNA which is the decisive target for inactivation and mutation induction at the UV range. The UV action spectrum for inactivation of *B. subtilis* spores spans over 7 orders of magnitude and declines exponentially with increasing wavelength at wavelengths longer than the peak wavelength around 260 nm[13,14]. Its spectral profile is similar to that averaged for affecting DNA (Figure 1).

On its way through the atmosphere, solar radiation is modified by scattering and absorption processes[15]. During the first 2.5 billion years of Earth's history, UV-radiation of wavelengths >200 nm reached the surface of the Earth due to lack of an effective ozone shield[16]. Following the rapid oxidation of the Earth's atmosphere about 2.1 billion years ago, an UV-absorbing ozone layer was built up in the stratosphere. Today, the

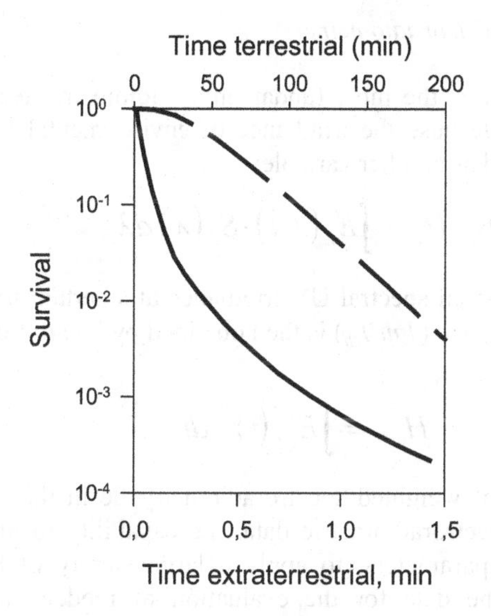

Figure 2. Survival of spores of *Bacillus subtilis* after exposure to extraterrestrial solar radiation (>190 nm) during the Spacelab 1 mission (solid line) or to global UV radiation at the surface of the Earth around noon in San Francisco area (37.5°N 122.2°W) on 10.8. 1985) (dashed line) (data modified from ref. 12).

stratospheric ozone layer effectively absorbs UV radiation at wavelengths shorter than 290 nm[17]. However, a decrease in total column ozone would shift the edge of the solar spectrum on Earth towards shorter wavelengths (Figure 1).

In order to determine the biological effectiveness of environmental UV radiation, spectral data are multiplied with an action spectrum of a relevant photobiological reaction[18]. The result is a biological effectiveness spectrum (Figure 1). It weights the environmental UV radiation according to the biological effectiveness at the different wavelengths. The biologically effective irradiance E_{eff} is then determined as follows:

$$E_{eff} = \int E_{\lambda}(\lambda) \cdot S_{\lambda}(\lambda)\, d\lambda \qquad (1)$$

with $E_{\lambda}(\lambda)$ = the solar spectral irradiance and $S_{\lambda}(\lambda)$ = the spectral sensitivity or action spectrum for a critical biological effect. E_{eff} is given in $(W/m^2)_{eff}$.

There are in general three different approaches to quantify a biologically effective solar irradiance. These are: (i) weighted spectroradiometry where the biologically weighted radiometric quantities are derived from physical solar spectral data by multiplication with an action spectrum of a relevant photobiological process according to equation (1); (ii) wavelength integrating chemical-based or physical broadband radiometers with spectral sensitivities similar to a biological response function; and (iii) biological dosimeters that directly weight the incident UV components of sunlight in relation to their biological effectiveness and to interactions between them. These different approaches of biologically weighted UV dosimetry are discussed in the following.

Biologically weighted spectroradiometry

Spectroradiometry is the most fundamental radiometric technique (see Seidlitz and Krins, this issue). Because the irradiance of environmental UV varies with time, time has to be considered as another variable:

$$E_{eff}(t) = \int E_{\lambda}(\lambda,t) \cdot S_{\lambda}(\lambda)\, d\lambda \qquad (2)$$

with $E_{\lambda}(\lambda,t)$ = the terrestrial spectral UV irradiance at a certain time. The biologically effective dose (BED) H_{eff} (in $(J/m^2)_{eff}$) is then obtained by integration over the time span of interest:

$$H_{eff} = \int E_{eff}(t) \cdot dt \qquad (3)$$

The advantages of weighted spectroradiometry lie in the principally attainable high accuracy of the spectroradiometric data, the capability to identify influences by various meteorological parameters, to apply a large variety of biological weighting functions and to use the data for the evaluation of model calculations and trend assessments. High demands are made on the instrument specifications, such as high accuracy - especially at the edge of the solar spectrum in the UV-B range -, high stray light suppression, high reproducibility and temperature stability. Frequent calibration with standard lamps and by use of field intercomparison campaigns with other spectroradiometers are indispensable[5,19,20]. On the other hand, because of the high demands on instrument specification and the relatively large investment costs, spectroradiometers are only installed at selected sites (e.g., UV-networks). To interpolate the discrete measurements made in typically 4 to 15 min intervals under unstable weather conditions, more or less correct assumptions have to be made using the information from additional continuously monitoring broadband radiometers. Yet, given even the utmost accuracy of physical spectral irradiances, due to experimental errors in the biological action spectra, the resulting calculated E_{eff} values may suffer from substantial uncertainties.

Errors may occur from inappropriate usage of action spectra, such as action spectra measured over too limited wavelength ranges. Although ozone depletion affects only the UV-B edge of sunlight, depending on the tail of the action spectrum, the responses to UV-A and even to visible light might be important to be included. Another source of errors lies in the measuring errors of the action spectra, which may arise especially in their long wavelength tails where the biological sensitivity decreases by orders of magnitude. Furthermore and most importantly, the method assumes simple additivity of the various wavelength components by incorporating a multiplicative constant appropriate for each wavelength. This is based on the fact that most action spectra have been developed with monochromatic radiation. If interactions occur - and they have been reported for various biological effects (reviewed in ref. 21) - simple additivity might be an insufficient basis for biologically weighted dosimetry and provides not a suitable indicator for the biological responses to the solar radiation reaching the surface of the Earth.

Action spectra as biological weighting functions

Action spectra play a pivotal role in characterizing biological effectiveness of environmental UV radiation[18,22]. They are most simply defined as the measurement of a biological effect as a function of wavelength. If cell killing is taken as biological endpoint, dose effects curves are obtained for different narrow band wavelength ranges and the biological sensitivity is derived either from the inactivation rate constant *k* or from the reciprocal value of the fluence, producing a given effect, e.g., $1/F_{10}$ for a dose producing a survival rate of 10^{-1} (Figure 3). The latter procedure is mostly used in case of non-linear dose effect curves (Figure 3).

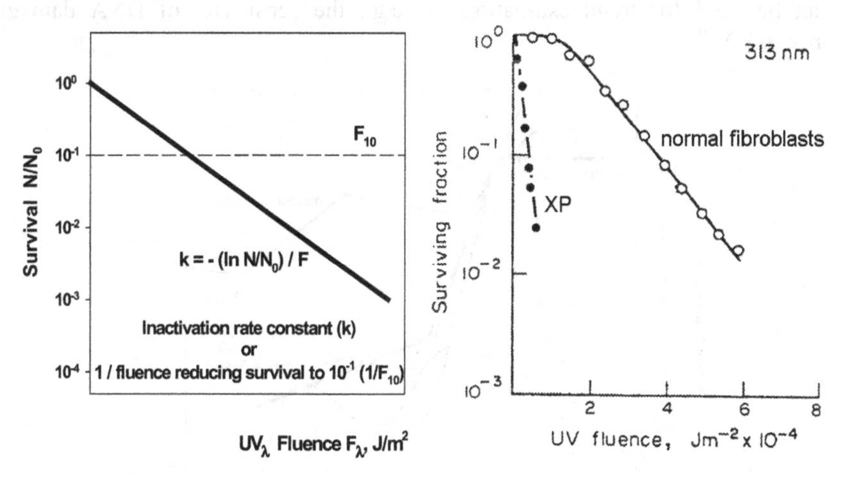

Figure 3. Typical exponential dose effect curve of cell inactivation (A) and experimental survival curves for normal human fibroblasts (open circles) and Xeroderma pigmentosum cells (closed circles) after irradiation with UV at 313 nm (B).(B from ref. 23).

Several criteria have to be met for the experimental determination of an action spectrum[21,22]:

- similar dose effect curves for all wavelengths;
- same quantum yield for all wavelengths;
- same absorption spectrum of the chromophore in vivo and in vitro;
- negligible or constant absorption or scattering in front of the chromophore;
- low absorption of UV by the sample in the wavelength range studied;
- reciprocity of time and dose rate.

Whereas these criteria hold relatively well when using small transparent cells or biomolecules in solution, extending such studies to tissues, larger organisms, or even ecosystems will inevitably lead to substantial absorption of UV by the layers surrounding the chromophore of interest.

Another practical difficulty arises from the need to extend the action spectroscopy to the UV-A range, which physically dominates in the solar spectrum but which is biologically less effective by several orders of magnitude than UV-B or UV-C radiation. This requires UV radiation sources of high intensity and also long exposure times. During these extended periods of irradiation with UV-A bands repair and restoration processes may occur that lead to a dose rate dependence. Furthermore, intermediate molecules. so-called photosensitizers, might be involved in genetic effects that absorb the incident UV-A photons and transfer the effect to the DNA[22].

In most cases, relative action spectra are used for weighted spectrometry, that are scaled to one at 300 nm (Figure 4). Depending on the biological endpoint under consideration, the action spectra may vary by orders of magnitude. Therefore, different biological effective irradiances and BEDs are obtained for different endpoints. These cannot be compared with each other, if relative action spectra have been used. However, they can be used for trend estimations as e.g., the sensitivity of DNA damaging irradiance to UV[30].

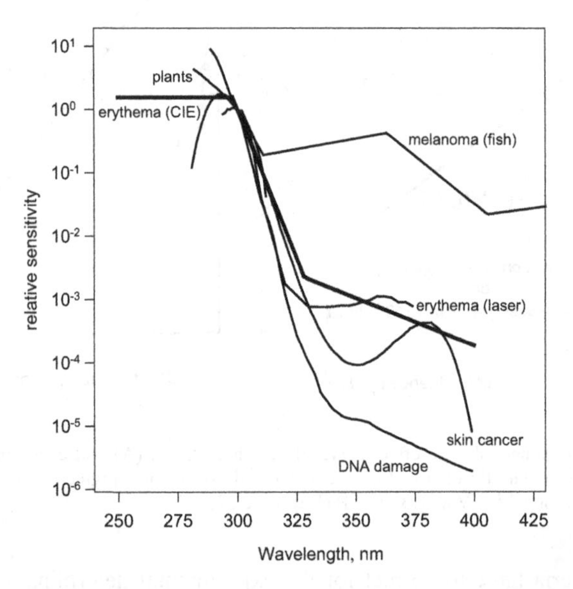

Figure 4. Action spectra for biological responses to UV utilized as biological weighting functions in weighted spectroradiometry: DNA damage (data from ref. 18,24), skin cancer (data from ref. 24,25), erythema (data from refs. 26, 27), malignant melanoma (data from ref. 28), and plant damage (data from ref. 29). Reference wavelength is $\lambda = 300$ nm.

Figure 5 shows, how relative action spectra can lead to erroneous assumptions. It shows absolute action spectra for normal fibroblasts and those of the DNA repair deficient strain Xeroderma pigmentosum (XP), obtained from the inactivation rate constant for different wavelengths as shown in Figure 3[23]. They clearly show that XP cells are more sensitive to UV-C and UV-B than normal cells - which is caused by the excision repair defect associated with XP cells -, however both cells have nearly identical sensitivities to UV-A. Probably other repair pathways take over in responses to UV-A which are equally efficient in both cells. In contrast, if both action spectra are expressed in relative units, e.g., scaled to one at the reference wavelength of 254 nm or 300 nm, both cell types, the normal as well as the XP one, appear to have the same sensitivity to UV-C and UV-B, whereas for longer wavelengths (300-365 nm) the cells of the normal type seem to become more sensitive than the XP cells, by up to one order of magnitude. This is certainly not the case, as demonstrated by the absolute action spectra. Much confusion could be avoided, if always absolute values of the original action spectra were available. Therefore, when using relative action spectra for comparing different biological UV responses or different strains for the same response, the absolute response at least to the reference wavelength has to be known for each endpoint or strain[7].

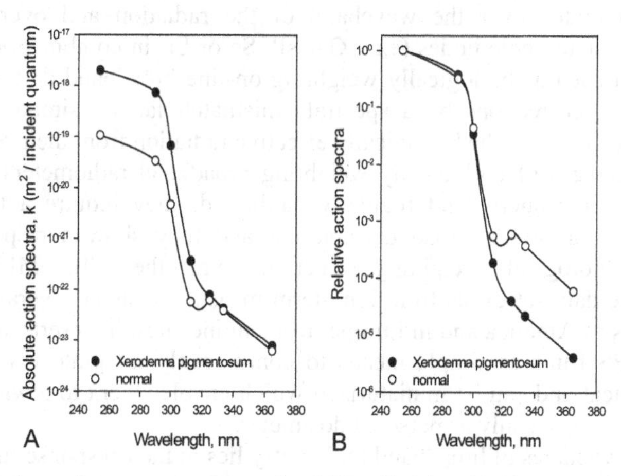

Figure 5. Absolute action spectrum of normal and XP cells (A) and relative action spectrum of the same cells scaled to one at 254 nm (B) (data from ref. 23).

For assessing risks from environmental radiation, the most widely used action spectra are that for DNA damage, for cancer induction, and for UV induced minimal erythema in human skin, which has been defined by the CIE (Commission International d'Éclairage) as erythema reference action spectrum (CIE MED spectrum) (Figure 4).

Weighting broadband radiometry

Different types of broadband radiometers have been developed that measure the irradiance of a defined UV band, e.g., UV-B or UV-A (see Häder, this issue). Several of them have spectral sensitivity similar to a specific biological response function, e.g., the CIE MED spectrum. However, because the spectral sensitivity curve of most biologically weighting broadband radiometers does not perfectly match with the most relevant action spectra, e.g. the CIE MED curve, their readings must be corrected with regard to the spectral irradiance of the UV emitting source and the action spectrum according to

$$H_{eff} = \frac{\int E_\lambda(\lambda,t) \cdot S_\lambda(\lambda) d\lambda dt}{\int E_\lambda(\lambda,t) \cdot \upsilon_\lambda(\lambda) d\lambda dt} \cdot F \qquad (4)$$

with H_{eff} = BED, $E_\lambda (\lambda,t)$ = the terrestrial spectral UV irradiance at a certain time or of an artificial UV source, $S_\lambda(\lambda)$ = the action spectrum for a critical biological effect, e.g., the CIE MED spectrum, $\upsilon_\lambda (\lambda)$ = the spectral response function of the sensor, and F = the equivalent dose of monochromatic radiation producing the same response of the detector. H_{eff} is given in $(J/m^2)_{eff}$ or MED.

The same requirement holds for photosensitive films with spectral sensitivities similar to photobiological responses (e.g., polysulphone (PS) films[31] or polycarbonate plastics[32]). The degree of deterioration of the detector is related to the incident UV dose. Their signal integrates over the waveband of the radiation and over the time of irradiation. Solid state photodiodes (e.g., GaAsP, Se or Si) in combination with optical filters have been used as biologically weighting on-line broadband detectors[33]. For all these radiometers, corrections by a spectral "mismatch factor" similar to Eq. 4 are required when determining the biologically effective radiation from their readings.

The advantages of biologically weighting broadband radiometers are that they are easy to use, if properly and regularly calibrated; they require a relatively low investment, which allows a broad distribution; and they show a response function similar to a biological weighting function, e.g. the CIE MED spectrum. A comprehensive data set exists from long-term measurements at several sites, e.g., in the United States of America and in Europe, from alpine sites, in deserts and in northern high altitudes. PS films as simple means to continuously integrate UV exposure are rugged, economical and can be miniaturized which enables them to a wide application in medical context, especially as personal dosimeter.

The disadvantages of broadband radiometry lies in their response function, which is similar, but not identical with any biological weighting function for the whole wavelength range, e.g., of solar UV radiation. Furthermore, substantial efforts are required concerning the precision of instrument calibration (e.g., long-term stability, sensitivity, wavelength coverage, temperature stabilization, power consumption, data logger capacity).

Biological UV dosimetry

The situation would be ideal, if the spectral response curve $\upsilon_\lambda (\lambda)$ of the UV detector would be identical to the photobiological effect under consideration. This is generally the case for biological dosimeters (Table 1) that automatically weight the incident UV components of sunlight in relation to the biological effectiveness of the different wavelengths and to interactions between them. In this case, in Eq. 4 $S_\lambda (\lambda) = \upsilon_\lambda (\lambda)$ and H_{eff} is given by the following simple relation:

$$H_{eff} = F \tag{5}$$

with F = the equivalent dose of monochromatic radiation producing the same response of the biodosimeter as the radiation under consideration. H_{eff} is given in $(J/m^2)_{eff}$[7]. In most cases, λ = 254 nm has been used as reference wavelength. This is based on the fact that this wavelength lies in the absorption maximum of DNA and on its availability as a sharp line from low pressure mercury lamps.

Table 1. Most commonly used sensory systems for biological UV dosimetry

Biological detector	Biological endpoint	Dosage unit	Application
Uracil[36]	Dimer formation	Dose to reduce O.D. by e^{-1}	Long-term monitoring
Vitamin D[37]	Isomerization to pre-D3	% conversion of 7 DHC	Environmental monitoring
DNA[38]	Inactivation, dimer formation	H_{eff} equivalent to 254 nm	Clear tropical marine water
Bacteriophage T7[34]	Inactivation of plaque formers	$(\|\ln(N/N_0\|)$	Long-term/ continuous monitoring; measurements in lakes, rivers, ocean
Bacterial cells (*E. coli* sp.)[39]	Inactivation, mutagenesis, role of repair processes, interactions	H_{eff} equivalent to 254 nm	Diurnal profiles, daily totals, vertical dose distribution in natural water
Bacterial spores (*B. subtilis* sp.)[35]	Inactivation; mutagenesis; spore photoproduct formation	ID or H_{eff} equivalent to 254 nm	Diurnal profiles, daily totals, long term monitoring, personal dosimetry
Biofilm[40]	Loss of biological activity	H_{eff} equivalent to 254 nm	Long term monitoring, personal dosimetry, trend estimation
RODOS: mammalian cells[41]	Inhibition of cell growth	H_{eff} from relative absorbance	Prototype

An alternative approach is to deduce the dosimetric quantity from the dose effect curves for a certain critical biological process. For exponential dose effect relationships, such as the inactivation curves of bacteriophage T7 or of repair deficient spores of *Bacillus subtilis*, the average number of lethal hits is given by the absolute value of the term $\ln(N/N_0)$, with N = number of viable individuals after irradiation and N_0 = number of viable individuals without irradiation[34,35]. This method allows a dosimetric characterization independent of a certain reference wavelength.

So far, in most cases simple test systems have been used as biological dosimeters (reviewed in[42]), such as the uracil molecule[36], DNA[38], bacteriophages[34], bacteria[35,39,40] or mammalian cells[41] (Table 1). Their action spectra agree fairly well with that for DNA damage[18] which has been actually obtained by averaging over responses in bacteria, bacteriophages and DNA. Induction rates for lethality, mutagenesis and dimerization have been used to determine directly the BED, either in terms of equivalent incident dose at 254 nm for DNA damaging capacity[38,39,40] or as number of lethal hits[34-36]. For ecological questions, more complex dosimetric systems, such as a motility test of flagellates[43] or survival of *Daphnia* have been introduced.

The advantage of a biological dosimeter stands or falls with the degree to which its action spectrum agrees with the spectral sensitivity of the photobiological phenomenon under consideration. Unlike the above mentioned chemical dosimeters, the biological dosimetry systems respond sensitively to small variations at the short-wavelengths edge of the solar spectrum. Although nearly all action spectra for biologically adverse effects of sunlight decrease with increasing wavelength, the rate of decline varies considerably (Figure 4). Especially the tail of the action spectra in the UV-A or visible waveband has to be checked carefully, since it may be different for different species or even strains, even for the same endpoint. This phenomenon reflects the complex interplay of a variety of cellular mechanisms for processing UV-induced damage (reviewed in[7]).

Most biological dosimeters are easy to use, at least by a trained technician. They are of low cost, which allows for a broad distribution. Some of them can be miniaturized so that they are suitable for personal dosimetry or to measure the distribution of the biologically effective UV in endangered ecosystems, cultivars, trees etc.

On the other hand, as every biological system, biological dosimeters may be sensitive to factors other than UV and their dynamic range may be restricted. In any case, a careful calibration and intercomparison with spectroradiometric data is needed.

3. Quantification of biologically effective dose using biological UV dosimeters

Biomolecules and viruses as biological UV dosimeters

Uracil dosimeter. The dimerization of uracil monomers in a polycrystalline layer is determined spectrophotometrically by a decrease in the absorption characteristics in the wavelength range 250-400 nm (Table 1). The effective dose for uracil bleaching is given in absolute values of H_{eff} (called H^U), which is equal to the exposure that decreases the optical density of the uracil layer to the e-th part of the difference between the original (non-irradiated) and the saturated values, i.e. on the average 1 hit per molecule[36]. Uracil dosimeters have been utilized at several sites for long-term UV monitoring.

DNA Dosimeter. The formation of cyclobutane pyrimidine dimers and other UV lesions in DNA (Table 1) is analyzed by a radiochromatographic assay[38], by PCR[44], or by an immunochemical reaction[45]. The BED is given in $(J/m^2)_{eff}$, equivalent to a dose of 254 nm producing the same effect. The DNA dosimeter was tested as solution in marine waters at different depths[38] and as dried film[44,45].

T7 Dosimeter. The inactivation of bacteriophage T7 (Table 1) is determined either by the plaque test or photometrically from the lysis of the host cells[34]. H_{eff} is given as H^{T7}, which equals to the absolute value of the term ln (N/N_0), with N = number of viable individuals after irradiation and N_0 = number of viable individuals without irradiation. The term $|\ln (N/N_0)|$ gives the average amount of UV damage in one bacteriophage particle[46]. T7 dosimeters have been utilized at different land sites and in different aquatic environments.[47].

Vitamin D Dosimeter. The endogenous synthesis of vitamin D under solar UV radiation is a widespread process and indicates one of the beneficial effects of solar UV-B radiation. The *in vitro* process of vitamin D synthesis by UV has been developed as a biological UV dosimeter (Table 1)[37]: 7-Dehydrocholesterol (7 DHC), the precursor of vitamin D3 in methanol, has been used as biomarker for UV exposure indicating the photoisomerisation of provitamin D3 (7 DHC) to previtamin D3 which is the initiating step of cutaneous vitamin D3 synthesis. Only low doses of UV are required at wavelengths below 315 nm. The photoconversion of 7 DHC to pre D3 is determined spectrophotometrically[48] or by HPLC[49]. This *in vitro* vitamin D dosimeter was utilized to determine the seasonal and latitudinal changes on the potential of sunlight to initiate cutaneous production of vitamin D3.

Cellular biological UV dosimeters

Bacteria as dosimeters. Bacterial systems have been frequently used as biodosimeters which measure lethality as the prime cellular UV effect that essentially reflects unrepaired DNA damage[18]. They include cells of *E. coli* in suspension[39,50-52] or spores of *B. subtilis* in suspension[53], as dry layers[35,54-57], or immobilised in a biofilm[7,40,58,59]. Due to their exponential inactivation kinetics repair deficient strains are especially suited for biological dosimetry. They respond highly sensitive to low doses of UV-light, because they accumulate the DNA photoproducts and consequently are killed at relatively low numbers of lesions in their DNA. The *E. coli* CSR 603 *uvrA6 recA1 phr-1* strain which is deficient in all dark repair as well as in photorepair is a kind of "worst case" indicator, because it quantifies the incidence of primary lethal damage by sunlight[51]. However, due to its extremely high UV sensitivity - it is inactivated by 3 orders of magnitude within 30 sec of insolation, it is too difficult to be used as field dosimeter. Its rec^{+} variant *E. coli* CSR 06 *uvrA6 phr-1*[39] and spores of *B. subtilis uvrA10 ssp-1*[35,50,54-56] or of *B. subtilis uvrA10 ssp-1 polA151*[57] are the most frequently used bacterial systems in biological dosimetry. The BED is determined either from the inactivation rate constant[14,35,50,54-57] as number of lethal hits[35,54-55] or as % lethality[39]. Because of their spectral responses[14], their long shelf-life, their resistance against other environmental parameters such as heat and humidity, the availability of repair-deficient mutants with exponential inactivation curves and the lack of a photoreactivation system, spores of *B. subtilis* especially meet many of the criteria for reliable biological dosimetry of sunlight. The spore dosimeter has been utilized at different latitudinal and seasonal conditions, in South America[50] in Japan[14,35,54,55,60], and in Antarctica[57] as well as in quality control of water disinfection by UV (254 nm)[61]. The *E. coli* cell dosimeter has mainly been used in aquatic environments in Antarctica[39] and in freshwater lakes in USA[62].

DLR-biofilm dosimeter. The DLR-Biofilm dosimeter consists of a monolayer of dry spores of *B. subtilis* immobilized in agarose on a transparent polyester plastic sheet[40]. After exposure to sunlight and calibration at 254 nm, the biofilm is incubated in nutrient broth medium and the proteins synthesized after spore germination and cell growth are stained. Figure 6 shows the DLR-Biofilm after insolation, calibration, and processing. The UV effect measured in this system is inhibition of biological activity, as expressed by the amount of protein synthesized inside the biofilm in a given incubation time. This amount of stained protein is determined by image analytic/photometric measurements. The BED H_{eff} in $(J/m^2)_{eff}$ is obtained according to Eq. 5 from the linear quadratic calibration curve determined at 254 nm from a calibration source. It relates the dose equivalent to that of the calibration source producing the same effect[63]. Its action spectrum deviates from the CIE MED curve by less than a factor of 2 over a wide range from UV-B to UV-A[7]. The response of the biofilm to UV is additive and follows the

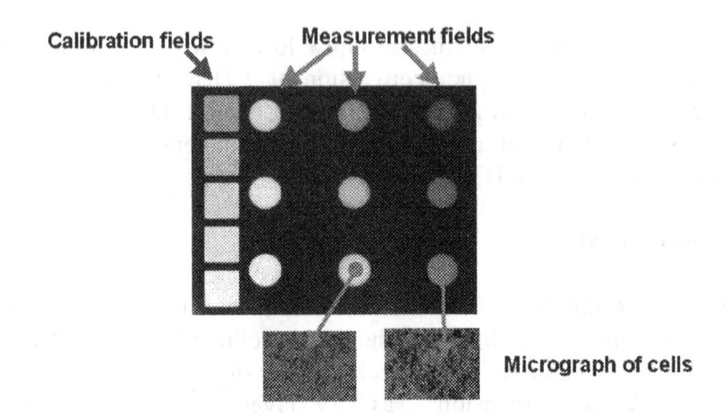

Figure 6. Biofilm with calibration fields and measurement fields after exposure, processing and staining and micrographs of the cells grown from spores during incubation.

reciprocity law. It is independent of temperature between -20°C and 70°C and humidity in the range of 20 % to 80 % relative humidity. The biofilm can be stored at room temperature for up to 9 months without detectable change in viability of the spores[40]. Hence, the biofilm technique combines both, the direct biological weighting of a biological dosimeter and the robustness and simplicity of a chemical film dosimeter. The biofilm has been used as UV dosimeter at different sites in Europe[46,64,65], in Arctic[33], in Antarctica[57], in space[63] and as personal dosimeter[66].

Euglena as bioindicator. In order to assess the potential hazards of enhanced solar UV-B radiation on phytoplankton productivity, sensitive test systems for freshwater (*Euglena*)[43,67,68] or marine habitats (dinoflagellates)[69] have been used as biomarkers for UV effects. A population of *Euglena* cells is significantly affected by unfiltered solar radiation after about 1-2 h of exposure. The test parameters include impairment of motility, photoorientation and growth. The biologically effective UV radiation is quantified in terms of decrease in growth rate[69], percentage of motile cells, speed or degree of phototactic orientation[43,67,68]. Phytoplankton test systems have been utilized in different field measurements in Germany and Portugal[67] and in marine waters[69].

Mammalian cells as bioindicator. Chinese hamster ovary cells grown on a UV transparent petri-PERM® foil have been used in the UV dosimeter „RODOS“ (rodent cells dosimeter). A special device allowed insolation of the cells and subsequent cellular growth under identical spatial arrangement of the cells. After insolation, incubation, fixation and staining, the DNA damaging effect of solar UV was determined by image analysis[41]. RODOS directly reflects the effects of solar UV radiation on health issues.

Criteria for biological UV dosimeters

Photobiological criteria. For biological dosimetry to be useful it is required that it is representative of a biologically significant process, such as UV effects on human health, ecosystem balance, or agricultural or fishery productivity. Figure 4 shows typical action spectra for biological UV effects of concern. Although nearly all action spectra for biologically adverse effects of sunlight decrease with increasing wavelength in the UV-B and UV-A range, the rate of decline may vary considerably. Even when considering the same endpoint, they may be different for different species or even strains[23,42,59]. This phenomenon reflects the complex interplay of a variety of cellular mechanisms for processing UV-induced damage.

Radiometric criteria. Besides the photobiological criteria, the general requirements on radiometric systems should be met in biological dosimetry. These latter criteria concern the reproducibility (standardized procedure, genetically defined homogenous material, robust against unspecific environmental stress parameters, long shelf life), the optical properties (identification curve, linearity of response, dynamic range, dose rate effect, action spectra using both, mono- and polychromatic radiation, additivity, dependence of response on temperature, relative humidity and angle of incidence). Furthermore the agreement of predicted with observed spectral responses (intercalibration with biologically weighted spectroradiometric data) should be verified. For applications as field or personal dosimeters the availability, suitability for routine measurements, easy handling, automatic registration, if possible, and low costs are important requirements.

Applications of biological UV dosimeters

Long-term trends of biologically effective solar radiation. Daily and annual profiles of biologically effective environmental UV radiation have been obtained using the T7 dosimeter[46,47,70], the spore dosimeter[35,50,64], and the biofilm dosimeter[46,58,65]. Using the spore dosimeter, Munakata showed in a 14 years study a trend of increase of the BED of solar UV radiation in Tokyo from 1980 to 1993[35]. Using the biofilm dosimeter in a more than 1 year lasting UV-monitoring campaign in Antarctica, Quintern et al. provided experimental proof of an enhanced level of biologically effective UV-B radiation during periods of stratospheric ozone depletion[58]. Figure 7 shows the annual profile of daily H_{eff} values separately for the UV-B and UV-A range, that were recorded at the Georg von Neumayer Station (70.4°S 80.3°W). Cut-off filters were used, to separate the effects at different UV ranges. After the polar night, when stratospheric ozone destruction led to values down to 150 DU, the BED for the UV-B range was elevated to values as high as those that were measured later on in mid summer. Figure 7 demonstrates that the enhancement in the UV-B dose coincided with the so-called ozone hole.

Contribution of UV-B and UV-A to the biologically effective dose H_{eff}. In order to monitor the biological effectiveness of distinct wavebands of the solar spectrum, cut off filters that largely remove the UV-B range of the solar spectrum have been used in biological dosimetry. Using the *E. coli* biodosimeter, Karentz and Lutze for the first time provided experimental proof of an enhanced UV-B/UV-A ratio during periods of reduced stratospheric ozone concentrations over Antarctica in 1988[39]. During periods of depleted stratospheric ozone the ratio UV-B/UV-A increased by a factor up to 10, from 0.3-0.4 to 2.5-3.4. This observation was subsequently confirmed with the spore dosimeter for the ozone depletion period in 1990[57] and with the biofilm for the

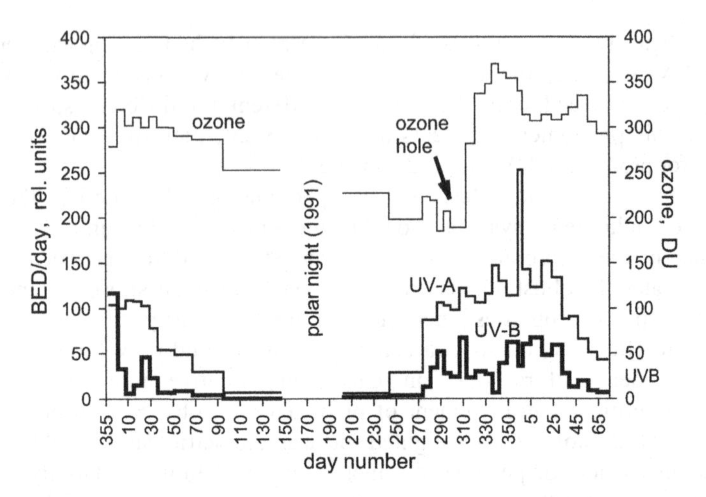

Figure 7. Ozone data and BED H_{eff} of solar UV-B and UV-A radiation determined with the biofilm dosimeter at the Georg von Neumayer Station (70°37'S, 8°22'W) in Antarctica (modified from ref. 58.

following period in 1991[58]. In the latter case, the ratio of biologically effective UV-B to biologically effective UV-A was raised by a factor of 2 during the ozone depletion period, from 0.1-0.3 to 0.5-0.6[58].

Sensitivity of the biologically effective solar irradiance to ozone. In a space experiment, extraterrestrial sunlight was used as natural radiation source, filtered through a set of different cut-off filters to simulate the terrestrial UV radiation climate at different ozone concentrations down to very low values[63]. The biologically effective irradiance E_{eff} as a function of simulated ozone column thickness was directly measured with the biofilm technique and compared with expected data from model calculations using the biofilm action spectrum. Figure 8 shows a dramatic increase in E_{eff} with decreasing (simulated) ozone concentration. The full spectrum of extraterrestrial solar radiation led to an increment of E_{eff} by nearly three orders of magnitude compared with the solar spectrum at the surface of the Earth for average total ozone columns. These data have been used to assess the history of the UV radiation climate of the Earth from the very beginning up to today[71].

Vertical attenuation of biologically effective UV irradiance in natural water. The spectral transmission of UV radiation in natural waters may vary substantially from one water body to another in response to the concentration of colored dissolved organic matter (CDOM) and other dissolved component. To assess the risks from enhanced UV-B radiation for aquatic ecosystems, it is important to determine the vertical profile of the biologically effective UV radiation within the water column, separately for each habitat of concern. Vertical attenuation coefficients of biologically effective UV irradiance have been determined for various freshwater and marine environments on the base of DNA, bacteriophage[47] or bacterial dosimeters. It is interesting to note that the biological attenuation coefficients, integrated over the whole spectral range of sunlight, correspond to spectroradiometrically determined attenuation coefficients in the range of 320 to 340 nm[62]. This suggests a substantial contribution of UV-A to the harmful effect

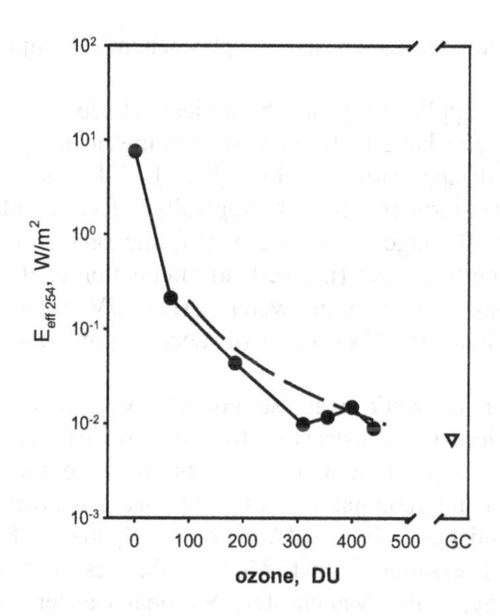

Figure 8. Biologically effective irradiance E_{eff} for different ozone column thicknesses (•) determined by the biofilm technique in a space experiment and the E_{eff} value measured in Köln on 6 August 1993 (∇). The dashed line shows the corresponding curve modeled for DNA damage. (modified from ref. 63).

of sunlight. During austral spring 1990, when stratospheric ozone concentrations fluctuated between 170 and 380 DU, the potential biological effect of in-water UV-B reaching depths down to 29 m was closely correlated to stratospheric ozone levels[72].

Personal UV dosimetry. To estimate the individual biologically effective UV dose, personal carry-along dosimeters are required, because they measure the dose in dependence of the individual behavior. Personal dosimeters, based on the uracil dosimeter[36], the spore dosimeter[73], and the DLR-Biofilm[66] have been applied at various instances, such as school children and different professional groups. Using the DLR-biofilm for personal dosimetry of astronauts on board of the space station MIR, it was concluded that long-term stay in space may lead to an underexposure and thereby to a Vitamin D deficiency[74]. In contrast to the general living module in space, the area behind a quartz window which was used by the astronauts for "sun-bathing" led to a severe overexposure, as measured by the DLR-Biofilm.

4. Conclusions

To assess the risks to human health and ecosystems from an enhanced UV-B radiation, accurate and reliable UV monitoring systems are required that weights the spectral irradiance according to the biological responses under consideration. Biological dosimetry meets these requirements by directly weighting the incident UV components of sunlight in relation to the biological effectiveness of the different wavelengths and to potential interactions between them. Bacteria, viruses and biomolecules have been developed as biological dosimeters. Their responses to environmental UV radiation, indicated as inactivation, mutagenesis or photochemical injury, reflect the UV sensitivity of DNA.

For assessing the applicability of a biological UV dosimeter, photobiological as well as radiometric criteria have to be met. If radiometrically properly characterized, there is a broad scope of applications of biological UV dosimeters, which include the determination of (i) long-term trends of biologically effective solar radiation; (ii) the contribution of the UV-B range to the BED; (iii) the sensitivity of the biologically effective solar irradiance to ozone; (iii) vertical attenuation coefficient of biologically effective solar irradiance in natural waters; (iv) UV tolerance and protection mechanisms; (v) the individual UV exposure of specific professional groups.

Acknowledgements. This work was supported by grants from the European Commission and the German Ministries of Research BMBF and Environment BMU. The work is based on cooperation with scientists from the following organizations: (i) characterization and intercomparison of biological dosimeters: Laboratory for Biophysics, HAS, Budapest; Inst. d'Aéronomie Spatiale, Brussels; Laboratory Atmospheric Physics, Thessaloniki; Inst. Medical Physics, Veterinary Univ. Vienna; Inst. Sci. & Technology, Univ. Manchester; National Cancer Center Research Inst. Tokyo; Inst. Physics, NAS, Kiev; Inst. Suisse de Rech. Exp. Cancer, Lausanne; JOBO Labortechnik, Gummerbach; Gigahertz-Optik, Puchheim; Fraunhofer Inst. Atmos. Res. Garmisch-Partenkirchen; Inst. Meteorol. & Climatology, Univ. Hannover; GSF, Neuherberg; Okazaki Large Spectrograph; (ii) personal dosimetry: Dermatology, Techn. Univ. Dresden; Dermatology, University Köln; CIE, TC6-53 Personal dosimetry of UV radiation; (iii) comparison of bioindicators: Inst. Botany & Pharm. Biol. Univ. Erlangen; Zoological Inst. Univ. München; Botanical Inst. Univ. Freiburg; (iv) UV-monitoring: BAS, Cambridge; AWI, Bremerhaven; INTA, Madrid.

References

1. Albritton, D.L., Aucamp, P.J., Megie, G. and Watson, R.T. (eds.) (1998) *Scientific Assessment of Ozone Depletion: 1998*, Report No. 44, WMO, Global Ozone Research and Monitoring Project, Geneva.
2. UNEP (1998) *Environmental Effects of Ozone Depletion: 1998 Asssessment*, ISBN92-607-1924-3.
3. Longstreth, J.D., de Gruijl, F.R., Kripke, M.L., Takazwa, Y. and van der Leun, J.C. (1995) Effects of increased solar UV radiation on human health. *AMBIO*, 24: 153-165.
4. Häder, D.-P. (ed.) (1997) *The Effects of Ozone Depletion on Aquatic Ecosystems*, R. G. Landes Company, Austin (TX), USA.
5. SCOPE (1992) *Effects of Increased Ultraviolet Radiation on Biological Systems*, SCOPE Report, SCOPE Secretariat, Paris.
6. Baumstark-Khan, C., Kozubek, S. and Horneck, G. (eds.) (1999) *Fundamentals for the Assessment of Risks from Environmental Radiation.*, NATO ASI Series, 2. Environmental Security – Vol. 55, Kluwer, Dordrecht.
7. Horneck, G. (1995) Quantification of the biological effectiveness of environmental UV radiation, *J. Photochem Photobio B:Biol.*, 31: 43-49.
8. Nicolet, M. (1989) Solar spectral irradiances and their diversity between 120 and 900 nm, *Planet. Space Sci.*, 37: 1249-1289.
9. Lean, J.L.Rottmann, G.J., Kyle, H.L., Woods, T.N., Hickey, J.R. and Puga, L.C. (1997) Detection and parameterization in solar mid- and near-ultraviolet radiation (200-400 nm), *J. Geophys. Res.*, 102: 29939-29956.
10. Schulze, R. (1970) *Strahlenklima der Erde*, Wissenschaftliche Forschungsberichte, Band 72, Dr. Dietrich Steinkopf Verlag, Darmstadt

11. Horneck, G., Bücker, H., Reitz, G., Requardt, H., Dose, K., Martens, K.D., Mennigmann, H.D. and Weber, P. (1984) Microorganisms in the space environment, *Science*, 225: 226-228.
12. Horneck, G. and Brack, A. (1992) Study of the origin, evolution and distribution of life with emphasis on exobiology experiments in Earth orbit. In *Advances in Space Biology and Medicine Vol. 2* (S.L. Bonting, ed.), JAI Press, Greenwich, CT, pp. 229-262.
13. Munakata, N., Saito, M. and Hieda, K. (1991) Inactivation action spectra of *Bacillus subtilis* spores in extended ultraviolet wavelengths (50-300 nm) obtained with synchrotron radiation, *Photochem Photobiol.*, 54: 761-768.
14. Munakata, N., Morohoshi, F., Hieda, K., Suzuki, H., Furusawa, Y., Shimura, H. and Ito, T. (1996) Experimental correspondence between spore dosimetry and spectral photometry of solar ultraviolet radiation, *Photochem. Photobiol.*, 63: 74-78.
15. Seidlitz, H.K., Thiel, S., Krins, A. and Mayer, H. (2001) Solar radiation at the Earth's surface. In *Sun Protection in Man* (P.U. Giacomoni, ed.), Elsevier, pp. 705-738.
16. Cockell, C.S. (2000) The ultraviolet history of the terrestrial planets – implications for biological evolution, *Planet. Space Sci.*, 48: 203-214.
17. Herman, J.R. and McKenzie, R.L. (1999) Ultraviolet radiation at the Earth's surface. In *Scientific Assessment of Ozone Depletion: 1998,* WMO Global Ozone Research and Monitoring Project-Report 44, WMO, Geneva, pp. 9.1-9.56.
18. Setlow, R.B. (1974) The wavelengths in sunlight effective in producing skin cancer: a theoretical analysis, *Proc. Natl. Acad. Sci. USA*, 71: 3363-3366.
19. McKenzie, R.L., Kotkamp, M., Seckmeyer, G., Erb, R., Roy, C.R., Gies H.P. and Toomey, S.J. (1993) First southern hemisphere intercomparison of measured solar UV spectra, *Geophys. Res. Lett.*, 20: 2223-2226.
20. Seckmeyer, G., Mayer B., Bernhard, G. (1998) *The 1997 Status of Solar UV Spectroradiometry in Germany: Results from the National Intercomparison of UV spectroradiometers, Garmisch-Partenkirchen, Germany,* IFU Band 55-1998, Fraunhofer Institut für Atmosphärische Umweltforschung, Garmisch-Partenkirchen.
21. Jagger, J. (1985) *Solar UV Actions on Living Cells*, Praeger, New York.
22. Coohill, T.P. (1991) Action spectra again? *Photochem. Photobiol.*, 54: 859-870.
23. Keyse, S.M., Moss, S.H. and Davies, D.J.G. (1983) Action spectra for inactivation of normal and Xeroderma pigmentosum human skin fibroblasts by ultraviolet radiations, *Photochem. Photobiol.*, 37: 307-312.
24. Madronich, S. and de Gruijl, F.R. (1994) Stratospheric ozone depletion between 1979 and 1992: implications for biologically active ultraviolet-B radiation and non-melanoma skin cancer incidence, *Photochem. Photobiol.*, 59: 541-546.
25. deGruijl, F.R., Sterenborg, H.J.C.M., Forbes, P.D., Davies, R.E., Cole, C., Kelfken, G., van Weelden, H., Slaper, H. and van der Leun, J.C. (1993) Wavelength dependence of skin cancer induction by ultraviolet radiation of albino hairless mice, *Cancer Res.*, 53: 53-60.
26. McKinley, A.F. and Diffey, B.L. (1987) A reference action spectrum for ultraviolet induced erythema in human skin, *CIE J.*, 6: 17-22.
27. Anders, A., Altheide, H.J., Knälmann, M. and Tronnier, H. (1995) Action spectra for erythema in humans investigated with dye lasers, *Photochem. Photobiol.*, 61: 200-205.
28. Setlow, R.B., Grist, E., Thompson, K. and Woodhead, A.D. (1993) Wavelengths effective in induction of malignant melanoma, *Proc. Natl. Acad. Sci. USA,* 90: 6666-6670.
29. Caldwell, M.M. (1971) Solar UV radiation and the growth and development of higher plants. In *Photophysiology Vol 6* (A.C. Giese, ed.), Academic Press, New York, pp. 131-177.
30. Madronich, S. (1993) The atmosphere and UV-B radiation at ground level. In *Environmental UV Photobiology* (R. Young, L.O. Björn, J. Moan and W. Nultsch, eds.), Plenum, New York, pp. 1-37.
31. Davis, A., Deane, G.H.W. and Diffey, B.L. (1976) Possible dosimeter for ultraviolet radiation, *Nature,* 261: 169-170.
32. Wong, C.F., Fleming, R. and Carter, S.J. (1989) A new dosimeter for ultraviolet-B radiation, *Photochem. Photobiol.*, 50: 611-615.
33. Cockell, C.S., Scherer, K., Horneck, G., Rettberg, P., Facius, R., Gugg-Helminger, A. and Lee, P. (2001) Exposure of Arctic field scientists to ultraviolet radiation evaluated using personal dosimeters, *Photochem. Photobiol.*, 74: 570-578.
34. Rontó, G., Gáspár, S. and Bérces, A. (1992) Phages T7 in biological UV dose measurements, *J. Photochem. Photobiol. B: Biol.*, 12: 285-294.

35. Munakata, N. (1993) Biologically effective dose of solar ultraviolet radiation estimated by spore dosimetry in Tokyo since 1980, *Photochem. Photobiol.*, 58: 386-392.
36. Gróf, P., Gaspar, S. and Rontó, G. (1996) Use of uracil thin layer for measuring biologically effective UV dose, *Photochem. Photobiol.*, 64: 800-806.
37. Galkin, O.N. and Terenetskaya, I.P. (1999) 'Vitamin D' biodosimeter; basic characteristics and potential applications, *J. Photochem. Photobiol. B: Biol.*, 53: 12-19.
38. Regan, J.D., Carrier, W.L., Gucinski, H., Olla, B.L., Yoshida, H., Fujimura, R.K. and Wicklund, R.I. (1992) DNA as a solar dosimeter in the ocean, *Photochem. Photobiol.*, 56: 35-42.
39. Karentz, D. and Lutze, L.H. (1990) Evaluation of biologically harmful ultraviolet radiation in Antarctica with a biological dosimeter designed for aquatic environments, *Limnol. Oceanogr.*, 35: 549-561.
40. Quintern, L.E., Horneck, G., Eschweiler, U. and Bücker, H. (1992) A biofilm used as ultraviolet dosimeter, *Photochem. Photobiol.*, 55: 389-395.
41. Baumstark-Khan, C., Hellweg, C.E., Scherer, K. and Horneck, G. (1999) Mammalian cells as biomonitors of UV-exposure, *Anal. Chim. Acta*, 387: 281-287.
42. Horneck, G. (1997) Biological UV dosimetry. In *The Effects of Ozone Depletion on Aquatic Ecosystems* (D.-P. Häder, ed.), R.G. Landes Company and Academic Press, Inc., pp. 119-142.
43. Häder, D.-P. and Worrest, R.C. (1991) Effects of enhaced solar ultraviolet radiation on aquatic ecosystems, *Photochem. Photobiol.*, 53: 717-725.
44. Yoshida, H. and Regan, J.D. (1997) UVB DNA dosimeters analyzed by polymerase chain reactions, *Photochem. Photobiol.*, 66: 82-88.
45. Ishigaki, Y., Takayama A., Yamashita S. and Nikaido O. (1999) Development and characterization of a DNA solar dosimeter, *J. Photochem. Photobiol. B: Biol.*, 50: 184-188.
46. Bérces, A., Fekete, A., Gáspár S., Gróf, P., Rettberg, P., Horneck, G. and Rontó, Gy. (1999) Biologial UV dosimeters in the assessment of biological hazards from environmental radiation, *J. Photochem. Photobiol. B: Biol.*, 53: 36-43.
47. Rontó, G., Gáspár, S., Gróf, P., Bérces A. and Gugolya, Z. (1994) Ultraviolet dosimetry in outdoor measurements based on bacteriophage T7 as a biosensor, *Photochem. Photobiol.*, 59: 209-214.
48. Terenetskaya, I.P. (1994) Provitamin D photoisomerisation as possible UVB monitor: kinetic study using a tuneable dye laser. *SPIE Proc Int. Conf. Biomed.Optics,* Vol. 2134 B, p.135.
49. Webb, A.R., Kline, L. and Hollick, M.F. (1988) Influence of season and latitude on the cutaneous synthesis of vitamin D_3: exposure to winter sunlight in Boston and Edmonton will not promote Vitamin D_3 synthesis in human skin. *J. Clin. Endocrin. Metabol.*, 67: 373-378.
50. Tyrrell, R. (1978) Solar dosimetry with repair deficient bacterial spores: action spectra, photoproduct measurements and a comparison with other biological systems, *Photochem. Photobiol.*, 27: 571-579.
51. Harm, W. (1979) Relative effectiveness of the 300-320 nm spectral region of sunlight for the production of primary lethal damage in *E. coli* cells, *Mut. Res.*, 60: 263-270.
52. Tyrrell, R.M. and Souza-Neto, A. (1981) Lethal effects of natural solar ultraviolet radiation in repair proficient and repair deficient strains of *Escherichia coli*: actions and interactions, *Photochem. Photobiol.*, 34: 331-337.
53. Tyrrell, R.M. (1980) Solar dosimetry and weighting factors, *Photochem. Photobiol.*, 31: 421-422.
54. Munakata, N. (1981) Killing and mutagenic action of sunlight upon *Bacillus subtilis* spores: a dosimetric system, *Mut. Res.*, 82: 263-268.
55. Munakata, N. (1989) Genotoxic action of sunlight upon *Bacillus subtilis* spores: monitoring studies at Tokyo, Japan, *J. Radiat. Res.*, 30: 338-351.
56. Wang, T.V. (1991) A simple convenient biological dosimeter for monitoring solar UVB radiation, *Biochem. Biophys. Res. Comm.*, 177: 48-53.
57. Puskeppeleit, M., Quintern, L.E., ElNaggar, S., Schott, J.U., Eschweiler, U., Horneck, G. and Bücker, H. (1992) Long-term dosimetry of solar UV radiation in Antarctica with spores of *Bacillus subtilis, Appl. Environm. Microbiol.*, 58: 2355-2359.
58. Quintern, L.E., Puskeppeleit, M., Rainer, P., Weber, S., ElNaggar, S., Eschweiler, U. and Horneck, G. (1994) Continuous dosimetry of the biologically harmful UV-radiation in Antarctica with the biofilm technique, *J. Photochem. Photobiol. B: Biol.*, 22: 59-66.
59. Quintern, L.E., Furusawa, Y., Fukutsu, K. and Holtschmidt, H. (1997) Characterization and application of UV detector spore films: the sensitivity curve of a new detector system provides good similarity to the action spectrum for UV-induced erythema in human skin, *J. Photochem. Photobiol. B:Biol.*, 37: 158-166.
60. Munakata N, (1995) Continual increase in biologically effective dose of solar UV radiation determined by spore dosimetry from 1980 to 1993 in Tokyo, *J. Photochem. Photobiol. B:Biol.*, 31: 63-68.

61. Sommer, R. and Cabaj, A. (1993) Evaluation of the efficiency of a UV plant for drinking water disinfection, *Wat. Sci. Tech.*, 27: 357-362.
62. Kirk, J.T.O., Hargreaves, B.R., Morris, D.P., Coffin, R.B., David, B., Frederickson, D., Karentz, D., Lean, D.R.S., Lesser, M.P., Madronich, S., Morrow, J.H., Nelson, N.B. and Scully, N.M. (1994) Measurements of UVB radiation in two freshwater lakes: an instrument intercomparison. *Arch. Hydrobiol. Beih.*, 43: 71-88.
63. Horneck, G., Rettberg, P., Rabbow, E., Strauch, W., Seckmeyer, G., Facius, R., Reitz, G., Strauch, K. and Schott, J.U. (1996) Biological dosimetry of solar radiation for different simulated ozone column thickness. *J. Photochem. Photobiol. B:Biol.*, 32: 189-196.
64. Munakata, N., Kazadzis, S., Bais, A.F., Hieda, K., Rontó, G., Rettberg, P. and Horneck, G. (2000) Comparisons of spore dosimetry and spectral photometry of solar UV radiation at four sites in Japan and Europe, *Photochem. Photobiol.*, 72: 739-745.
65. De la Torre, R., Horneck, G., Rettberg, P., Luccini, E., Vilaplana, J.M. and Gil, M. (2001) Monitoring of biologically effective UV irradiance at El Arenosillo (INTA), Andalucia, in Spain, *Adv. Space Res.*, 26: 2015-1019.
66. Rettberg, P. and Horneck, G. (1998) Messung der UV-Strahlenbelastung mit dem DLR-Biofilm. *Derm*, 4: 180–184.
67. Häder, D.-P. and Häder, M.A. (1988) Inhibition of motility and phototaxis in the green flagellate, *Euglena gracilis*, by UV-B radiation, *Arch. Microbiol.*, 150: 20-25.
68. Häder, D.-P., Worrest, R.C., Kumar, H.D. and Smith, R.C. (1998) Effects on aquatic ecosystems, *J. Photochem. Photobiol. B: Biol.*, 46: 53-68.
69. Nielsen, T., Björn, L.O. and Ekelund, N.G.A. (1995) Impact of natural and artificial UVB radiation on motility and growth rate of marine dinoflagellates, *J. Photochem. Photobiol. B: Biol.*, 27: 73-79.
70. Rontó, G., Gróf, P. and Gáspár, S. (1995) Biological UV dosimetry - a comprehensive problem, *J. Photochem. Photobiol. B:Biol.*, 31: 51-56.
71. Cockell, C.S. and Horneck, G. (2001) The history of the UV radiation climate of the Earth – theoretical and space-based observations, *Phtochem. Photobiol.*, 73: 447-451.
72. Karentz, D. and Spero, H.J. (1995) Response of natural *Phaecocystis* population to ambient fluctuations of UVB radiation caused by Antarctic ozone depletion, *J. Plankton Res.*, 17: 1771-1789.
73. Munakata, N., Ono, M. and Watanabe, S. (1998) Monitoring of solar UV exposure among school children in five Japanese cities using spore dosimeter and UV-coloring labels, *Jpn. J. Canc. Res.*, 89: 235-245.
74. Rettberg, P., Horneck, G., Zittermann, A. and Heer, M. (1998) Biological dosimetry to determine the UV radiation climate inside the MIR station and its role in vitamin D biosynthesis. *Adv. Space Res.*, 22: 1643-1652.

USE AND EVALUATION OF BIOLOGICAL SPECTRAL UV WEIGHTING FUNCTIONS FOR THE OZONE REDUCTION ISSUE

MARTYN M. CALDWELL AND STEPHAN D. FLINT
Utah State University
Logan, Utah 84322-5230 USA

1. Introduction

1.1. ROLES OF BIOLOGICAL SPECTRAL WEIGHTING FUNCTIONS IN THE OZONE REDUCTION ISSUE

Biological spectral weighting functions (BSWF) are used in evaluating the stratospheric ozone problem in several ways. Best known is their use as the basis for radiation amplification factors (RAF) that predict the relative increment in "biologically effective" UV radiation for each step of ozone depletion [1]. They are used to evaluate latitudinal gradients of solar UV-B [2] and to calculate how much supplemental UV to supply in experiments with UV-B-emitting lamp systems [3]. Outdoor experiments with filters intended to reduce natural solar UV have also been conducted, though they are less common than the UV-supplementation experiments. Thus, for assessing the importance of ozone column reduction, the BSWF's play several critical roles.

1.2. WAVELENGTH SENSITIVITY OF BSWF'S

In the solar UV spectrum, the absorption coefficient of ozone plays a decisive role in the wavelength composition of sunlight. Other atmospheric agents of radiation attenuation are important, but they have much less wavelength specificity compared with the absorption by ozone (Figure 1).

Because ozone absorption is so wavelength specific, the ground-level solar spectrum drops by orders of magnitude in a mere 30 nm of wavelength. But, often action spectra, and therefore BSWF, can also increase by orders of magnitude in the same general wavelength range as the solar UV spectral irradiance is decreasing (Figure 2).

F. Ghetti et al. (eds.), Environmental UV Radiation: Impact on Ecosystems and Human Health and Predictive Models, 71–84.

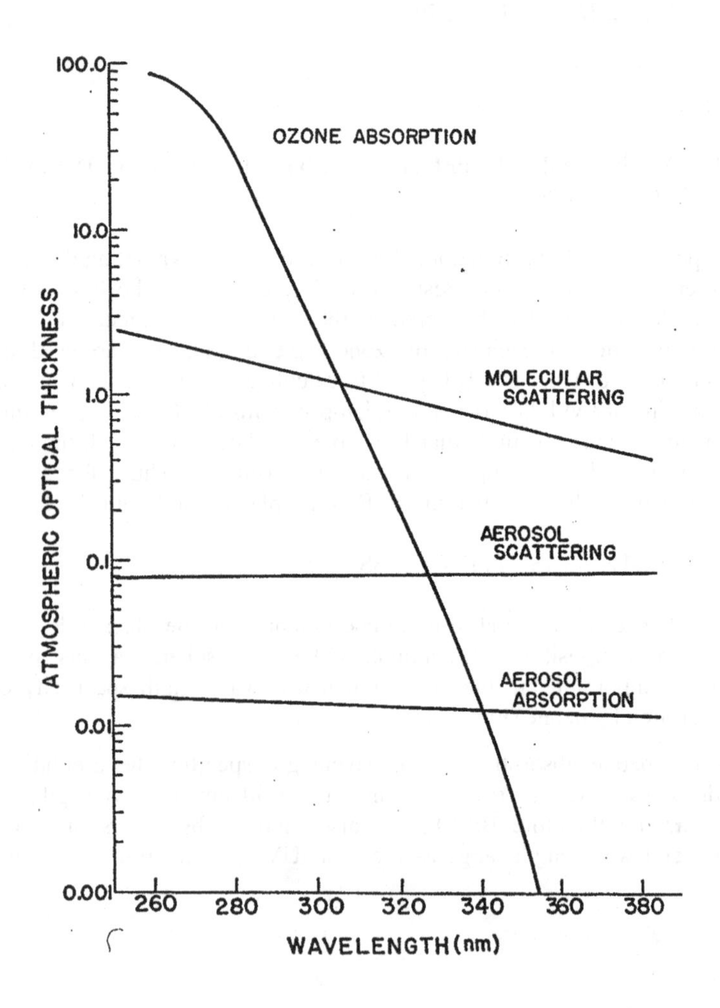

Figure 1. Solar radiation attenuation processes in the atmosphere as a function of wavelength. Each of the four processes that attenuate direct beam solar radiation are represented for a standard atmospheric thickness as would be encountered under cloudless sky conditions with minimal atmospheric aerosols. Ozone absorption is the product of the ozone absorption coefficient [4] and a standard atmospheric ozone thickness of 0.300 atm cm as is typical of temperate latitudes during summer. Scattering by air molecules and aerosols, as well as aerosol absorption, are represented from the data presented by Dave and Halpern [5]. From Caldwell [6].

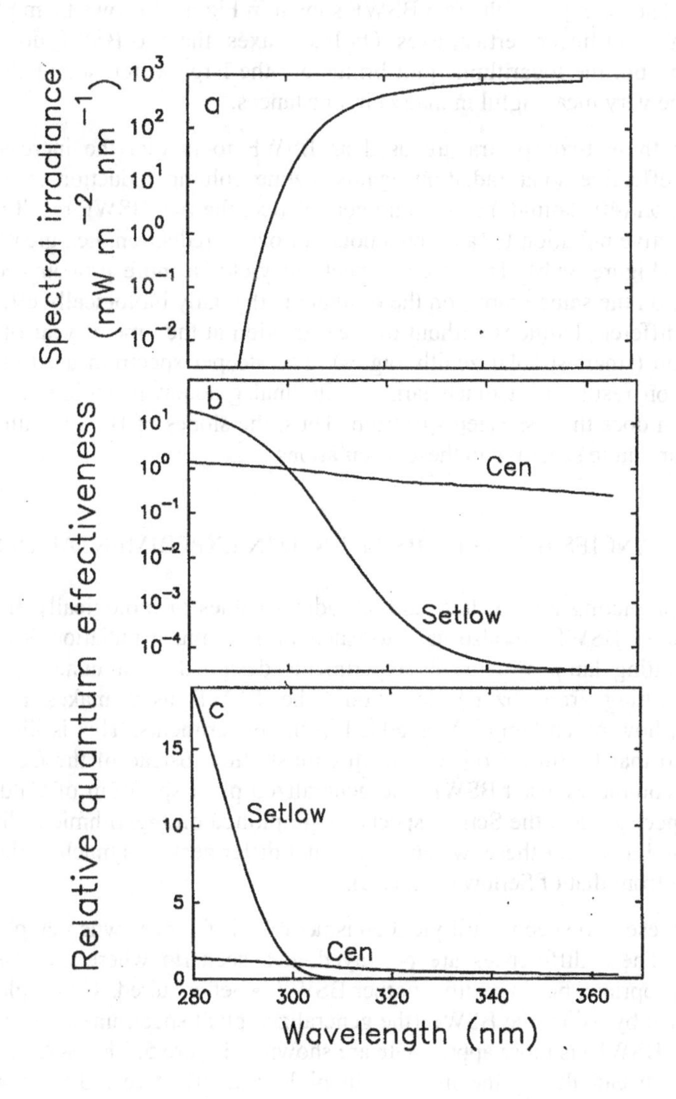

Figure 2. (a) Solar spectral irradiance predicted by a model for temperate-latitude conditions in summer; (b, c) two UV action spectra used as BSWF for sensitivity analysis: The Cen and Björn spectrum for enhancement of ultraweak luminescence in leaves of Brassica napus [7], the Setlow composite spectrum for DNA damage [8]. Spectra have been normalized at 300 nm and are shown plotted against log (b) and linear (c) axes.

1.3. BSWF PREDICTIONS RELATED TO OZONE REDUCTION

Most action spectra and BSWF decrease in relative effectiveness with increasing wavelength. The example of the two BSWF shown in Figure 2 shows them plotted both on logarithmic and linear vertical axes. On linear axes, the two BSWF do not look so very different, but the logarithmic plot brings out the large difference in slopes which turns out to be very meaningful in many circumstances.

When these two spectra are used as BSWF to predict the increase in daily biologically effective solar radiation against ozone column reduction at a temperate latitude location (40° latitude) at the summer solstice, the two BSWF result in different levels of effective radiation for a given amount of ozone reduction, i.e., greatly different RAF values (Figure 3a,b). The steeper spectrum yields a more pronounced increase. Also, plotted on the same graphs on the ordinate is the daily biologically effective solar radiation at different latitudes without ozone reduction at the time of year of maximum solar radiation (smallest solar zenith angles). The steeper spectrum used as BSWF in this calculation results in a much larger latitudinal gradient in biologically effective solar UV than does the less steep spectrum. Thus, the slopes of BSWF with increasing wavelength are quite sensitive in these calculations.

1.4. DISCREPANCIES BETWEEN BSWF USED IN EXPERIMENTAL DESIGN

Apart from predicting RAF values and latitudinal values of biologically effective UV (as in Figure 3), BSWF are also used to calculate how much radiation is to be added with UV-emitting lamp systems in experiments designed to simulate the extra UV radiation resulting from ozone depletion. The BSWF used makes a substantial difference in how much lamp UV is added in the experiments. This is illustrated in a brief scenario that follows. However, in this illustration instead of the Cen and Björn spectrum, a commonly used BSWF, the generalized plant spectrum of Caldwell [9] is used. This spectrum and the Setlow spectrum are plotted on logarithmic ordinate scales in Figure 4 and it is clear these two spectra do not differ nearly as much as the BSWF of Cen & Björn from that of Setlow (Figure 2).

Yet, these two spectra still yield considerable differences when employed in the calculations. These differences are portrayed in a scenario where one assumes one BSWF is appropriate, but in reality another BSWF is better suited. This is plotted as the "errors" caused by using one BSWF (the generalized plant spectrum) when another (the Setlow DNA BSWF) is more appropriate are shown in Figure 5. The "RAF error" is that associated with calculating the increase in biological effective radiation with ozone column reduction (essentially, the RAF as shown in Figure 3).

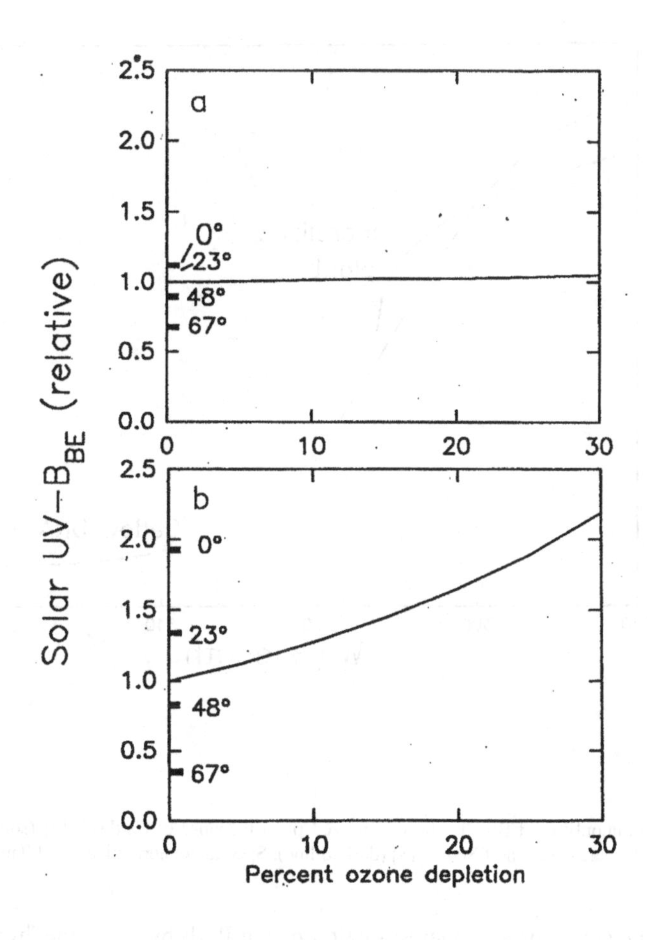

Figure 3. The relationship between ozone reduction and UV-B calculated with the two BSWF of Figure 2. (a) Cen and Björn [7]; (b) Setlow [8]. These represent relative total integrated daily biologically effective UV-B on the summer solstice at 40° latitude. Also shown are the relative daily integrated UV-B values for current ozone conditions at different latitudes at the time of year of maximum solar irradiation. From [10].

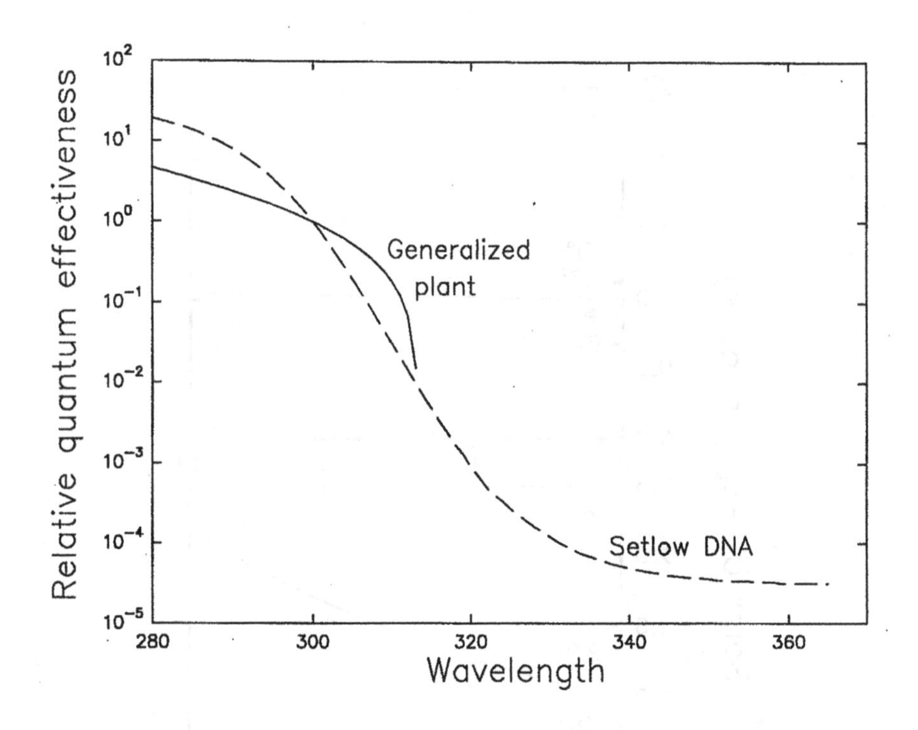

Figure 4. Two commonly used BSWF: the generalized plant response of Caldwell [9] (solid line) and the composite DNA damage function of Setlow [8] (dashed line). Spectra are normalized at 300 nm.

This error is negative, or an underestimation of the RAF by using the "wrong" BSWF and the error increases in magnitude as the ozone layer is diminished to greater extent. The "enhancement error" is the error in determining how much supplemental UV-B should be supplied from lamp systems to simulate a given level of ozone reduction in field experiments. To predict the amount of lamp UV needed, several factors come into play including the RAF itself, the differences in the solar spectral irradiance with and without ozone reduction, and the spectral irradiance coming from the lamp system. The details of these calculations are given elsewhere [11].

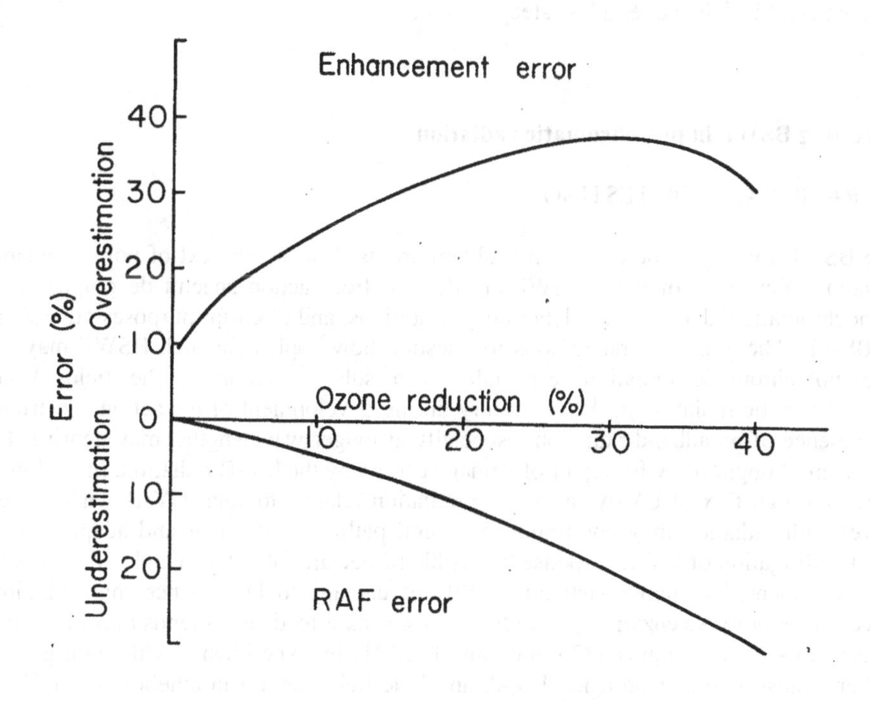

Figure 5. Relative RAF and enhancement errors as a function of ozone reduction for field UV supplementation experiments under temperate-latitude conditions in the summer (40° latitude at the summer solstice). The scenario for these errors is that the RAF values and the enhancements were calculated following the generalized plant action spectrum when the DNA damage spectrum was indeed more appropriate. From Caldwell et al. [11].

The enhancement error is in the opposite direction from the RAF error in this scenario (Figure 5). Even at low levels of simulated ozone reduction, the error is appreciable. This error increases at higher ozone depletion levels, but not in the same trajectory as the RAF error. The so-called "enhancement error" arises because the spectral composition of UV coming from the lamps is different from that of sunlight with ozone reduction; it contains proportionately more shortwave UV-B radiation [11]. (If the lamp could supply an increment of UV-B radiation with exactly the same wavelength composition as the increased solar UV-B radiation resulting from ozone reduction, then this "enhancement error" would not exist.) Thus, the shape of BSWF's are critical when they are used in experiments and in experimental simulations of solar radiation with ozone reduction. If the two BSWF in Figure 2 had been used in portraying this scenario of RAF and enhancement errors, the errors would be much larger.

In general, a BSWF that falls more steeply with increasing wavelength will yield a greater RAF. But, in experiments involving fluorescent lamp systems designed to supply added UV in a manner simulating reduced ozone, a steeper BSWF results in less supplemental UV than does a less steep BSWF.

2. Testing BSWF in polychromatic radiation

2.1. RATIONALE FOR TESTING

The BSWF for the ozone reduction problem are used in the context of polychromatic radiation. Yet, most of these BSWF are derived from action spectra developed with monochromatic radiation under laboratory conditions, and often for purposes other than as BSWF. There are several reasons to question how applicable such BSWF may be with polychromatic radiation, especially with solar radiation in the field. With monochromatic radiation in the laboratory during development of the action spectrum, the absence of broadband radiation, especially at longer wavelengths, may deprive the organism of opportunity for repair of damage caused by the UV-B radiation. In sunlight, there is a high flux of UV-A and visible radiation relative to solar UV-B. This longer wavelength radiation may contribute to several pathways of repair and adaptation to UV-B. Mitigation of UV-B response by visible (especially blue light) and UV-A flux is best documented in photoreactivation (PR) of damage to DNA since this radiation drives the photolyase enzyme. For higher plants, peak effectiveness tends to occur at the longer UV-A wavelengths (375-400 nm) [12,13]. In experiments with light-grown higher plants, both supplemental UV-A and blue light can aid in ameliorating UV-B-caused chlorosis [14]; this may or may not involve PR.

In general, for plant growth and morphological response to UV, several phenomena may interact: DNA damage [13, 15, 16], free radical formation [7, 17, 18], and the action of photoreceptors such as flavins [19,20,21]. The manner in which UV-A and PFD interact with UV-B can also involve several possibilities. For example, UV-A and blue light can drive the photolyase enzyme to repair a common lesion of DNA caused by UV-B [22,23], in the process of photoreactivation (PR); UV-B can induce PR and other DNA repair systems [13, 24]; UV-A can cause growth delay (at least in bacteria) allowing more time for dark DNA repair [25], but UV-B and UV-A at high flux rates can damage DNA repair systems [26]. For free radical damage, UV-A can stimulate carotenoid quenching [26]. The blue light receptor appears to be important in mitigating UV-B damage [27]. Interactions between the UV-B, UV-A/blue, and phytochrome receptors have also been demonstrated [28, 29]. Thus, there are many potential pathways by which radiation of different wavelengths may interact. In the laboratory with monochromatic radiation the plant is exposed to radiation of only one wavelength at a time and this removes the possibility of the various wavelength interactions that might be expected for plants exposed to sunlight in nature. Also, the time frame of exposures during action spectrum development with monochromatic radiation in the laboratory usually involves hours, or even fractions of an hour. Yet, organisms in nature are exposed to the full spectrum of solar radiation for days, weeks or longer. Thus, BSWF derived from such action spectra need to be tested in polychromatic radiation. Ideally, this should include longer wavelength radiation at rather high flux, as occurs in sunlight.

2.2. SOME EXAMPLES OF TESTS OF BSWF

There are data sets in the literature and experiments still in progress in which the explicit goal was to test BSWF under polychromatic radiation. There are also a number of data sets that involve tests of BSWF with polychromatic radiation, even though this was not the primary purpose of the work. A collection of data sets are briefly reviewed here and the salient results summarized. The polychromatic radiation was from lamps, solar radiation or a combination of these. A reasonable test should include polychromatic combinations of some rather different wavelength compositions.

Hunter et al. [30] compared the effects of solar UV on anchovy larvae survival with that from filtered fluorescent sunlamps in the laboratory. The two dose response relationships were considerably different when plotted with dose as total unweighted UV-B radiation (Figure 6a). However, if the UV from the two sources was weighted with a BSWF based on an action spectrum they developed for larval survival, the two dose-response relationships overlapped (Figure 6b). This was not the primary goal of their work, but serves to suggest that this BSWF provided a realistic biologically effective radiation calculation for radiation sources of very disparate spectral composition.

In a similar fashion using combinations of polychromatic radiation from lamp systems, from filtered sunlight, or both, several tests of BSWF have been reported. Plant stem elongation was tested under different spectral distributions of radiation from a xenon arc lamp in a growth chamber and in the field under a combination of sunlight and filtered fluorescent UV lamps [31]. The generalized plant BSWF fit well at shorter wavelengths (<310 nm) but underestimated the response at longer UV-B and UV-A wavelengths. Mazza et al. [32] used different combinations of filtered sunlight to establish that induction of UV-B-absorbing compounds (e.g., flavonoids) was largely the result of exposure to shorter wavelengths (<315 nm) and was reasonably consistent with the generalized plant spectrum. Similar results were reported with different species by Krizek et al. [33].

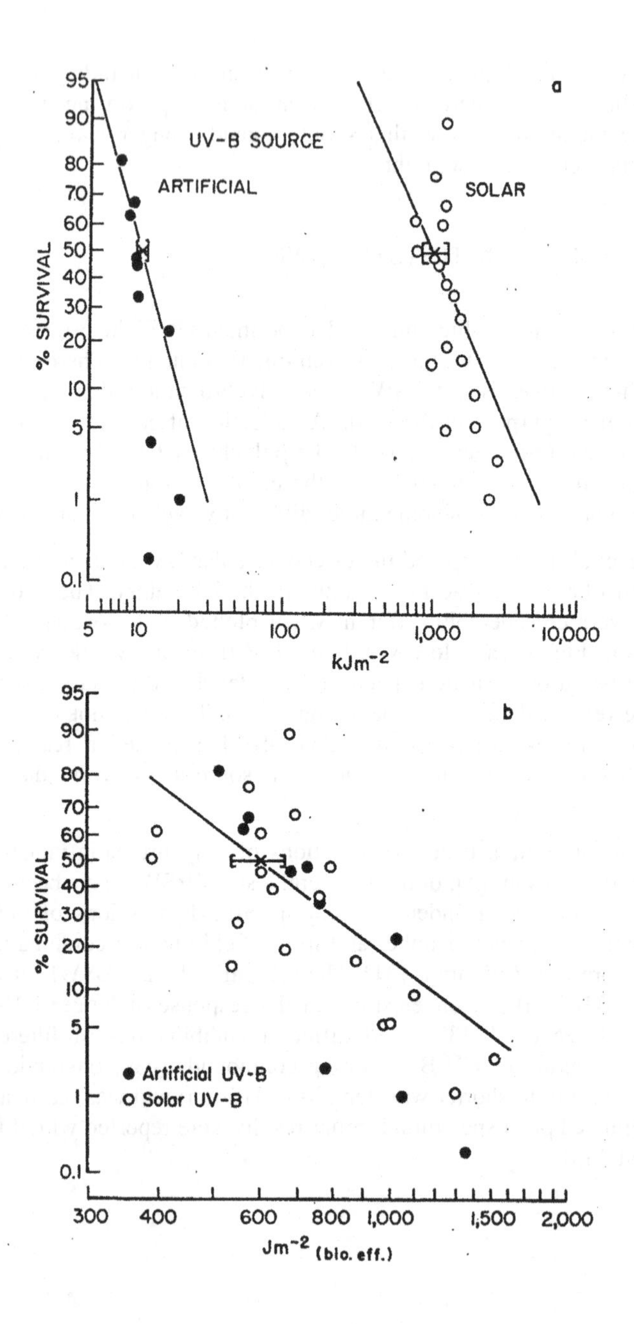

Figure 6. (a) Relation between unweighted UV-B dose cumulated over 12 days and survival of larval anchovy exposed to artificial (solid circle) and solar (open circle) radiation. The lines represent the regression of survival against $\log_{10}$ dose; (b) Relation between weighted UV-B dose using an action spectrum they developed for survival as a BSWF cumulated over 12 days for both artificial and solar exposures. From Hunter et al. [30].

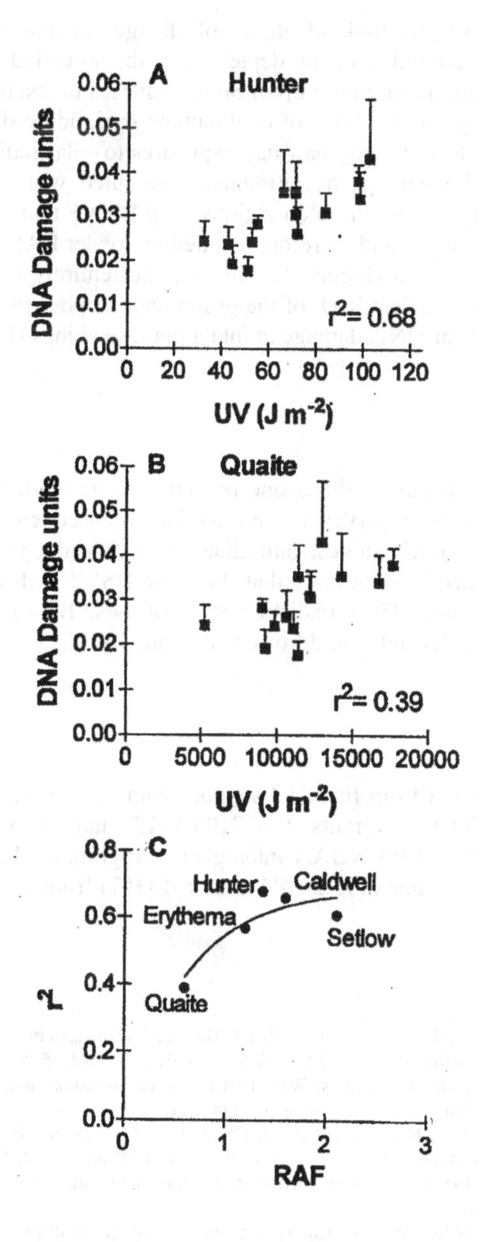

Figure 7. Leaf DNA damage (in form of cyclobutane pyrimidine dimers) of *Gunnera magellanica* exposed for half-day periods to different intensities and spectral combinations of solar radiation in Tierra del Fuego. (a) When the solar spectral irradiance was weighted according to the BSWF developed by Hunter et al. [34] for anchovy larvae survival. (b) When weighted according to the BSWF of Quaite et al. [16] for DNA damage in alfalfa seedlings. (c) Goodness of fit (as coefficient of determination, r^2) of five different BSWF as a function of the RAF that result from these BSWFs. From [34].

Rousseaux et al. [34] took advantage of changes in solar spectral irradiance as the Antarctic vortex containing ozone depleted air, the so-called "ozone hole", passed over Tierra del Fuego at the southern tip of South America to test different BSWF. They measured DNA damage in the form of cyclobutane pyrimidine dimers extracted from intact leaves in the field following half-day exposures to solar radiation. In this manner they tested how well solar spectral irradiance weighted with five different BSWF predicted DNA damage under these conditions. The BSWF that declined more steeply with increasing wavelength (and therefore resulted in greater RAF values) generally had the better fit with these data (Figure 7). This was not entirely expected since the one BSWF tested with the smallest RAF of the group had the poorest fit and yet was based on an action spectrum for DNA damage in intact plant seedlings (Figure 7).

3. Conclusion

The use of BSWF in assessing the ozone reduction issue is clearly highly important. Even small differences among BSWF can have large consequences when employed in various predictions of RAF values, latitudinal gradients and in experimental design involving lamp systems. It is unlikely that the same BSWF will be appropriate for all responses to UV radiation. Thus, realistic testing of these BSWF under polychromatic radiation, especially under field conditions, is critical.

Acknowledgements

This work derives in part from financial support from the United States Department of Agriculture (CSRS/NRICG grants 95-37100-1612 and 98-35100-6107) and the NSF/DOE/NASA/USDA/EPA/NOAA Interagency Program on Terrestrial Ecology and Global Change (TECO) grants (95-24144 and 98-14357) from the US National Science Foundation.

References

1. Madronich, S; McKenzie, RL; Björn, LO; Caldwell, MM (1998) Changes in biologically active ultraviolet radiation reaching the Earth's surface. *J. Photochem. Photobiol. B: Biol.* 46, 5-19.
2. Caldwell, MM; Robberecht, R; Billings, WD (1980) A steep latitudinal gradient of solar ultraviolet-B radiation in the arctic-alpine life zone. *Ecology* 61, 600-611.
3. Caldwell, MM; Gold, WG; Harris, G; Ashurst, CW (1983) A modulated lamp system for solar UV-B (280-320 nm) supplementation studies in the field. *Photochem. Photobiol.* 37, 479-485.
4. Inn, ECY; Tanaka, Y (1953) Absorption coefficient of ozone in the ultraviolet and visible regions. *J. Opt. Soc. Amer.* 43, 870-873.
5. Dave, JV; Halpern, P (1976) Effect of changes in ozone amount on the ultraviolet radiation received at sea level of a model atmosphere. *Atmos. Environ.* 10, 547-555.
6. Caldwell, MM (1981) Plant response to solar ultraviolet radiation in: Lange, OL; Nobel, PS; Osmond, CB; Ziegler, H (eds.) Encyclopedia of Plant Physiology, Vol. 12A Physiological Plant Ecology I. Responses to the Physical Environment. Springer, Berlin, pp. 169-197.
7. Cen, YP; Björn, LO (1994) Action spectra for enhancement of ultraweak luminescence by UV radiation (270-340 nm) in leaves of *Brassica napus*. *J. Photochem. Photobiol. B: Biol.* 22, 125-129.

8. Setlow, RB (1974) The wavelengths in sunlight effective in producing skin cancer: a theoretical analysis. *Proc. Nat. Acad. Sci.* 71, 3363-3366.
9. Caldwell, MM (1971) Solar ultraviolet radiation and the growth and development of higher plants. in: Giese, AC (ed.) Photophysiology. Vol. 6. Academic Press, New York, pp. 131-177.
10. Caldwell, MM; Flint, SD (1997) Uses of biological spectral weighting functions and the need of scaling for the ozone reduction problem. *Plant. Ecol.* 128, 66-76.
11. Caldwell, MM; Camp, LB; Warner, CW; Flint, SD (1986) Action spectra and their key role in assessing biological consequences of solar UV-B radiation change. in: Worrest, RC; Caldwell, MM (eds.) Stratospheric Ozone Reduction, Solar Ultraviolet Radiation and Plant Life. Springer, Berlin, pp. 87-111.
12. Ikenaga, M; Kondo, S; Fujii, T (1974) Action spectrum for enzymatic photoreactivation in maize. *Photochem. Photobiol.* 19, 109-113.
13. Pang, Q; Hays, JB (1991) UV-B-inducible and temperature-sensitive photoreactivation of cyclobutane pyrimidine dimers in *Arabidopsis thaliana. Plant Physiol.* 95, 536-543.
14. Adamse, P; Britz, SJ (1996) Rapid fluence-dependent responses to ultraviolet-B radiation in cucumber leaves: the role of UV-absorbing pigments in damage protection. *J. Plant. Physiol.* 148, 57-62.
15. Beggs, CJ; Schneider-Ziebert, U; Wellmann, E (1986) UV-B radiation and adaptive mechanisms in plants. in: Worrest, RC; Caldwell, MM (eds.) Stratospheric Ozone Reduction, Solar Ultraviolet Radiation and Plant Life. Springer, Berlin, pp. 235-250.
16. Quaite, FE; Sutherland, BM; Sutherland, JC (1992) Action spectrum for DNA damage in alfalfa lowers predicted impact of ozone depletion. *Nature* 358, 576-578.
17. Panagopoulos, I; Bornman, JF; Björn, LO (1990) Effects of ultraviolet radiation and visible light on growth, fluorescence induction, ultraweak luminescence and peroxidase activity in sugar beet plants. *J. Photochem. Photobiol. B: Biol.* 8, 73-87.
18. Balakumar, T; Gayathri, B; Anbudurai, PR (1997) Oxidative stress injury in tomato plants induced by supplemental UV-B radiation. *Biologia Plantarum* 39, 215-221.
19. Ballaré, CL; Barnes, PW; Kendrick, RE (1991) Photomorphogenic effects of UV-B radiation on hypocotyl elongation in wild type and stable-phytochrome-deficient mutant seedlings of cucumber. *Physiol. Plant.* 83, 652-658.
20. Ballaré, CL; Barnes, PW; Flint, SD (1995) Inhibition of hypocotyl elongation by ultraviolet-B radiation in de-etiolating tomato seedlings. I. The photoreceptor. *Physiol. Plant.* 93, 584-592.
21. Wilson, MI; Greenberg, BM (1993) Specificity and photomorphogenic nature of ultraviolet-B-induced cotyledon curling in *Brassica napus* L. *Plant Physiol.* 102, 671-677.
22. Jagger, J; Stafford, RS; Snow, JM (1969) Thymine-dimer and action-spectrum evidence for indirect photoreactivation in *Escherichia coli. Photochem. Photobiol.* 10, 383-395.
23. Takeuchi, Y; Murakami, M; Nakajima, N; Kondo, N; Nikaido, O (1998) The photorepair and photoisomerization of DNA lesions in etiolated cucumber cotyledons after irradiation by UV-B depends on wavelength. *Plant Cell Physiol.* 39, 745-750.
24. Menezes, S; Tyrrell, RM (1982) Damage by solar radiation at defined wavelengths: involvement of inducible repair systems. *Photochem. Photobiol.* 36, 313-318.
25. Jagger, J (1981) Near-UV radiation effects on microorganisms. Photochem. Photobiol. 34, 761-768.
26. Webb, RB (1977) Lethal and mutagenic effects of near-ultraviolet radiation. *Photochemical and Photobiological Reviews* 2, 169-261.
27. Adamse, P; Britz, SJ; Caldwell, CR (1994) Amelioration of UV-B damage under high irradiance. II. Role of blue light photoreceptors. *Photochem. Photobiol.* 60, 110-115.
28. Mohr, H (1986) Coaction between pigment systems. in: Kendrick, RE; Kronenberg, GHM (eds.) Photomorphogenesis in Plants. Martinus Nijhoff, Dordrecht, pp. 547-564.
29. Gaba, V; Black, M (1987) Photoreceptor interaction in plant photomorphogenesis: the limits of experimental techniques and their interpretations. *Photochem Photobiol.* 45, 151-156.
30. Hunter, JR; Kaupp, SE; Taylor, JH (1981) Effects of solar and artificial ultraviolet-B radiation on larval northern anchovy, *Engraulis mordax. Photochem. Photobiol.* 34, 477-486.
31. Flint, SD; Caldwell, MM (1996) Scaling plant ultraviolet spectral responses from laboratory action spectra to field spectral weighting factors. *J. Plant. Physiol.* 148, 107-114.
32. Mazza, CA; Boccalandro, HE; Giordano, CV; Battista, D; Scopel, AL; Ballare, CL (2000) Functional significance and induction by solar radiation of ultraviolet-absorbing sunscreens in field-grown soybean crops. *Plant Physiol.* 122, 117-125.

33. Krizek, DT; Britz, SJ; Mirecki, RM (1998) Inhibitory effects of ambient levels of solar UV-A and UV-B radiation on growth of cv. New Red Fire lettuce. *Physiol. Plant.* 103, 1-7.
34. Rousseaux, MC; Ballare, CB; Giordano, CV; Scopel, AL; Zima, AM; Szwarcberg-Bracchitta, M; Searles, PS; Caldwell, MM; Diaz, SB (1999) Ozone depletion and UVB radiation: Impact on plant DNA damage in southern South America. *Proc. Nat. Acad. Sci.* 96, 15310-15315

RESPONSE TO UV-B RADIATION: WEIGHTING FUNCTIONS AND ACTION SPECTRA

FRANCESCO GHETTI, COSTANZA BAGNOLI AND GIOVANNI CHECCUCCI
CNR Istituto di Biofisica
Via G. Moruzzi 1, I-56100 Pisa, Italy

1. Introduction

In the last twenty years, the presence in the atmosphere of ozone-depleting substances (CFCs, HCFCs, halons, carbon tetrachloride, etc.) has been reducing the ozone concentration in the stratosphere over high and mid-latitudes of both hemispheres. Notwithstanding ozone concentration in the atmosphere is very low, the stratospheric layer, which contains approximately 90% of the total ozone, has the function of a protective filter for the Earth's surface, cutting-off solar radiation under 280 nm and greatly reducing UV-B radiation (280-315 nm).

The reduction of stratospheric ozone has been recognized as the main cause of the increase of UV-B irradiance at the Earth's surface. This increase has been estimated in the range 6-14% in the last twenty years. Also spectral measurements performed at various sites located in Europe, North and South America, Antarctica and New Zealand have correlated UV-B radiation increase with the ozone decline.[1]

UV-B has various direct adverse effects on human health (skin cancer, immunosuppression, eye disorders), terrestrial plants and aquatic organisms (DNA alterations, photosynthesis inhibition, reduced growth). Moreover, due to the differences in UV-B sensitivity and adaptation among the various species, shifts in species composition may occur as a consequence of increased UV-B radiation, thus leading indirectly to alterations in ecosystems.[2-4]

As the enhancement of solar UV-B is highly wavelength specific and increases when the wavelength decreases, action spectroscopy plays a central role in assessing the effects of ozone depletion on the biosphere.[5,6]

2. Action spectroscopy

Action spectroscopy is a non-destructive technique for studying *in vivo* the absorption properties and, in some cases, the primary reactions of photoreceptors involved in triggering biological photoresponses. Action spectroscopy can be used in the investigation of any light-dependent phenomenon for which a standard response can be defined. It consists in the quantitative analysis of the response of the system as a function of the wavelength of the stimulating light and the outcome of this analysis, the action spectrum, is a measure of the relative efficiency of light of different wavelengths in inducing a defined effect on the examined biological system.

F. Ghetti et al. (eds.), Environmental UV Radiation: Impact on Ecosystems and Human Health and Predictive Models, 85–93.

A prerequisite for the determination of an action spectrum is the definition of a biological parameter for quantifying the photoresponse, such as, for example, the length or the weight of a plant, the concentration of a gas or an enzyme, etc..

An action spectrum is the plot of the effectiveness of the light in inducing the observed photoresponse as a function of wavelength. The effectiveness of the light in inducing the photoresponse can be quantified, for example, as the reciprocal of the amount of light required by the system to produce a fixed level of response. A simple procedure to determine an action spectrum could be to choose a standard level of response and to measure at each wavelength the radiation intensity inducing that response. However, a necessary prerequisite of action spectroscopy is the proportionality between magnitude of the response and intensity of the stimulus; therefore, to avoid distortion of the action spectrum, a succession of graded responses should be measured, for each wavelength, to construct an intensity-response curve and to verify that the chosen standard level of response fall on the linear part of the curve.

Other wavelength-dependent distortions of the action spectrum could be due to the molecular environment where the target molecule is embedded. In fact, the presence of other pigments can cause screening or reflection effects, in such a way that the effectiveness of a particular spectral range on the target molecule could be reduced or enhanced, respectively. Moreover, in a molecular environment with many scattering centres, the efficiency in eliciting the photoresponse of the shorter wavelengths can be apparently decreased, because of the dependence of the diffusion on the inverse of the wavelength. The sample should also not absorb too much radiation at the tested wavelengths, so that, at least ideally, every molecule involved in the photoprocess could be exposed approximately to the same amount of photons and, consequently, have the same probability of responding.

An important criterion to determine if the intensity-response curves at the different wavelengths are indicative of the same mechanism of action is to check if they can be reasonably superimposed, by multiplying each of them by an appropriate factor, thus showing that they have the same "shape". In other words, given any pair of intensity-response curves, the ratio of two fluence rates corresponding to the same level of response should be approximately a constant for any level of response and, consequently, the difference of their logarithms should be also a constant. This implies that the curves plotted on a logarithmic abscissa should appear parallel. If a fluence rate-response curve has been determined with a sufficient number of data points, it can be fitted with an appropriate mathematical function. This, besides allowing the calculation of the fluence rates for any standard level of response, makes it possible to check more precisely if the curves have the same shape.

3. Environmental UV action spectroscopy

Action spectroscopy can also provide significant information in the field of ecophysiological plant research and, more generally, in the field of environmental photobiology. In fact, action spectroscopy is an useful tool for investigating the biological effects of ultraviolet radiation, and, in particular, of the UV-B wavelengths (280-315 nm), on terrestrial plants and phytoplankton organisms.[6]

The effectiveness spectrum ($ES(\lambda)$) is the product of an action spectrum ($AS(\lambda)$) for a biological effect with a given solar irradiance spectrum $E(\lambda)$, originated by a particular ozone-depletion scenario; it gives indication on the biological significance of the change in solar UV-B.

$$ES(\lambda) = AS(\lambda)E(\lambda)$$

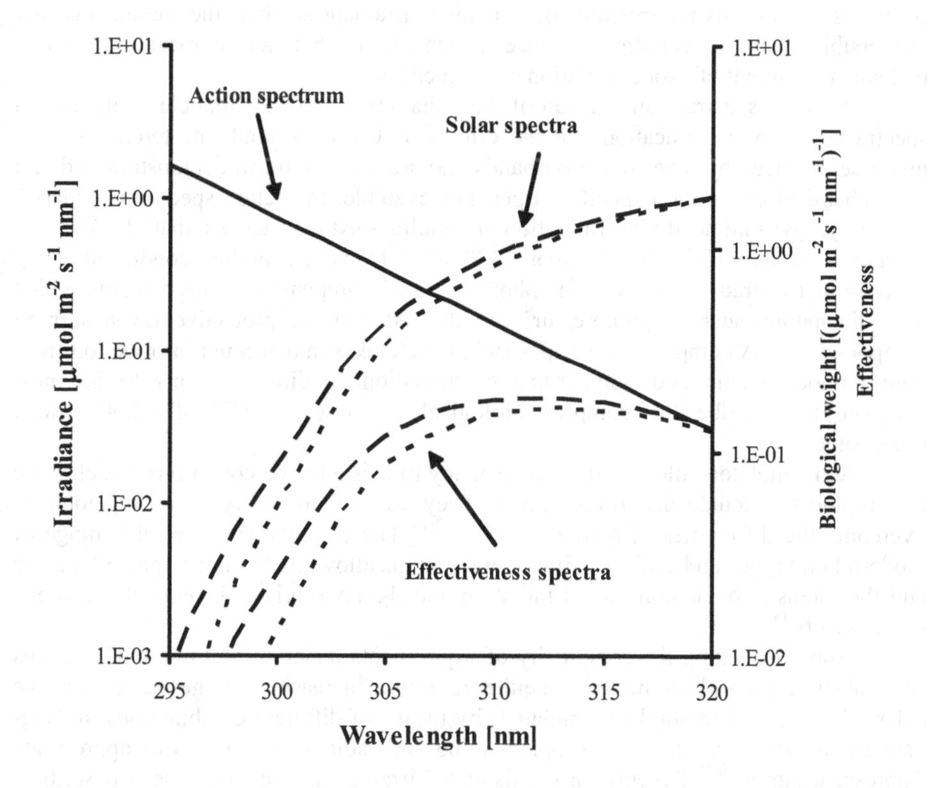

Figure 1. Effectiveness spectra. Two different solar spectra multiplied by an action spectrum (in which the biological weight is expressed in reciprocal units of irradiance) give two different effectiveness spectra. The areas under the two effectiveness spectrum curves are an useful measure of the overall effect of the two different irradiation conditions.

The overall effect of a particular irradiation condition can be estimated by means of the biologically active irradiance (E_{BA}), defined as the area under the ES curve:

$$E_{BA} = \int AS(\lambda)E(\lambda)d\lambda$$

Ecologically relevant studies, performed in order to determine action spectra, require the careful consideration of various experimental features. Treatment duration should be adequate, because short exposures may not allow to observe the overall response of the organism; the examined UV effects should be tested on whole organisms, because *in vitro* studies can have limitations for extrapolation to the natural performance of the whole organism; the presence of reciprocity effects should be carefully tested, when performing experiments with different duration at different wavelengths.[7]

An important aspect to consider is the use of irradiation conditions not too far from present levels of natural radiation or those under predictable ozone depletion scenarios. Using extraterrestrial or so high irradiances that the organisms are irreversibly damaged certainly produce a clear effect, but not so meaningful for a realistic assessment of ozone depletion consequences.

Also the spectral composition of the radiation is crucial. Monochromatic action spectra can provide indications on the effects of UV-B without interference due to processes induced by other radiation bands and are very useful to demonstrate a direct photochemical effect on a specific target. For example, the action spectrum for DNA cyclobutyl pyrimidine dimer formation in alfalfa seedlings shows that the damage extends well into UV-A up to about 370 nm.[8] However, studies conducted using polychromatic irradiation allow for photoregulated compensatory mechanisms which occur in nature, such as photorepair[9] and the induction of protective UV-absorbing compounds.[10,11] A complementary approach, which takes into account information from both monochromatic and polychromatic irradiation conditions, seems to be most adequate to describe the complex biological responses to UV of whole, intact organisms.

Sun simulators allow to reproduce at any time irradiation conditions and climatic factors and to exclude disturbing agents. They are therefore very promising tools to overcome the difficulties of outdoor studies.[12-14] The employment of a wide range of modern lamp types and different filter combinations allow to obtain the spectral quality and the intensity of natural global radiation and the appropriate shaping of the short-wave cut-off.[15,16]

Notwithstanding the complexity of experiments under natural radiation, various spectral studies have been performed either removing increasingly larger portions of the solar UV spectrum (natural or simulated) by means of different combinations of long-pass filters (see Fig. 2),[17-19] or supplementing UV radiation by means of appropriate fluorescent lamps.[20,21] Recently, a sophisticated irradiation system has been described, which uses both filters and supplementary UV lamps and can provide computer-controlled modulated changes in any radiation band between about 250 and 730 nm.[22-24]

Two different approaches has been proposed for extracting an action spectrum from experimental data obtained with polychromatic exposures in which UV irradiance is varied by means of filter combinations with progressively decreasing cut-off wavelength. Differential action spectroscopy (DAS) takes into account the spectral bands given by the difference in irradiance of two filter combinations with successive cut-off wavelength. Assuming that different spectral bands supplement one another in producing the observed effect, it is possible to estimate the weighted contribution of each band by dividing the differences in response to the treatment by the corresponding differential irradiances. Finally, the action spectrum is derived by plotting the weighted effects as a function of the average wavelength of the corresponding band. As UV action spectra tends to exponentially increase with decreasing wavelength, the main limit of this method is given by the potentially large variability of the response in the range of each differential band, which could significantly alter the shape of the resulting action spectrum.[25] This limit, however, is present also for monochromatic action spectroscopy, when using currently available interference filters. The differential method can be improved by using several irradiance levels for each differential band, thus reducing statistical errors and allowing a more correct shaping of the action spectrum.[7]

Figure 2. Polychromatic action spectroscopy. Increasingly larger portions of a simulated solar spectrum are removed by means of different combinations of long-pass filters, in order to determine an action spectrum of the UV inhibition of photosynthesis in microalgae.

If DAS reduces the problem to the monochromatic situation, the other approach proposed by Rundel[25] is a deconvolution technique in which it is assumed that the action spectrum (which in this case can be defined as "biological weighting function", BWF[26]) is an analytical continuous function $\varepsilon(\lambda)$ and the predicted system response R_i to the i-th irradiation regime $E_i(\lambda, t)$ can be expressed as depending on the biologically active irradiance integrated over the irradiation time:

$$R_i = F[\iint_{t,\lambda} \varepsilon(\lambda) E_i(\lambda, t) d\lambda dt]$$

Once a functional form for $\varepsilon(\lambda)$ has been chosen in dependence of a given number of parameters, the latter can be determined by means of iterative calculations to minimize the differences between each R_i and the corresponding measured response. The limit of this method is that the *a priori* choice of both the form of $\varepsilon(\lambda)$ and the functional relationship F between response and exposure is critical for the correct determination of the action spectrum. However, the use of multivariate statistical analysis (such as, for example, principal component analysis) allows to obtain more detailed BWFs.[26,27]

The method proposed by Rundel[25] was used to retrieve a biological weighting function of the UV inhibition of photosynthesis in the green microalga *Dunaliella salina*, after exposure to simulated solar radiation for 30 minutes.[12]

In this work, the combination of two PAR (Photosynthetic Active Radiation) levels and two UV levels allowed to obtain four different basic radiation regimes: high UV and high PAR (UVPAR), low UV and high PAR (uvPAR), high UV and low PAR (UVpar), low UV and low PAR (uvpar). For each of these four regimes ten different irradiation conditions were applied, ranging from a full simulator spectrum to an almost complete exclusion of UV radiation. The first was obtained by means of an UV transparent quartz plate and the others by using various colour filter glasses (WG 295, WG 305, WG 320, WG 335, WG 345, WG 360, GG 385, GG 400, GG 420). The inhibition of photosynthesis was evaluated by measuring the chlorophyll fluorescence parameter, ΔF/Fm', which provides a reliable estimate of photosynthetic efficiency.

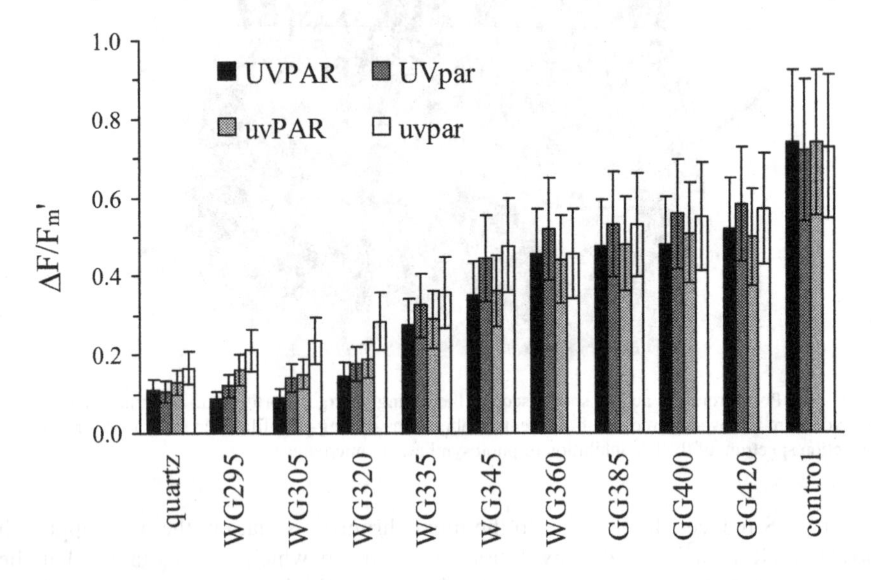

Figure 3. Chlorophyll fluorescence parameter **ΔF/Fm'** measured after the 30 min exposure to the forty different irradiation conditions of simulated solar radiation (redrawn from reference 12).

A biological weighting function was retrieved from the 40 values of the fluorescence data (see Fig. 3) assuming that the value P of ΔF/Fm' after irradiation could be calculated by means of the following model function:

$$P = P_0 \, [1/(1 + \Sigma\varepsilon_{UV}(\lambda)E_{UV}(\lambda)\Delta\lambda + \varepsilon_{PAR}E_{PAR})]$$

where P_0 is the control value; $E_{UV}(\lambda)$ is the spectral photon flux density in the UV and E_{PAR} is the integrated photon flux density in the PAR range; $\varepsilon_{UV}(\lambda)$ and ε_{PAR} (in reciprocal photons per square meter and per second) are the spectral biological weighting function for UV and the average biological weighting factor for PAR, respectively. As previously determined biological weighting functions showed a marked difference in slope in the UV-B as compared to UV-A, the analytical continuous function representing $\varepsilon_{UV}(\lambda)$ was chosen formed by two different exponential functions: the first one from 285 nm to a wavelength λ_0 and the second from λ_0 to 400 nm, the wavelength λ_0 (285 nm < λ_0 < 400 nm) being a free parameter of the fitting procedure. To avoid discontinuity in the UV, the constraint that the two functions had the same value at λ_0 was used. The best fit, obtained by means of an iterative process based on the least-squares method, returned the value of 331 nm for λ_0 and yielded the following components for the biological weighting function (see Fig. 4):

$$\varepsilon_{UV}(\lambda) = 6.1164\times10^{15}\times e^{-0.12\times\lambda} \qquad 285\text{ nm} < \lambda < 331\text{ nm}$$

$$\varepsilon_{UV}(\lambda) = 3.235\times10^{3}\times e^{-0.0346\times\lambda} \qquad 331\text{ nm} < \lambda < 400\text{ nm}$$

$$\varepsilon_{PAR} = 3.17\times10^{-4} \qquad 400\text{ nm} < \lambda < 700\text{ nm}$$

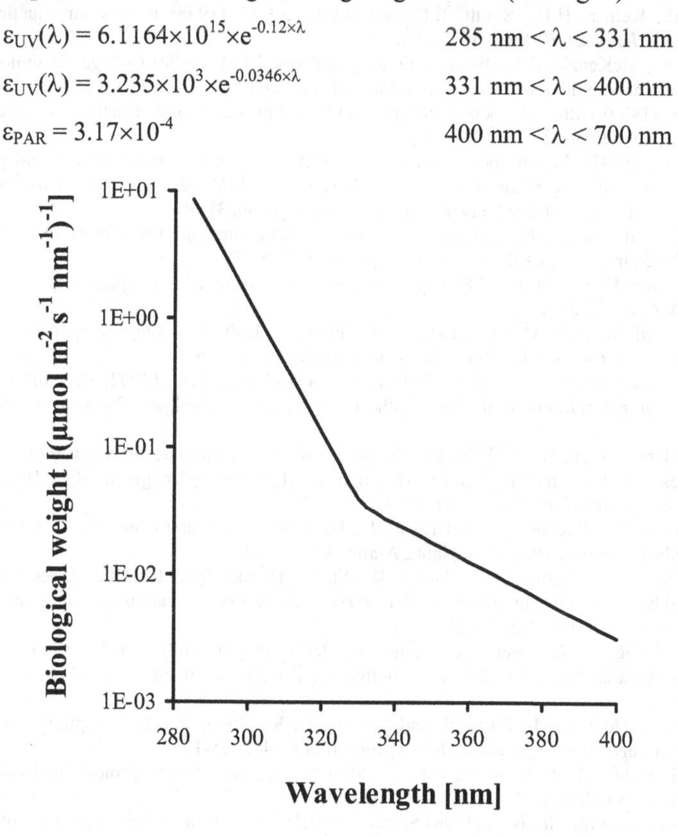

Figure 4. Biological weighting functions for the inhibition of the chlorophyll fluorescence parameter ΔF/Fm' in *D. salina* (redrawn from reference 12).

This biological weighting function for the UV inhibition of photosynthesis in *D. salina* shows quite a good agreement, both in shape and absolute magnitude, with that determined for the microalga *Phaeodactylum*.[27]

Acknowledgement. This work was supported by the EC Project EUVAS (Contract ENV4-CT97-0538)

References

1. United Nations Environment Program (UNEP), World Meteorological Organization (WMO) (2002) *Scientific assessment of ozone depletion: 2002 - Executive summary*, http://www.unep.org/ozone/sap2002.shtml.
2. Longstreth, J., de Gruijl, F.R., Kripke, M.L., Abseck, S., Arnold, F., Slaper, H.I., Velders, G., Takizawa, Y. and van der Leun, J.C. (1998) Health risks, *J. Photochem. Photobiol. B: Biol.*, 46: 20-39.
3. Caldwell, M.M., Björn, L.O., Bornmann, J., Flint, S.D., Kulandaivelu, G., Teramura, A.H. and Tevini, M. (1998) Effects of increased solar ultraviolet radiation on terrestrial ecosystems, *J. Photochem. Photobiol. B: Biol.*, 46: 40-52.
4. Häder, D.-P., Kumar, H.D., Smith, R.C. and Worrest, R.C. (1998) Effects on aquatic ecosystems, *J. Photochem. Photobiol. B: Biol.*, 46: 53-68.
5. Madronich, S., McKenzie, R.L., Björn, L.O. and Caldwell, M.M. (1998) Changes in biologically active UV radiation reaching the Earth's surface, *J. Photochem. Photobiol. B: Biol.*, 46: 5-19.
6. Coohill, T.P. (1989) Ultraviolet action spectra (280 to 380 nm) and solar effectiveness spectra for higher plants, *Photochem. Photobiol.* 50: 451-457.
7. Holmes, M.G. (1997) Action spectra for UV-B effects on plants: monochromatic and polychromatic approaches for analysing plant responses. In *Plant and UV-B: responses to environmental change* (P.J. Lumsden, ed.), Cambridge University Press,Cambridge, pp. 31-50.
8. Quaite, F.E., Sutherland, B.M. and Sutherland, J.C. (1992) Action spectrum for DNA damage in alfalfa lowers predicted impact of ozone depletion, *Nature*, 358: 576-578.
9. Sinha, R.P. and Häder, D.-P. (2002) UV-induced DNA damage and repair: a review, *Photochem. Photobiol. Sci.*, 1: 225-236.
10. Rozema, J. and Björn, L.-O. (2002) (Eds.), Evolution of UVB absorbing compounds in aquatic and terrestrial plants, *J. Photochem. Photobiol. B: Biol.*, Special Issue, 66.
11. Häder, D.-P., Lebert, M., Sinha, R.P., Barbieri, E. and Helbling, E.W. (2002) Role of protective repair mechanisms in the inhibition of photosynthesis in marine microalgae, *Photochem. Photobiol. Sci.*, 1: 809-814.
12. Ghetti, F., Hermann, H., Häder, D.-P. and Seidlitz, H.K. (1999) Spectral dependence of the inhibition of photosynthesis under simulated global radiation in the unicellular green alga *Dunaliella salina*, *J. Photochem. Photobiol. B:Biol.*, 48: 166-173.
13. Ries, G., Heller, W., Puchta, H., Sandermann Jr., H., Seidlitz, H.K and Hohn, B. (2000) Elevated UV-B radiation reduces genome stability in plants, *Nature* 406: 98-101.
14. Ibdah, M., Krins, A., Seidlitz, H. K., Heller, W., Strack, D. and Vogt, T. (2002) Spectral dependence of flavonol and betacyanin accumulation in *Mesembryanthemum crystallinum* under enhanced UV radiation, *Plant, Cell and Environment*, 25: 1145-1154.
15. Thiel, S., Döhring, T., Köfferlein, M., Kosak, A., Martin, P. and Seidlitz, H. K. (1996) A phytotron for plant stress research: how far can artificial lighting compare to natural sunlight?, *J. Plant Physiol.*, 148: 456-463.
16. Döhring, T., Köfferlein, M., Thiel, S. and Seidlitz, H. K. (1996) Spectral shaping of artificial UV-B irradiation for vegetation stress research, *J. Plant Physiol.* 148: 115-119.
17. Searles, P.S., Caldwell, M.M. and Winter, K. (1995) The response of five tropical dicoltyledon species to solar ultraviolet-B radiation, *Am. J. Bot.*, 82: 445-453.
18. Boucher, N.P. and Prézelin, B.B. (1996) Spectral modeling of UV inhibition of *in situ* Antarctic primary production using a field-derived biological weighting function, *Photochem. Photobiol.* 64: 407-418.
19. Hermann, H., Häder, D.-P. and Ghetti, F. (1997) Inhibition of photosynthesis by solar radiation in *Dunaliella salina*: relative efficiencies of UV-B, UV-A and PAR, *Plant, Cell and Environment*, 20: 359-365.
20. Caldwell, M.M, Gold, W.G., Harris, G. and Ashurst, C.W. (1983) A modulated lamp system for solar UV-B (280-320 nm) supplementation studies in the field, *Photochem. Photobiol.*, 37: 479-485.
21. McLeod, A.R. (1997) Outdoor supplementation systems for studies of the effects of increased UV-B radiation, *Plant. Ecol.*, 128: 78–92.

22. Holmes, M.G. (2002) An outdoor multiple wavelength system for the irradiation of biological samples: analysis of the long-term performance of various lamps and filter combinations, *Photochem. Photobiol.*, 76: 158-163.
23. Cooley, N.M., Truscott, H.M.F., Holmes, M.G. and Attrige, T.H. (2000) Outdoor ultraviolet polychromatic action spectra for growth responses of *Bellis perennis* and *Cynosurus cristatus*, *J. Photochem. Photobiol. B: Biol.*, 59: 64-71.
24. Holmes, M.G. and Keiler, D.R. (2002) A novel phototropic response to supplementary ultraviolet (UV-B and UV-A) radiation in the siliquas of oilseed rape (*Brassica nspus* L.) grown under natural conditions, *Photochem. Photobiol. Sci.*, 1: 890-895.
25. Rundel, R.D. (1983) Action spectra and estimation of biologically effective UV radiation, *Physiol. Plant.*, 58: 360-366.
26. Cullen, J.J. and Neale, P.J. (1997) Biological weighting functions for the effects of ultraviolet radiation on aquatic systems, In *The effects of ozone depletion on aquatic ecosystems*, (D.-P. Häder, ed.), Academic Press, R.G. Landes Company, Austin, 97-118.
27. Cullen, J.J., Patrick, J.N. and Lesser, M.P. (1992) Biological weighting function for the inhibition of phytoplankton photosynthesis by ultraviolet radiation, *Science*, 258: 646-650.

ELDONET – EUROPEAN LIGHT DOSIMETER NETWORK

DONAT-P. HÄDER AND MICHAEL LEBERT
Friedrich-Alexander-Universität, Institut für Botanik und Pharmazeutische Biologie, Staudtstr. 5, D-91058 Erlangen, Germany

1. Abstract

A network of three channel dosimeters has been installed in Europe and other continents to continuously and automatically monitor solar radiation in the UV-B (280 – 315 nm), UV-A (315 – 400 nm) and PAR (photosynthetic active radiation, 400 – 700 nm) wavelength ranges to follow long-term and short-term changes in the light climate. The instruments are housed in rugged cases to withstand extreme environments from the polar circle to the tropics. The entrance optic is based on an integrating sphere and the wavelength selection is done by appropriate filter and photodiode combinations. Software packages have been developed to poll the data, display them graphically and store them. Quality control is warranted by careful and frequent calibration of the instruments as well as national and international intercalibrations. The data are sent to a central server in Pisa from where they can be downloaded on the Internet free of charge. The network has been in operation for the last five years and has constantly been growing in numbers. The data are being used to extract important information on changing light climate conditions and the development of stratospheric ozone.

2. Background

All forms of the biota on our planet are affected by solar radiation. Scientists investigating these effects need precise instrumentation to measure the irradiation. Starting from sporadic measurements, systematic monitoring of solar radiation has been established to satisfy these needs. Distinct wavelength bands induce different responses, so that it became important to measure spectrally resolved data. Following the discovery of stratospheric ozone depletion caused by anthropogenic pollution, scientists became concerned about the effects of the resulting increasing UV-B radiation (280 - 315 nm, C.I.E. definition) on life[1]. One of the first study areas was an increase in UV-induced human skin cancers[2-3]. Furthermore, UV-B is suspected to induce cataracts, impair the human immune system and to be responsible for other health effects in humans and animals[4-6].

Other concerns deal with both wild and crop plants where UV-B causes adverse effects, including decreased productivity and reduced food quality[7-10]. Solar UV radiation inhibits photosynthesis[11-12], stomatal movement[13-14], growth and development in about 50 % of all plants investigated so far[15-18].

F. Ghetti et al. (eds.), Environmental UV Radiation: Impact on Ecosystems and Human Health and Predictive Models, 95–108.

Aquatic ecosystems are the third area of research[1,19-22] since they account for about 50 % biomass productivity on our planet. They also play a decisive role in global carbon fluxes[23]. Due to their essential need for solar energy the primary producers inhabit the top layers of the water column (the euphotic zone) where they are simultaneously exposed to solar ultraviolet radiation to which they have been found to be very sensitive[24-26].

To estimate the deleterious potential of various wavelength bands of solar radiation, several networks monitoring solar radiation have been installed or are in the planning stage. One is the Brewer network using spectroradiometers[27,28]. The Umweltbundesamt and the Bundesamt für Strahlenschutz have installed spectroradiometers at four stations in Germany (Offenbach, Schauinsland, Neuherberg, Zingst), which measure solar radiation at high spectral resolution (0.5 - 5 nm) as well as the integrated total UV spectrum. A major drawback is that, though in principle the data should be available to interested scientists, currently there is only limited personnel to evaluate the data, and the bureaucratic obstacles to actually obtain and use these data are difficult to overcome.

Other European activities include the SUVDAMA (spectral UV data and management) project initiated by Seckmeyer (Hannover, Germany) and the UV-radiative transfer using a set of well-defined spectroradiometers (Webb, Reading, England). Another project is based on the intercomparison of existing spectroradiometers (Bais, University of Thessaloniki, Greece).

Some networks specifically target the UV band to follow stratospheric ozone depletion. The Robertson-Berger (R-B) network has been measuring UV irradiance at eight stations since 1974[29,31]. The spectral sensitivity (290 - 330 nm) covers the wavelength range defined for erythemal sensitivity, but this does not coincide with the CIE (Commission Internationale d`Eclairage) definition of UV-B (280 - 315 nm). There was a long discussion why the R-B meter network indicated a decrease in solar UV radiation with time, while satellite data showed a gradual ozone depletion[32-33]. Eventually, this contradiction was solved by the observation that most R-B meters were deployed at meteorological stations in the neighborhood of airports, and the increasing atmospheric pollution due to higher air traffic more than cancelled the increase in UV-B reaching the surface[34]. Furthermore, a significant temperature sensitivity as well as aging were found over the years resulting in drifts in wavelength accuracy and amplitude which offset the factual increases in UV-B[35].

Other worldwide activities are located in New Zealand[36], the European Alps[37-38] and the Antarctic[39]. All these projects use spectrally resolved radiation data[27].

The rationale for the design of the European light dosimeter network (ELDONET) was to install many stations throughout Europe (and on other continents) to cover a wide land area. This approach excluded the installation of double monochromator spectroradiometers because of their considerable costs and demanding maintenance. Instead, rugged, accurate and reliable, but low-cost, three-channel dosimeters have been developed with filter functions corresponding to the UV-B (280 - 315 nm, CIE definition), UV-A (315 - 400 nm) and PAR (photosynthetically active radiation, 400 - 700 nm). The physically measured values can be convoluted with any biological filter function (e.g. DNA or plant sensitivity) to determine the biological

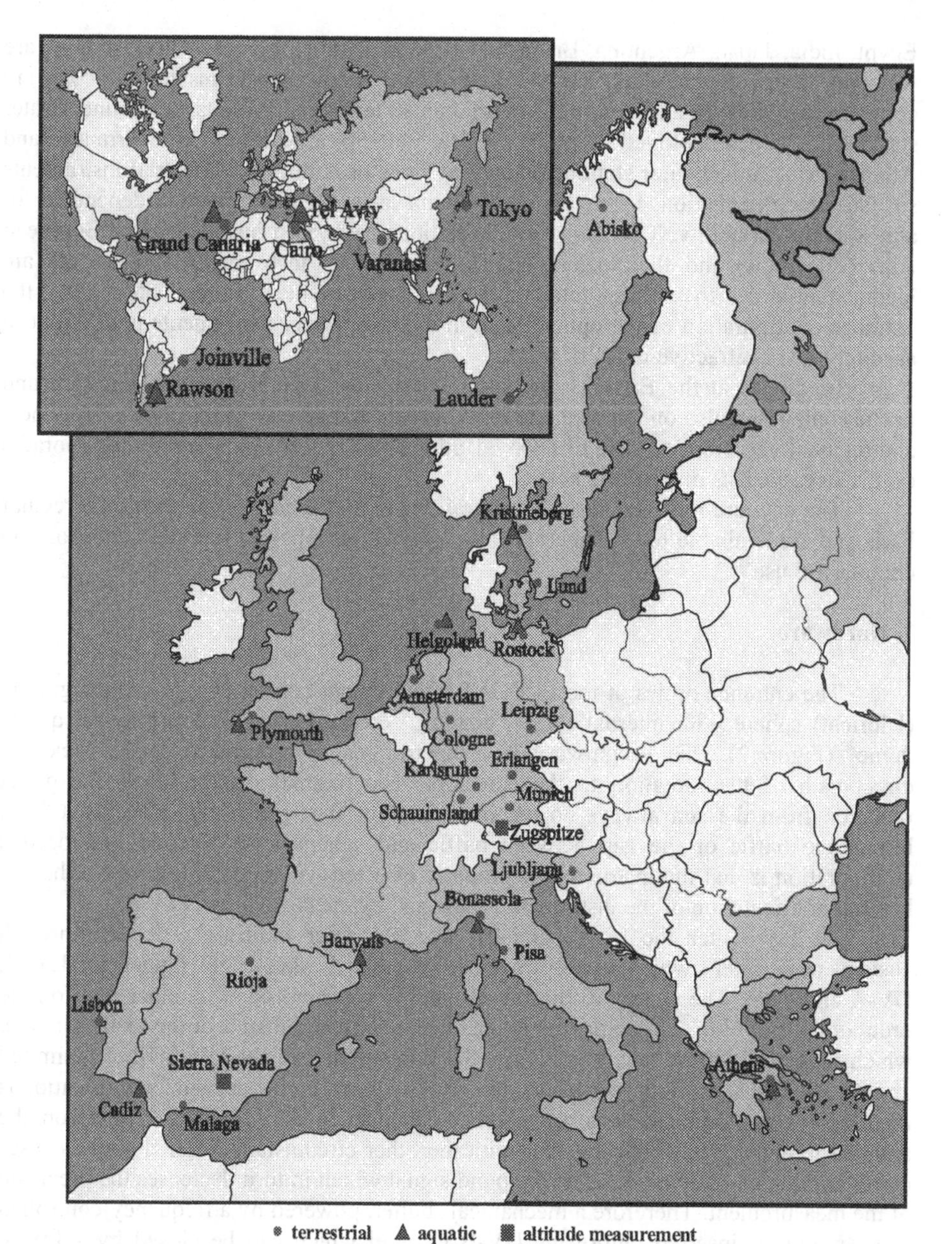

Figure 1. Location of the ELDONET stations. Squares: high altitude stations; triangles: underwater instruments.

from North Sweden (Abisko, 69° N) to the Canary Islands (Gran Canaria, 28° N) and extend east to west from Greece (Corinth, 24° E) to the Canary Islands (16° W) to monitor the major light climate areas in Europe[40]. Further stations have been installed in

Egypt, India, Japan, Argentina, Brazil and New Zealand (Figure 1). Other stations are welcome to join the network. For this purpose the hardware and installation costs are comparatively low. In addition to the more than 40 terrestrial stations, seven underwater stations have been deployed in the North Sea, Baltic Sea, Kattegat, Mediterranean and Atlantic. Most underwater stations operate in close proximity of terrestrial instruments so that the attenuation in the water column can be determined independent of atmospheric variability. Two high altitude stations have been installed in the European Alps (Zugspitze) and the Spanish Sierra Nevada. All instruments and sensors are identical in order to warrant intercomparison between the different sites. Only the entrance configuration of the optical head in the underwater instruments is different to account for the refractive index of water.

Irradiance at the Earth's surface does not only depend on solar elevation and ozone content but also on atmospheric transparency and albedo which shows large local and regional variation[34,37]. The ELDONET allows us to follow local trends and sporadic occurrences such as mini ozone holes.

The data are transferred to the central server of the network in Pisa on a regular basis and are available on the internet (http://www.eldonet.org/) free of charge for non commercial use[41].

3. Hardware

The entrance optics of the ELDONET instrument consist of a 10 cm integrating (Ulbricht) sphere with internal $BaSO_4$ coating[42] covered with a hemispherical quartz dome[40] (Figure 2). This design warrants a superior cosine response. A baffle blocks the direct path of the radiation to the detectors. The internal design forces the direct radiation from the sun always on the inner wall of the sphere and prevents it from hitting the baffle or the gap between baffle and sphere. One precondition for this configuration is that the instrument is properly oriented with respect to true North. The horizontal orientation of the instrument is adjusted by means of a level gauge.

The light rays enter the Si photodiode detectors after multiple reflection through custom-made filters which closely match the selected wavelength ranges for UV-B, UV-A and PAR. The energy in the UV-B band in solar emission is rather low (on the order of up to 2 W m^{-2}) as compared to the other wavelength bands or the total radiation, which exceeds 1000 W m^{-2}. In addition, the transmission of the UV-B filter is limited and the spectral sensitivity of the receptors low, so that a high electronic amplification is required. Si photodiodes have an inherent dependence of their dark current on the ambient temperature, which together with the other circumstances listed above causes considerable fluctuations of the background signal which in turn decreases the precision of the measurement. Therefore a mechanical shutter, powered by a frequency-controlled servo motor, is inserted before the UV-B channel which can be closed by software control to determine the dark value at regular intervals in order to

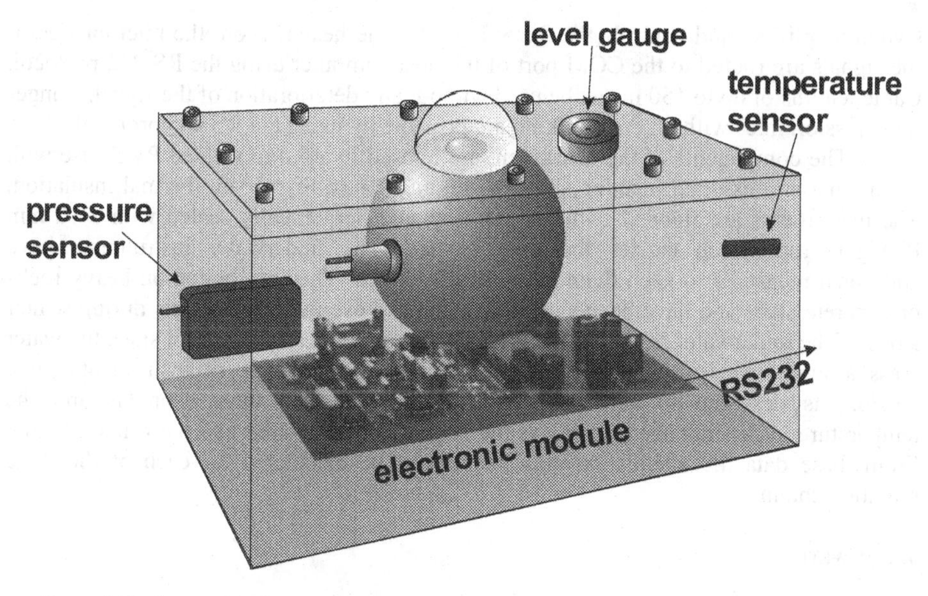

Figure 2. Hardware of the ELDONET instrument.

eliminate temperature-related drifts and potential aging. This preventive measure is not necessary for the other two channels since their dark values are stable within one digital unit over a wide range of temperatures. The shutter function is controlled by the built-in microprocessor using switches which signal the end positions of the shutter movement. The signals of the three photodiodes are routed to separate operation amplifiers with precision field effect transistor (FET) inputs featuring very low offset voltages and minimal temperature dependence of offset and amplification. All signals from each stage are low-pass filtered to minimize high frequency noise. External noise is further reduced by careful shielding in a Faraday cage. All supply voltages are highly stabilized. The last stage of each amplification channel is a switch operation amplifier which can be switched to 1 time, 10 times or 100 times amplification, depending on the amplitude of the incoming signal, under software control. This measure increases the dynamic range of the radiation measurements to 1:409500. In addition to the three radiation channels the instruments have separate temperature sensors for the external and internal temperature. Underwater instruments possess a pressure sensor to monitor the water depth. All signals are routed to a 12 bit analogue/digital (A/D) converter with 8 parallel channels.

Since the instruments are installed over a wide variety of geographic locations from polar regions to the tropics, they experience a wide range of ambient temperatures from –50° to +50°C. In order to limit this range, the interior is maintained at an elevated temperature of 35° C controlled by a heater operated by a fuzzy logic. Thus the internal temperature can fluctuate only in a limited range of about 15° C.

All internal functions are controlled by a microprocessor. It generates the pulse width modulation for the shutter servo, it polls the A/D channels and averages the data over 256 measurements. Furthermore, it automatically adjusts the amplification of the switch amplifiers and controls the fuzzy logic for the heater. From the microprocessor the signals are routed to the COM port of the host computer using the RS 232 protocol. Cable lengths of up to 150 m can be used without any deterioration of the signal. Longer transmission lines without cross talking are possible by using the RS 485 protocol.

The components of the instrument are housed in a 6 mm welded PVC case with an internal and external metallic shielding and a 12 mm Styrofoam thermal insulation. The instruments are flooded with dry nitrogen and hermetically sealed with a 10 mm Plexiglas top which carries the optical head. The underwater instruments have additional measures to keep them hermetically sealed. They are bolted on heavy rocks or concrete slabs and have been shown to withstand even strong currents during winter storms. The underwater instruments have no internal temperature control since the water mass around them limits the temperature range. A third type of instruments (dive version) is designed to determine the light gradients in three channels and the temperature in dependence of depth as the instrument is lowered into the water column. From these data the attenuation coefficients can be calculated for each of the three radiation channels.

4. Software

The software package WinDose 2000 for the ELDONET instruments has been developed in Visual Basic (Microsoft, Redmond, WA, USA) and runs under the Windows platform. When the program is started, it tests the available serial ports for the presence of an ELDONET instrument. The instrument answers by sending back its unique code number which also defines whether it is a terrestrial or aquatic instrument or a dive version. Before the first measurements commence at a new location the geographic coordinates for longitude and latitude and the site name are entered. The calibration requires the knowledge of the solar angle. Therefore the program determines the solar angle from the longitude and latitude data and the time and date provided by the host computer. In order to warrant high precision measurements the time should be as accurate as possible. The computer hardware clock can be automatically updated by programs on the net when the computer is connected to the Internet. Internally the software calculates in Universal time and Windows provides the offset to the local time as well as the shift from daylight saving to winter time and back, so that the local time can be displayed on screen. The measurements start in the morning when the sun is 12° below the horizon and end when it sinks 12° below the horizon. This condition may not be met when the instrument is located near one of the poles. The geographic data, the individual number of the instrument, the name of the location and the calibration data of the specific instrument are stored in an individual initialization (.INI) file.

The microprocessor within the instrument sends updated values once every second to the host computer where they are displayed in numerical and graphical form. In addition, it sends averages over one-minute intervals which are stored in a disk file which is updated every 15 min. The main screen shows the radiation and external temperature data in graphical as well as in numerical form together with the doses accumulated since the start of the program for the day. For the aquatic instruments the water depth is displayed. A toolbar allows to display selected data, print them or store them to disk in graphical form. The status bar at the bottom of the screen informs about sunrise, sunset, system time, active window, path and filename of the current data file, available disk space and the COM-port used by the ELDONET instrument. An icon indicates the quality of the serial connection.

Menus allow printer setup and printing or copying the current graph. If WinDose 2000 detects an error such as communication problems or data exceeding the maximal value the program issues visual and acoustic warnings to alert the user. Advanced options allow indoor measurements not restricted by sunrise and sunset, dark current measurements used during calibration of the instrument. In this mode the measured data are recorded and stored in digital units. The depth readings can be set to zero in the aquatic version to adjust for the air pressure at the measurement site.

All data are stored in ASCII files which facilitates further analysis of the data by the analysis program developed for the project (WinData 2000) or by spreadsheet programs such as EXCEL. The final record of the day contains a header with the location information and shows the accumulated doses in the three channels for the day.

5. Calibration

Each radiation channel in every instrument is accurately calibrated against a NIST traceable 1000 W quartz halogen calibration lamp with a constant spectral composition operated with a highly stabilized power supply. All calibration factors are stored in the individual .INI file. Furthermore, every instrument is tested under all solar angles under blue sky in a location with free view of the horizon. The absolute calibration was derived from an intercomparison of several spectroradiometers and the ELDONET instrument in September 1997 in Garmisch-Partenkirchen (Southern Germany). Instrument intercomparison is a vital tool for quality control and documentation[43-44]. All instruments should be recalibrated at least every year. Frequently ELDONET instruments are compared with double monochromator spectrophotometers by running them side by side in parallel several days at different geographic locations under different weather and irradiation conditions to warrant absolute solar calibration.

In addition to cloud-related changes the solar spectral emission changes over the day with less contribution of short wavelengths in the early morning and late evening hours because of the different path lengths of the sun rays through the atmosphere. The filter transmissions of the three channels are not perfectly square in the selected wavelength bands, therefore the measured values need to be corrected in dependence of the solar angle. This is done by direct comparison with a double monochromator spectroradiometer for all possible solar angles during the daily cycle on clear days. This check against the sun is carried out for all instruments, a procedure which is now standard for most instrument manufacturers. The PAR and UV-A channels did not deviate from the spectroradiometer values, but for the UV-B channel a correction function was determined which is built into the program. The ELDONET data show only minimal deviation from the spectroradiometer values. However, it is recommended that the calibration procedure should be repeated at least every year.

Quality control and documentation is an indispensable prerequisite of high-precision measurements of solar radiation. The resolution and detection limits of the instrument are 1 mW m^{-2} for PAR, 0.1 mW m^{-2} for UV-A and 5 μW m^{-2} for UV-B. The error of measurement with respect to a calibrated spectroradiometer is less than 10 % for the UV-B, 5 % for the UV-A and 2 % for the PAR channel.

The specific data of each individual instrument are recorded in a log sheet and updated during each repair or maintenance. During their lifetime up to now - some instruments have been in use for over 6 years - the reproducibility was found to be 1.4±1.0 % for the PAR channel, 1.9±1.5% for UV-A and 3.8±3.8% for UV-B based on a total of 41 instruments, indicating a long-term stability of the instruments[40].

6. Central server

The central server is installed in Pisa, Italy and is now accessible for everybody interested in irradiance or dose data (http://www.eldonet.org/). The server is hosted on an IBM Power/PC and the software consists of two main programs written in Perl 5. One program is devoted to generate the graphs in GIF format from the submitted data using the PBMPLUS image manipulation library. Data are sent to the server either by automatic FTP, by modem or as e-mail attachments. The incoming files are first checked for plausibility and completeness and then transferred into the directory for the specific station.

After entering the home page, the user can select a station either on the interactive map or from the stations list. There is a short description of the ELDONET project and network as well as for the individual stations. The data are selected from a calendar and can be viewed in graphical form. The availability of the data is limited to 2 months, but all existing data for the whole lifetime of the project can be downloaded by FTP in numerical form and then displayed with a data sheet program such as EXCEL.

7. Data

Figure 3 shows a measurement at Ljubljana (Slovenia, 46° N and 15° E) on 11 August 1999. In addition to some scattered clouds, the solar eclipse with an occlusion of 96.6 % at around 12:48 h is clearly visible in all spectral channels. The monthly doses for the southernmost terrestrial European station of Gran Canaria form a broad peak during the summer months reaching 320 MJ m^{-2} in the PAR channel, 48 MJ m^{-2} in the UV-A and 1.3 MJ m^{-2} in the UV-B channel. The UV-B irradiance strongly depends on the path length of the solar rays through the atmosphere and thus on the solar angle directly related with the time of the year. The other decisive parameter is the total ozone column. To eliminate the effect of the changing cloud cover, the ratio of UV-B over UV-A can be calculated. The UV-B/UV-A ratio peaks in early Autumn rather than in the Summer which is due to lower ozone values in Autumn, measured by the TOMS instrument on the NASA Earth Probe satellite, or the predicted ozone concentration calculated on the basis of Green's model[45].

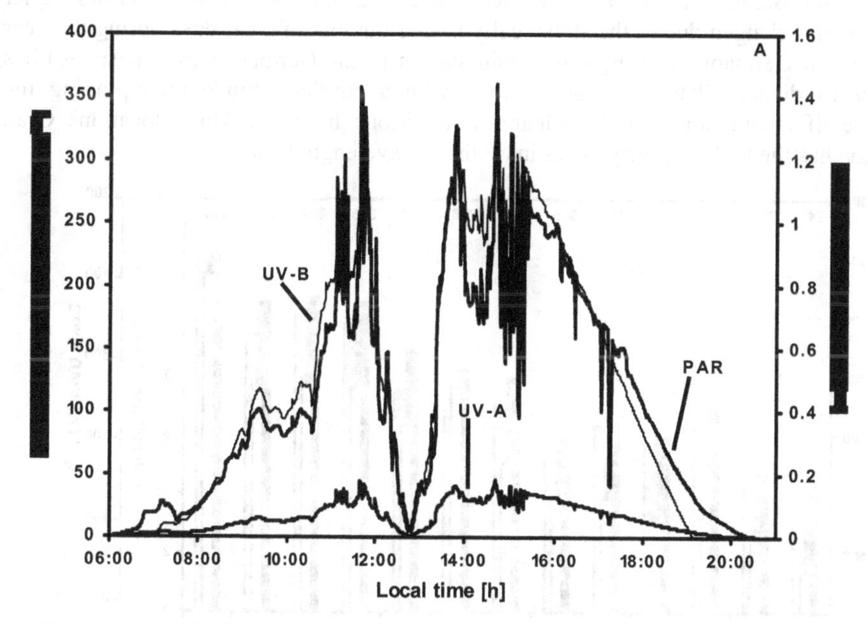

Figure 3. ELDONET measurements at Ljubljana (Slovenia) during the solar eclipse on 11 August 1999.

Erlangen (Germany, 50° N, 11° E) represents a typical central European location. The monthly doses peak at about 244 MJ m^{-2} for PAR in the summer. July had more cloud cover than June or August. Also UV-A and UV-B irradiances were maximal in June which corresponds with the highest solar angles. The same is true for the ratio UV-B/UV-A while the ratios UV-B/PAR and UV-A/PAR have their maximum in July. Plotting the UV-B/PAR ratios on a daily basis also shows a clear maximum in July which is due to the antagonistic functions of the decreasing total ozone column values and the decreasing solar angles in autumn.

The northernmost station of the network (Abisko, North Sweden, 69° N, 19° E) is located above the Nordic polar circle and also has maximal irradiances in June. During winter the sun does not rise above the horizon resulting in extremely low irradiances. In summer the PAR peak doses exceed those measured in Erlangen corresponding to the fact that in this month the sun does not sink below the horizon so that the doses are effectively accumulated over 24 h periods. While UV-B/PAR and UV-B/UV-A ratios peak in July, the UV-A/PAR has a maximum in December and January which is due to the higher amount of diffuse skylight during the winter months in comparison to the direct radiation which increases with longer wavelengths.

Figure 4 summarizes the yearly doses for 1998 for all European stations at which measurements were performed on more than 250 days of the year arranged in a North-South gradient. While there is a clear dependence on the latitude one can detect a few remarkable exceptions. Helgoland (island in the German bight, 54° N, 8° E) shows comparatively higher doses than Rostock with a comparable latitude (on the Southern coast of the Baltic Sea, 54° N, 12° E). Karlsruhe (Germany, 49° N, 8° E) shows higher doses than Erlangen due to the statistically higher number of clear days during the year in Western Germany. The high mountain station in the German Alps (Zugspitze) has higher irradiances than low-land stations with a similar latitude underpinning the altitude effect and more abundant clear sky conditions in pristine Alpine locations. Gran Canaria has the highest yearly doses in all three wavelength bands.

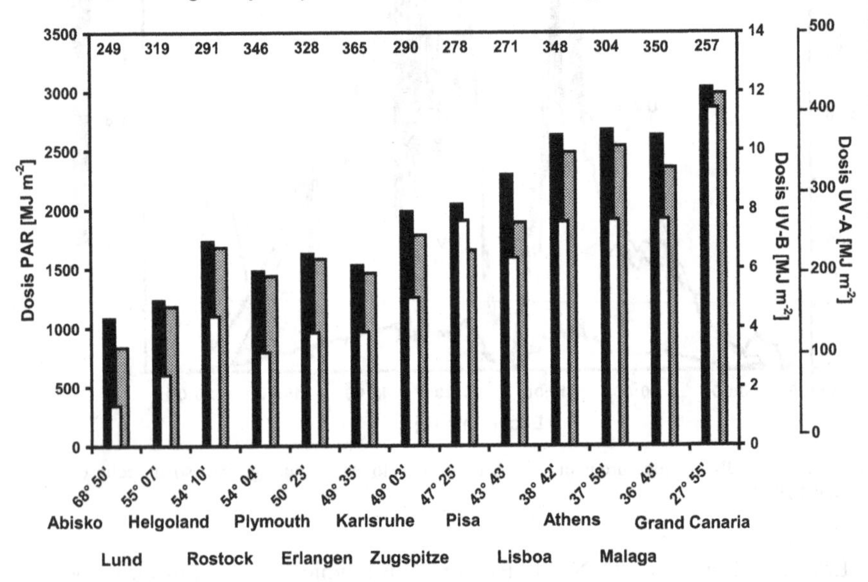

Figure 4. Yearly doses for several European stations.

8. Correlating ELDONET data with ozone values

Ozone data measured by satellites (TOMS, Earth Probe) can be obtained from the Internet (toms.gsfc.nasa.gov) on a daily basis including stratospheric ozone concentration, reflectivity and aerosol. Erythemal doses are estimated from the extraterrestrial solar irradiance (298 nm – 400 nm), the Earth-Sun distance, the biological action spectrum for erythemal damage[46], the irradiation time, the cloud cover and the optical cloud thickness, the solar zenith angle and the total column ozone. The procedure is described in detail on the TOMS data home page (ftp://toms.gsfc.nasa.gov/pub/eptoms/data/erynotes.pdf). We developed a program to automatically extract the relevant ozone values from the data sets which are organized by latitude (in 1 degree steps) and longitude (in 1.25 degree steps). The corresponding ozone concentrations show the typical seasonal high variability in early and mid spring of each year caused by mini ozone holes which originate from the ozone hole detected frequently over the Arctic in early spring and drift over North and mid Europe. This variability is not revealed by the monthly doses because of their short-term occurrence but can easily be seen in the UV-B over PAR and UV-B over UV-A ratios. Changes in the ozone concentration are most effective in the UV-B readings and have a much smaller impact on the UV-A irradiances. While the spring UV-B increase can be observed at all European stations to a different extent, no obvious trend can be extracted over the four-year measurement period.

There is a weak correlation between the UV-B dose and the ozone concentration for a given location (correlation coefficient: 0.15). A stronger correlation exists between reflectivity and measured doses in all measured bands (correlation coefficient$_{UV\text{-}B}$: 0.52, correlation coefficient$_{UV\text{-}A}$: 0.61, correlation coefficient$_{PAR}$: 0.67). Reflectivity is controlled by the cloud cover and other reflecting substances in the atmosphere. Thus, local weather conditions have a significant influence on the measurements. In Sierra Nevada, Spain (latitude: 37.67° N) for many days the correlation holds, but significant differences occur from end of May when fog or clouds cover the mountain top. TOMS data are only measured once a day so that sudden changes of the reflectivity are not visible in the data sets.

The correlation between ozone depletion and increasing UV-B irradiances is now well established. The current UV-B doses are made available to the public e.g. in Australia, but the awareness of possible health threats of increased UV-B radiation is still low in Europe. The ELDONET network offers a low-cost and readily available source of information for local information and prediction. The spectral sensitivity of the UV-B sensor and filter combination used in the ELDONET instrument closely resembles the CIE erythema action spectrum so that the erythemally weighted dose can be easily determined from the UV-B readings of the instruments and from these the mean erythemal doses (MED). For a Caucasian skin type II one MED is equivalent to 220 J biologically weighted radiation according to the German DIN norm.

Up to now only seven stations measure solar radiation in the water column and they have not all been active for the whole measuring period, but there are sufficient reliable data. Penetration of solar radiation into the top layers of the water column is strongly controlled by the turbidity of the water. In the coastal waters off Helgoland solar UV-B does not penetrate below 1 m of water, but the penetration is much stronger in the Atlantic station in Gran Canaria, especially during the winter months when the transparency is very high.

9. Conclusions and outlook

To the present day, the ELDONET project has been very successful in monitoring solar radiation. These data are of high quality and even allow to detect subtle, transient changes in the light climate. One of the advantages of the instrument is the low cost of about 2300 US dollar, so that the number of instruments in the network can easily be increased. Interested stations are invited to join the network and make their data available on the central server to the interested scientific community and the public. Another advantage is that the measured values can be convoluted with a given biological weighting function.

Work is in progress to develop a computer simulation to predict the irradiances and doses of solar radiation for each location for the whole day and each day over the year based on existing models[47-48]. The intention is to automatically compare the modeled data with the actually measured values. This can be used to alert the user in the case of significant deviations. This procedure is straight forward for clear days. In the next step, it is planned to use satellite data to model also the irradiance during cloudy days.

Acknowledgments. This work was supported by the european union (Environment programme, EV5V-CT94-0425; DG XII). The authors thank M. Schuster and H. Wagner for their excellent technical assistance

References

1. Häder, D.-P., Kumar, H.D., Smith, R.C. and Worrest, R.C. (1998) Effects on aquatic ecosystems. In *UNEP Environmental Effects Panel Report*, UNEP, Nairobi, pp. 86-112.
2. van der Leun, J.C. (1988) Ozone depletion and skin cancer, *J. Photochem. Photobiol. B: Biol.*, 1: 493-496.
3. van der Leun, J.C. (1990) Introduction: photobiology and the ozone layer, *J. Photochem. Photobiol. B: Biol.*, 8: 113-118.
4. De Fabo, E.C., Noonan, F.P. and Frederick, J.E. (1990) Biologically effective doses of sunlight for immune suppression at various latitudes and their relationship to changes in stratospheric ozone, *Photochem. Photobiol.*, 52: 811-817.
5. Noonan, F.P. and De Fabo, E.C. (1990) Ultraviolet-B dose-response curves for local and systemic immunosuppression are identical, *Photochem. Photobiol.*, 52: 801-810.
6. Longstreth, J., de Gruijl, F.R., Kripke, M.L., Abseck, S., Arnold, F., Slaper, H.I., Velders, G., Takizawa, Y. and van der Leun, J.C. (1998) Health risks, *J. Photochem. Photobiol. B: Biol.*, 46: 20-39.
7. Björn, L.O. (1989) Consequences of decreased atmospheric ozone: Effects of ultraviolet radiation on plants. In *Atmospheric Ozone Research and its Policy Implications* (T. Schneider, ed.), Elsevier Science, Amsterdam, pp. 261-267.
8. Panagopoulos, I., Bornman, J.F. and Björn, L.O. (1989) The effect of UV-B and UV-C radiation on *Hibiscus* leaves determined by ultraweak luminescence and fluorescence induction, *Physiol. Plant.*, 76: 461-465.
9. Bornman, J.F. (1991) UV radiation as an environmental stress in plants, *J. Photochem. Photobiol. B:Biol.*, 8: 337-342.
10. Caldwell, M.M., Björn, L.O., Bornman, J.F., Flint, S.D., Kulandaivelu, G., Teramura, A.H. and Tevini, M. (1998) Effects of increased solar ultraviolet radiation on terrestrial ecosystems, *J. Photochem. Photobiol. B: Biol.*, 46: 40-52.
11. Strid, A., Chow, W.S. and Anderson, J.M. (1990) Effects of supplementary ultraviolet-B radiation on photosynthesis in *Pisum sativum*, *Biochem. Biophys. Acta*, 1020: 260-268.
12. Renger, G., Rettig, W. and Gräber, P. (1991) The effect of UVB irradiation on the lifetimes of singlet excitons in isolated photosystem II membrane fragments from spinach, *J. Photochem. Photobiol. B: Biol.*, 9: 201-210.
13. Negash, L. (1987) Wavelength-dependence of stomatal closure by ultraviolet radiation in attached leaves of *Eragrostis tef*: Action spectra under backgrounds of red and blue lights, *Plant Physiol. Biochem.*, 25: 753-760.
14. Negash, L., Jensen, P. and Björn, L.O. (1987) Effect of ultraviolet radiation on accumulation and leakage of $^{86}Rb^{+}$ in guard cells of *Vicia faba*, *Physiol. Plant.*, 69: 200-204.
15. Cen, Y.-P. and Bornman, J.F. (1990) The response of bean plants to UV-B radiation under different irradiances of background visible light, *J. Exp. Bot.*, 41: 1489-1495.
16. Teramura, A.H., Sullivan, J.H. and Lydon, J. (1990) Effects of UV-B radiation on soybean yield and seed quality: a 6-year field study, *Physiol. Plant.*, 80: 5-11.
17. Teramura, A.H., Sullivan, J.H. and Ziska, L.H. (1990) Interaction of elevated ultraviolet-B radiation and CO_2 on productivity and photosynthetic characteristics in wheat, rice, and soybean, *Plant Physiol.*, 94: 470-475.
18. Rozema, J., Tosserams, M. and Magendans, E. (1995) Impact of enhanced solar UV-B radiation on plants from terrestrial ecosystems. In *Climate Change Research: Evaluation and Policy Implications* (S. Zwerver, R.S.A.R. v. Rompaey, M.T.J. Kok and M.M. Berk, eds.), Elsevier Science, Amsterdam, pp. 997-1004.

19. Smith, R.C., Prezelin, B.B., Baker, K.S., Bidigare, R.R., Boucher, N.P., Coley, T., Karentz, D., MacIntyre, S., Matlick, H.A., Menzies, D., Ondrusek, M., Wan, Z. and Waters, K.J. (1992) Ozone depletion: ultraviolet radiation and phytoplankton biology in Antarctic waters, *Science*, 255: 952-959.
20. Häder, D.-P., Worrest, R.C., Kumar, H.D. and Smith, R.C. (1995) Effects of increased solar ultraviolet radiation on aquatic ecosystems, *AMBIO*, 24: 174-180.
21. Gerber, S., Biggs, A. and Häder, D.-P. (1996) A polychromatic action spectrum for the inhibition of motility in the flagellate *Euglena gracilis*, *Acta Protozool.*, 35: 161-165.
22. Jimenez, C., Figueroa, F.L., Aguilera, J., Lebert, M. and Häder, D.-P. (1996) Phototaxis and gravitaxis in *Dunaliella bardawil*: Influence of UV radiation, *Acta Protozool.*, 35: 287-295.
23. Sarmiento, J.L. and Le Quere, C. (1996) Oceanic carbon dioxide uptake in a model of century-scale global warming, *Science*, 274: 1346-1350.
24. Cullen, J.J. and Lesser, M.P. (1991) Inhibition of photosynthesis by ultraviolet radiation as a function of dose and dosage rate: results for a marine diatom, *Mar. Biol.*, 111: 183-190.
25. Raven, J.A. (1991) Responses of aquatic photosynthetic organisms to increased solar UVB, *J. Photochem. Photobiol. B: Biol.*, 9: 239-244.
26. Häder, D.-P. and Worrest, R.C. (1997) Consequences of the effects of increased solar ultraviolet radiation on aquatic ecosystems. In *The Effects of Ozone Depletion on Aquatic Ecosystems* (D.-P. Häder, ed.), Acad. Press, R.G. Landes Company, Austin, pp. 11-30.
27. Diffey, B.L. (1996) *Measurement and trends of terrestrial UVB in Europe*, OEMF spa., Milano.
28. Webb, A.R. (1998) *UVB instrumentation and applications*, Gordon and Beach Science Publishers.
29. Scotto, J., Cotton, G., Urbach, F., Berger, D. and Fears, T. (1988) Biologically effective ultraviolet radiation: surface measurements in the United States, 1974 to 1985, *Science*, 239: 762-764.
30. Frederick, J.E. (1992) An assessment of the Robertson-Berger ultraviolet meter and measurements: Introductory comments, *Photochem. Photobiol.*, 56: 113-114.
31. Kennedy, B.C. and Sharp, W.E. (1992) A validation study of the Robertson-Berger meter, *Photochem. Photobiol.*, 56: 133-141.
32. Kerr, J.B. (1997) Observed dependencies of atmospheric UV radiation and trends. In *Solar Ultraviolet Radiation Modelling, Measurement and Effects, Nato ASI Series I, Vol. 52* (C.S. Zerefos and A.F. Bais, eds.), Springer Verlag, Berlin, pp. 259-266.
33. Lubin, D. and Jensen, E.H. (1997) Satellite mapping of solar ultraviolet radiation at the earth's surface. In *Solar Ultraviolet Radiation Modelling, Measurement and Effects, Nato ASI Series I, Vol. 52* (C.S. Zerefos and A.F. Bais, eds.), Springer Verlag, Berlin, pp. 95-118.
34. Dickerson, R.R., Kondragunta, S., Stenchikov, G., Civerolo, K.L., Doddridge, B.G. and Holben, B.N. (1997) The impact of aerosols on solar ultraviolet radiation and photochemical smog, *Science*, 278: 827-830.
35. Johnson, B. and Moan, J. (1991) The temperature sensitivity of the Robertson-Berger sunburn meter, model 500, *J. Photochem. Photobiol. B: Biol.*, 11: 277-284.
36. McKenzie, R.L., Paulin, K.J. and Kotkamp, M. (1997) Erythemal UV irradiances at Lauder, New Zealand: relationship between horizontal and normal incidence, *Photochem. Photobiol.*, 66: 683-689.
37. Feister, U. and Grewe, R. (1995) Spectral albedo measurements in the UV and visible region over different types of surfaces, *Photochem. Photobiol.*, 62: 736-744.
38. Blumthaler, M. (1997) Broad-band detectors for UV-measurements. In *Solar Ultraviolet Radiation Modelling, Measurement and Effects, Nato ASI Series I, Vol. 52* (C.S. Zerefos and A.F. Bais, eds.), Springer Verlag, Berlin, pp. 175-185.
39. Booth, C.R. and Lucas, T.B. (1994) UV spectroradiometric monitoring in polar regions. In *Stratospheric Ozone Depletion/UV-B Radiation in the Biosphere, NATO ASI Series I, Vol. 18* (R.H. Biggs and M.E.B. Joyner, eds.), Springer Verlag, Berlin, pp. 237.
40. Häder, D.-P., Lebert, M., Marangoni, R. and Colombetti, G. (1999) ELDONET - European Light Dosimeter Network: hardware and software, *J. Photochem. Photobiol. B: Biol.*, 52: 51-58.
41. Marangoni, R., Barsella, D., Gioffré, D., Colombetti, G., Lebert, M. and Häder, D.-P.: ELDONET – European light dosimeter network. Structure and function of the ELDONET server, *J. Photochem. Photobiol. B: Biol.*, 58: 178-184.
42. Khanh, T.C. and Dähn, W. (1998) Die Ulbrichtsche Kugel. Theorie und Anwendungsbeispiele in der optischen Strahlungsmeßtechnik, *Photonik*, 4: 6-9.
43. Seckmeyer, G., Thiel, S., Blumthaler, M., Fabian, P., Gerber, S., Gugg-Helminger, A., Häder, D.-P., Huber, M., Kettner, C., Köhler, U., Köpke, P., Maier, H., Schäfer, J., Suppan, P., Tamm, E. and Thomalla, E. (1994) Intercomparison of spectral-UV-radiation measurement systems, *Appl. Optics*, 33: 7805-7812.

44. Leszczynski, K., Jokela, K., Ylianttila, L., Visuri, R. and Blumthaler, M. (1998) Erythemally weighted radiometers in solar UV monitoring: results from the WMO/STUK intercomparison, *Photochem. Photobiol.*, 67: 212-221.
45. Green, A.E.S. and Chai, S.-T. (1988) Solar spectral irradiance in the visible and infrared regions, *Photochem. Photobiol.*, 48: 477-486.
46. McKinlay, A.F. and Diffey, B.L. (1987) A reference action spectrum for ultraviolet induced erythema in human skin, *CIE J.*, 6: 17-22.
47. Björn, L.O. and Murphy, T.M. (1985) Computer calculation of solar ultraviolet radiation at ground level, *Physiol. Veg.*, 23: 555-561.
48. Stamnes, K. (1997) Transfer of ultraviolet light in the atmosphere and ocean: a tutorial review. In *Solar Ultraviolet Radiation Modelling, Measurement and Effects, Nato ASI Series I, Vol. 52* (C.S. Zerefos and A.F. Bais, eds.), Springer Verlag, Berlin, pp. 49-64.

GENETIC AND MOLECULAR ANALYSIS OF DNA DAMAGE REPAIR AND TOLERANCE PATHWAYS

B. M. SUTHERLAND
Brookhaven National Laboratory,
Upton, NY, 11973, USA

1. Introduction

Radiation can damage cellular components, including DNA. Organisms have developed a panoply of means of dealing with DNA damage. Some repair paths have rather narrow substrate specificity (e.g. photolyases, which act on specific pyrimidine photoproducts in a specific type (e.g., DNA) and conformation (double-stranded B conformation) of nucleic acid. Others, for example, nucleotide excision repair, deal with larger classes of damages, in this case bulky adducts in DNA.

A detailed discussion of DNA repair mechanisms is beyond the scope of this article, but one can be found in the excellent book of Friedberg et al. [1] for further detail. However, some DNA damages and paths for repair of those damages important for photobiology will be outlined below as a basis for the specific examples of genetic and molecular analysis that will be presented below.

2. DNA Damages

The major photoproduct formed in DNA by UV radiation is the cyclobutyl pyrimidine dimer (CPD); although [6-4] pyrimidine-pyrimidone adducts are induced at lower level, they are also significant in production of biological damage. Other pyrimidine photoproducts (e.g., hydrates) are formed by UV, but purine photoproducts are induced only at much lower levels.

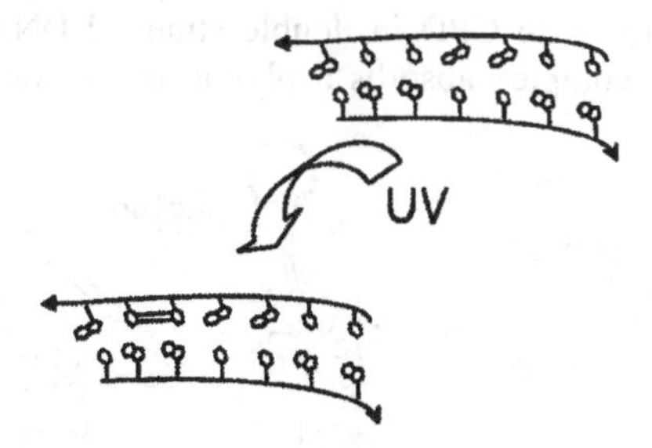

Figure 1. Induction of cyclobutyl pyrimidine dimer in DNA by UV radiation.

In addition to DNA damages formed by UV, all organisms living in an oxygen environment suffer from oxidative damage to DNA. They also carry out replication

F. Ghetti et al. (eds.), Environmental UV Radiation: Impact on Ecosystems and Human Health and Predictive Models, 109–120.

and other cellular metabolic processes that result in formation of single and double strand breaks. Thus, in addition to the UV-induced damages discussed above, cells must cope with oxidized purines, oxidized bases and abasic sites (sites of base loss), as well as single and double strand breaks. To deal with these UV-induced and other DNA alterations, cells have developed several distinct yet in some cases overlapping repair pathways.

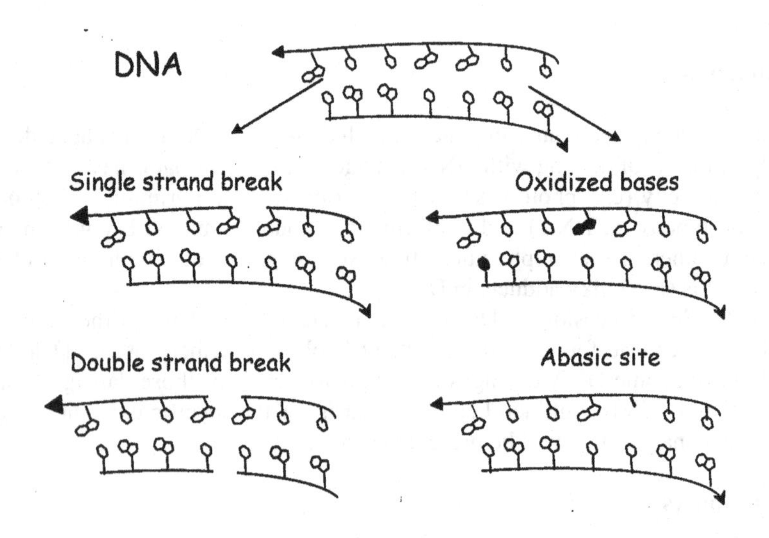

Figure 2. DNA damages induced by normal cellular metabolism in air.

3. Photorepair

Pyrimidine dimer photoreactivation is a one-enzyme repair path. The enzyme photolyase binds specifically to CPD in double stranded DNA, forming an enzyme-substrate complex. That complex absorbs a photon in the long UV or visible range,

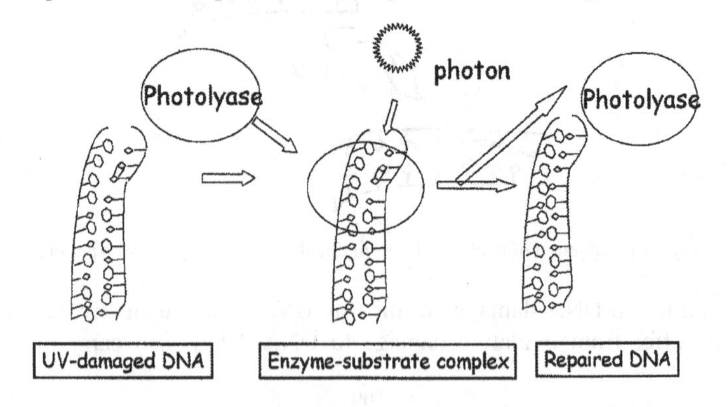

Figure 3. Mechanism of photorepair of CPD in DNA by photolyase.

upon which the dimer is monomerized and the enzyme is released to seek another dimer. It is important to note that the formation of the enzyme-substrate complex is temperature-dependent, and also dependent on the concentration of the enzyme (number of photolyases per cell) and concentration of the substrate (CPD in DNA). The photolytic reaction depends on the wavelength and concentration (intensity) of the second substrate, the photon, but is virtually independent of temperature. Further, the enzyme-substrate complex is remarkably stable in the dark, allowing dissection of the two component reactions, i.e., enzyme-substrate complex formation and dimer photolysis.

4. Nucleotide Excision Repair

Nucleotide excision repair deals with a wide variety of bulky lesions in DNA, including CPD. Several enzymes are required (including a nuclease, polymerase, ligase and other

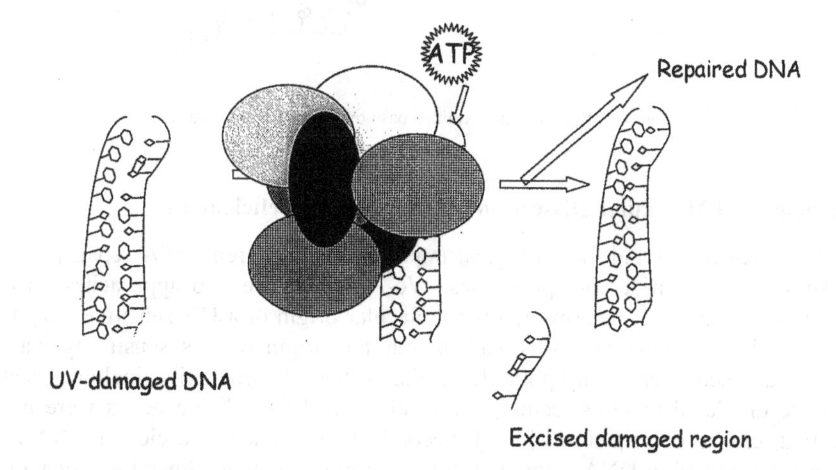

Figure 4. Repair of CPD in DNA by nucleotide excision repair.

proteins), and nucleoside triphosphates are used as an energy source. In addition to UV damages, this path deals with DNA alterations induced by some chemicals and even by ionizing radiation.

5. Base Excision Repair

Base excision repair deals with altered bases in DNA through a glycosylase activity, which removes the altered base leaving an abasic site, followed by a lyase-mediated cleavage of the phosphodiester backbone, and removal of the abasic site and restoration of the DNA by some of the proteins involved in nucleotide excision repair. The glycosylase and lyase activities may reside in the same protein or may involve separate enzymes. Abasic sites (sites of base loss) are dealt with by endonucleases that recognize and cleave the DNA at abasic sites. In human cells, an important enzyme in human cells for repair of abasic sites in human cells is Apel, a 5' endonuclease that specifically cleaves DNA (which is then repaired by nucleotide excision repair).

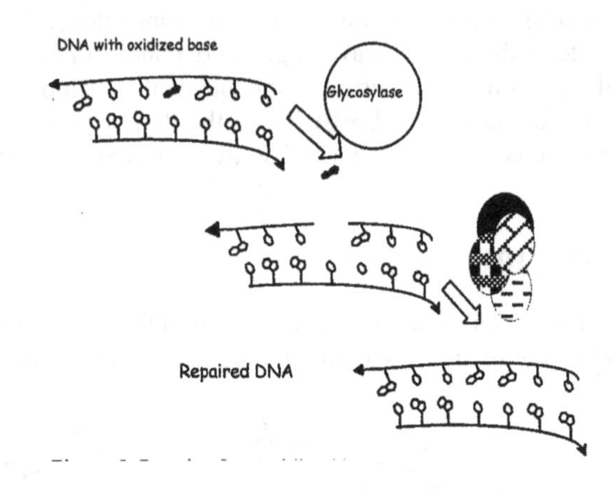

Figure 5. Repair of an oxidized base by base excision repair.

6. Principles of Molecular Dissection of DNA Repair Deficiencies

The molecular origin and the biological importance of different DNA repair paths can be explored by a variety of approaches. We consider here two approaches, in quite different systems. First, determining the molecular origin of a UV sensitive rice strain, the UV radiation sensitivity was known, but the origin of this sensitivity was not understood. There were many possible origins of the UV sensitivity, including several that were unrelated to DNA damage or repair, and the candidate genes were neither identified, cloned or sequenced [2, 3]. Second, the sequence of a cloned DNA repair gene was examined in DNA samples from 124 normal humans. Since the biochemistry of the resulting protein was well understood, the portions of the gene whose alteration was expected to alter the function of the enzyme was known, and the potential impact of any polymorphisms could be anticipated and compared with actual activity of the altered gene product when expressed in *Escherichia coli* cells [4]. We will examine the approaches used in both these cases.

7. UV Sensititive Rice: Norin 1

Rice cultivar Norin 1 is an ancestor of many Japanese commercially-important rice strains.

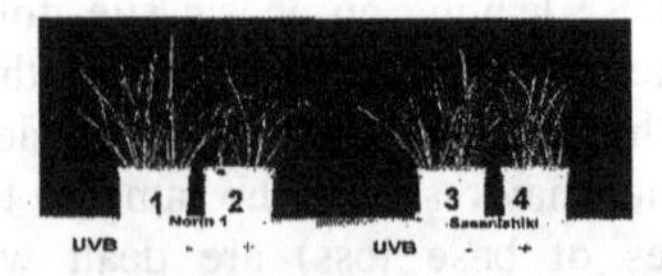

Figure 6. UV sensitive Norin 1 grown in the absence (1) and presence (2) of supplemental UVB radiation, and UV resistant Sasaniskiki grown in the absence (3) and presence (4) of UVB.

Investigators determining the effects of UVB supplementation on various rice strains found that Norin 1 was exquisitely sensitive to added UVB, compared with closely related strains.

Further, the effect of supplemental UVB on other close relatives, e.g., Sasanishiki, could be ameliorated by the addition of additional white light. However, the detrimental effect of UVB could not be effectively prevented by white light supplementation. These data suggested that Norin 1 might be UV sensitive due to some deficiency in protection against UV, perhaps due to increased DNA damage or decreased DNA repair. The lack of improved UV-resistance under conditions of increased white light further suggested that Norin 1 might be deficient in light-mediated DNA repair or photorepair, the principal repair path of UV-induced DNA damage in most plants. [Background studies had already shown that Norin 1 did not differ significantly in non-repair physiological responses from its relatives.]

The first possibility was that Norin 1 was deficient in UV-protecting pigments or other cellular components, so that more damage would be formed per unit UV exposure than in its UV resistant relatives. The level of CPD per unit dose was thus determined in Norin 1 and Sasanishiki; the data clearly indicate that similar levels of CPD are formed in seedlings of both strains. Thus, equal exposures of UVB to these

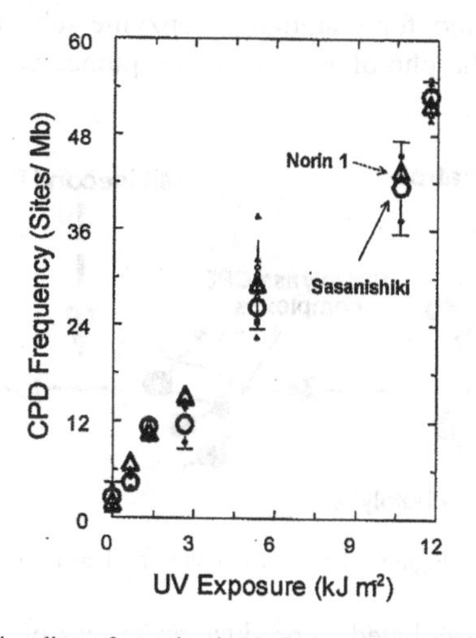

Figure 7. Equal pyrimidine dimer formation in Norin 1 and Sasanishiki rice seedlings.

strains induce equal frequencies of DNA damage [3]. This shows that the UV sensitivity of Norin 1 did not stem from induction of more damage than in the UV resistant Sasanishiki.

Although these data suggested that Norin 1 might be defective in some repair path of UV damage, it did not indicate which repair path was affected. Hidema et al. evaluated nucleotide excision repair and photorepair in seedlings of Norin 1 and of Sasanishiki [3]. They found that UVB-irradiated seedlings of Norin 1 carried out photorepair much more slowly than Sasanishiki, whereas (within the time examined) the excision properties were similar.

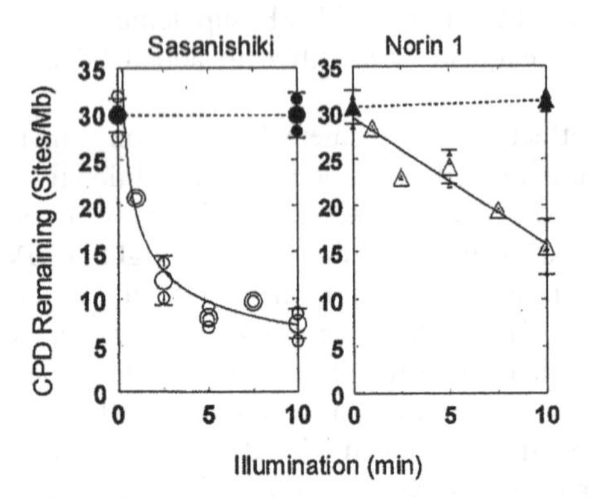

Figure 8. Photorepair (open symbols) and exision repair (closed symbols) in Sasanishiki and Norin 1.

These data implicated a photorepair deficiency in Norin 1, but did not indicate if it resulted from a regulatory mutation (producing fewer normal photolyase molecules) or from a structural mutation (resulting in normal numbers of less-effective photolyases).

A powerful method for quantitating enzyme-substrate complex formation in photolyase reactions is the photoflash technique, pioneered by W. Harm, C.S. Rupert and H. Harm [5].

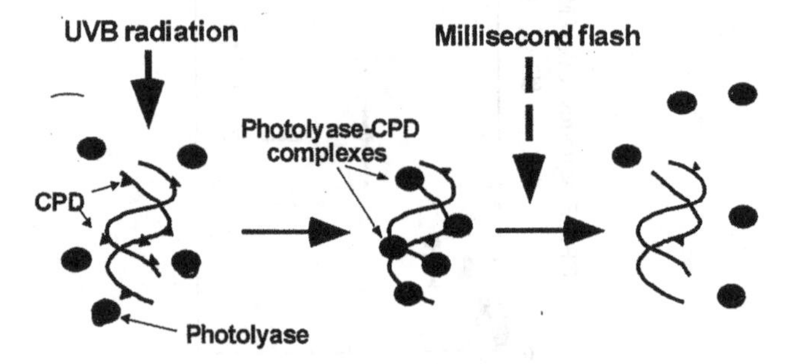

Figure 9. Principles of photoflash analysis.

In this method, DNA is irradiated to produce an excess of pyrimidine dimers. Time is allowed for all available photolyase molecules to bind to a dimer, forming an enzyme-substrate complex that is stable in the absence of light. Then one intense flash of light is applied, of sufficient intensity to photolyze all the existing photolyase-dimer complexes, and of short enough time duration that no photolyase molecules can locate and bind to a second dimer. Measurement of the number of dimers repaired then gives a direct count of the number of active enzyme-substrate complexes, and thus of the number of active photolyases.

Hidema and colleagues probed the nature of this photorepair deficiency in Norin 1 in two series of experiments using photoflash approaches to "count" the number of active photolyase molecules in a cell, to determine the rate of association of the enzyme-substrate complexes, and its stability. In the first series, the rate of association of the enzyme-substrate complexes and the final levels of these complexes were determined: If Norin 1 had a regulatory mutation in photolyase production, the final level of E-S complexes in Norin 1 would always be less than in Sasanishiki, no matter how long was allowed for complex formation. However, if it had a structural mutation

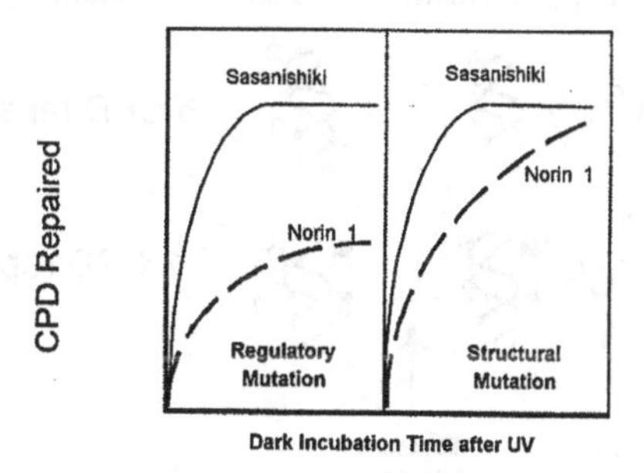

Figure 10. Theoretical expectations for photolyase-dimer complex formation in regulatory mutants and in structural mutants of photolyase.

in the photolyase gene, the normal levels of altered photolyase might have lowered substrate affinity (manifested by slower E-S complex formation), but given long enough time, could form the same number of E-S complexes as its UV-resistant relative.

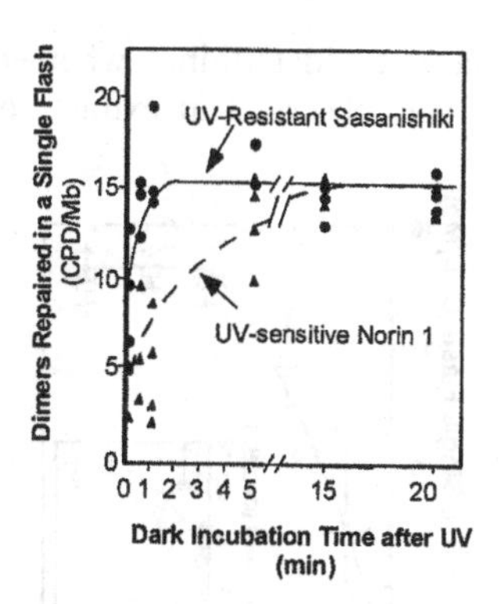

Figure 11. Photoflash analysis of photolyase-dimer complex formation in Norin 1 and Sasanishiki suggests a structural mutation in photolyase.

The data clearly showed that the latter situation was the case, prevailed, strongly suggesting that Norin 1 contained a structural mutation in photolyase that affected its function, not a regulatory mutation that decreased the number of normal photolyase molecules in each cell.

Hidema et al. wanted to confirm that this was the case, and to exclude the possibility that some non-repair factor that might slow E-S complex formation in Norin 1. One way to probe this possibility was to determine the properties of the E-S complex formed in both strains. If the slower photolyase binding (see Figure 11) was ue to an extrinsic factor, the complexes, once formed, should be similar to those formed in Sasanishiki. However, if the slower binding reflected lowered affinity of Norin 1 for dimers in DNA, the E-S complexes of Norin 1 might be less stable than those of Sasanishiki. Complex stability can be probed by their thermal stability: E-S complexes would be allowed to form, then challenged by incubation at a series of temperatures before administration of a photoflash. Decreased E-S stability would be reflected by

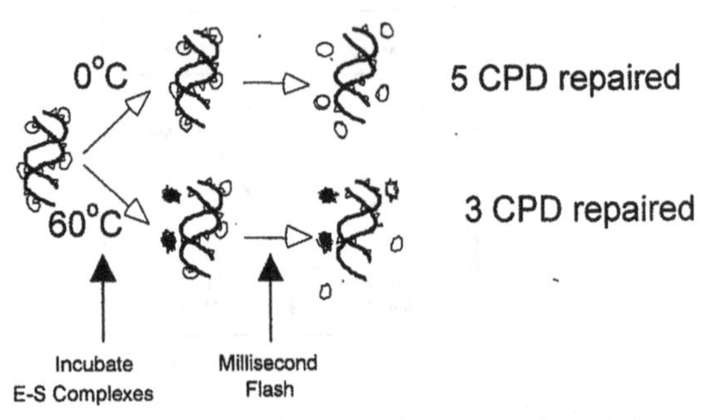

Figure 12. Measurement of photolase-dimer complex stability by photoflash analysis.

lower photorepair in the flash reaction, whereas unaltered E-S complex stability would be reflected in similar properties of the complexes in both strains. However, since artifacts could result from heating intact seedlings, it would be preferable to use extracts of the plants.

First, it was necessary to determine whether the apparent photolyase deficiency observed *in planta* was observed in extracts of the seedlings. They made

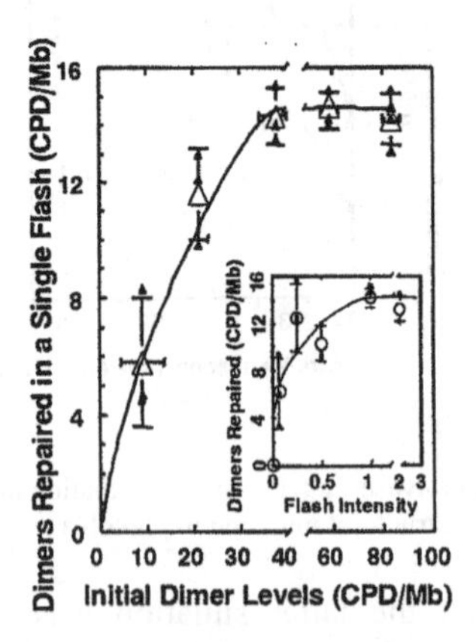

Figure 13. Photolyase deficiency in Norin 1 is observed in vivo and in vitro

extracts of seedlings of the two strains, and tested their photorepair activity *in vitro*, and compared the results with those for dimer repair *in planta*. The results are shown in Figure 13. The data show that extracts of Norin 1 seedlings also had lower photolyase activity than did Sasanishiki, just as is observed in the intact plants.

They then hypothesized that if the altered Norin 1 photolyase formed stable E-S complexes more slowly, the complexes that were eventually formed might be less stable than those from Sasanishiki. They added cell extracts from Norin 1 or from Sasanishiki to UV-irradiated DNA in vitro, and incubated the mixes in the dark to allow

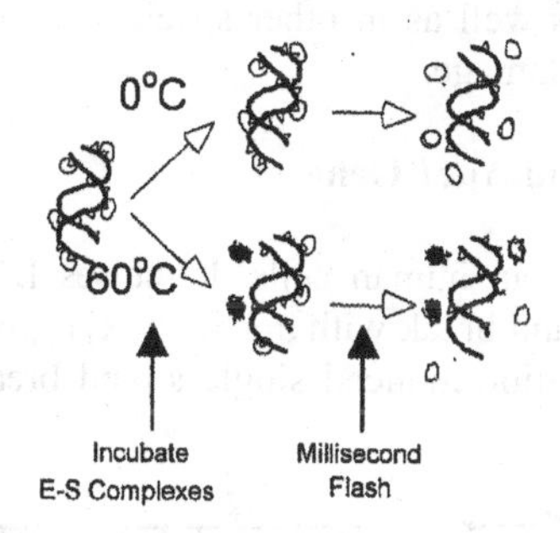

Figure 14. Measurement of photolyase-dimer complex stability by photoflash analysis.

complexes to form. They then incubated aliquots of the complexes at different temperatures, and as a function of time after complex formation, administered one saturating photoflash to allow photolysis of all existing photolyase-CPD complexes.

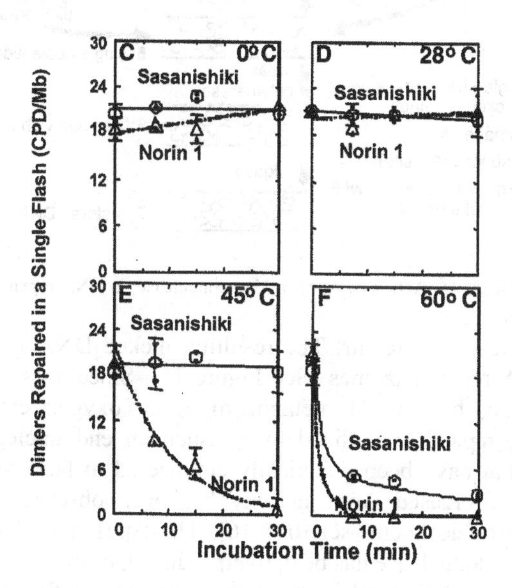

Figure 15. Determination of photolyase-dimer complex stability in extracts of Norin 1 and Sasanishiki at 0, 28, 45 and 60°C.

The resulting data clearly showed that the E-S complexes formed by the Norin photolyase were less thermostable than those from Sasanishiki, strongly suggesting that the Norin 1 photolyase gene has a mutation, most probably a mutation in its structural gene that affects photolyase function.

Photolyase genes from Norin 1 and from Sasanishiki are currently being cloned and sequence to determine the exact location(s) of any putative mutations and study their effect on photolyase function. Since photorepair is the primary DNA repair paths in plants, these studies should form the basis for determining whether other UV sensitive cultivars of rice as well as in other species contain mutations in this critical repair path for UV-induced damage.

8. Polymorphisms of Human Apel Gene

Apel is a 5'endonuclease from human cells. It incises DNA at the 5' side of abasic sites, producing a single strand break with a 3' hydroxyl group and a 5' abasic terminus. It can also function at radiation-induced single strand breaks, or at sites of base loss

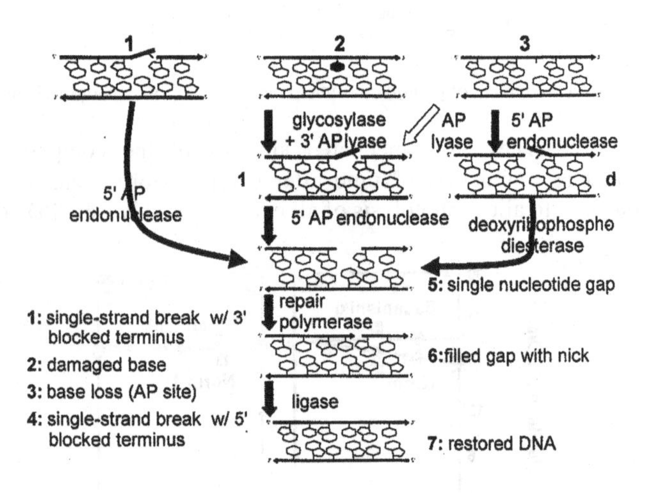

Figure 16. Action of Apel, a 5′ endonuclease, in DNA repair.

resulting from base excision repair. The resulting nicked DNA is then repaired by a series of other DNA repair enzymes (See Figure 16). Since abasic sites are produced by ionizing radiation, by normal metabolism in an oxygen environment, and by glycosylases during repair of oxidized bases, such an endonuclease is essential for normal life. Mice that have been genetically engineered to lack Apel do not survive embryogenesis and decreased Apel activity has been observed in brain tissue of patients with amyotrophic lateral sclerosis [6]. The Apel gene has been cloned and sequenced, and the endonuclease has been overproduced, characterized, and its structure determined by X-ray crystallography. The Apel gene has two functional regions, a refl domain and a nuclease domain. David Wilson and Harvey Mohrenweiser and their

Figure 17. Human Apel Gene with Refl and Nuclease Domains.

groups at Lawrence Livermore National Laboratory took advantage of high throughput DNA sequencing with the knowledge base of Ape1 structure and function to look for sequence changes (polymorphisms) among normal humans in the sequence of their Ape1 genes (see Figure 18).

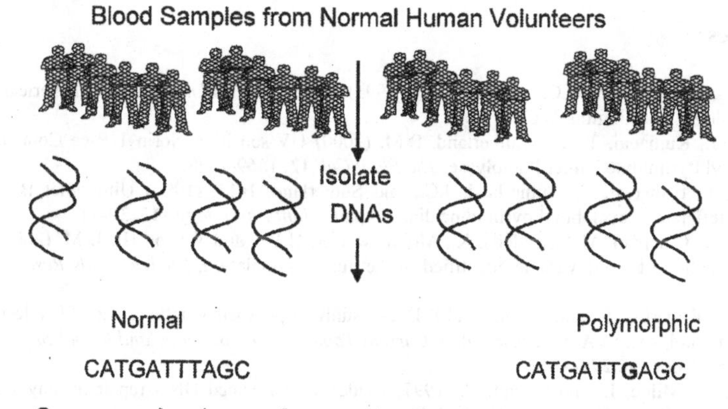

Figure 18. Scheme for determining DNA sequence polymorphisms in Ape1 gene in human volunteers.

They sequenced the Ape1 gene from over 128 normal, unrelated people, and detected sequence changes in the repair domain of this gene from DNA in seven individuals.

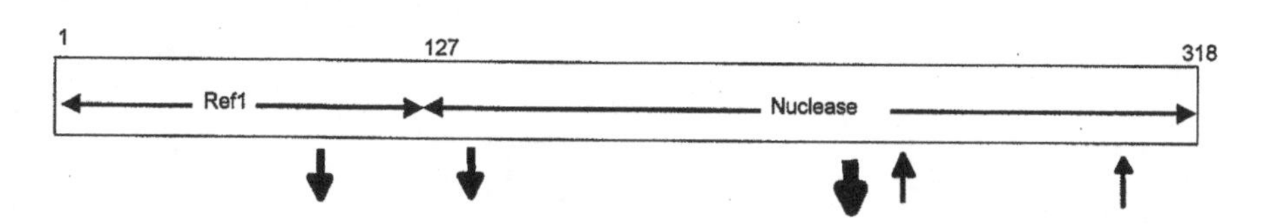

Figure 19. Human Ape1 gene, showing the Ref1 and nuclease domains, and locations of polymorphisms.

Of these mutations, Ape1 activity was decreased in 3 (shown as bold, downward-pointing arrows), of those, one was very strongly reduced (heavy bold, down-pointing arrow), increased in 2 (dotted, upward–pointing arrow) and unaffected in 2 (not shown). These mutations may affect human health, including cancer proneness.

9. Conclusions

Alterations of the structural genes coding for DNA repair enzymes can have severe effects on the organisms, whether a simple bacterium, or a higher plant or person. Modern biochemical and molecular biological approaches allow determination of the molecular origin of radiation sensitivity, including those resulting from deficiencies in DNA repair.

Acknowledgments

Research supported by the Office of Biological and Environmental Research of the US Department of Energy.

References

1. Friedberg, E.C., Walker, G.C., and Siede, W. (1995) *DNA repair and mutagenesis*, American Society for Microbiology, Washington, D.C.
2. Hidema, J., Kumagai, T., and Sutherland, B.M. (2000) UV-sensitive Norin 1 Rice Contains Defective Cyclobutyl Pyrimidine Dimer Photolyase, *The Plant Cell* **12,** 1569-1578.
3. Hidema, J., Kumagai, T., Sutherland, J.C., and Sutherland, B.M. (1996) Ultraviolet B-sensitive rice cultivar deficient in cyclobutyl pyrimidine dimer repair, *Plant Physiology* **113,** 39-44.
4. Hadi, M.Z., Coleman, M.A., Fidelis, K., Mohrenweiser, H.W., and Wilson III, D.M. (2000) Functional characterization of Ape1 variants identified in the human population, *Nucleic Acids Research* **20,** 3871-3879.
5. Harm, W., Rupert, C.S., and Harm, H. (1971) The study of photoenzymatic repair of UV lesions in DNA by flash photolysis., in A. C. Giese (eds.), *Current Topics in Photobiology and hotochemistry*, Academic Press, New York, pp. 279-324.
6. Kisby, G.E., Milne, J., and Sweatt, C. (1997) Evidence of reduced DNA repair in amyotrophic lateral sclerosis brain tissue, *Neuroreport* **8,** 1337-40.

UV-B AND UV-A RADIATION EFFECTS ON PHOTOSYNTHESIS AT THE MOLECULAR LEVEL

COSMIN SICORA, ANDRÁS SZILÁRD, LÁSZLÓ SASS, ENIKŐ TURCSÁNYI, ZOLTÁN MÁTÉ AND IMRE VASS
Institute of Plant Biology, Biological Research Center
P.O.Box 521, Szeged, H-6701 Hungary

1. Abstract

Ultraviolet radiation is a well known damaging factor of plant photosynthesis. Here we studied the mechanism of damage induced by the UV-B and UV-A spectral regions to the light energy converting Photosystem II (PSII) complex, which is the origin of electron flow for the whole photosynthetic process. Our results show that the primary UV damage occurs at the catalytic Mn cluster of water oxidation, which is most likely sensitized by the UV absorption of Mn(III) and Mn(IV) ions ligated by organic residues. The presence of visible light enhances the photodamage of PSII, but has no synergistic interaction with UV radiation. UV-induced damage of PSII can be repaired via *de novo* synthesis of the D1 and D2 reaction center protein subunits. This process is facilitated by low intensity visible light, which thereby can protect against UV-induced damage. However, the photodamage induced by visible light at high intensity (above 1000 $\mu Em^{-2}s^{-1}$) cancels the protective effect. The protein repair of PSII is also retarded by the lack of DNA repair as shown in a photolyase deficient cyanobacterial mutant.

2. Introduction

Solar radiation is not only the vital source of energy for the biosphere on our planet, but also acts as an adverse environmental factor for various forms of life. The efficiency of light-induced damage increases with decreasing wavelength. Thus, from natural sunlight that reaches the Earth the UV-B (280-315 nm) spectral range has the highest damaging potential. Therefore, recent reductions in the stratospheric ozone layer, which enhances UV-B intensity on the surface of Earth and ecologically significant depths of the ocean have initiated extensive research efforts to elucidate molecular mechanisms regulating responses of various organisms to UV-B radiation[1].

Photosynthetic organisms entirely depend on sunlight as the ultimate source of energy for their survival. As a consequence, their exposure to UV radiation is unavoidable, which poses a potentially adverse impact leading to decreased oxygen evolution and CO_2 fixation[2-5], reduction in dry weight, secondary sugars, starch, and total chlorophyll[6-7]. In cells of non-photosynthetic organisms, DNA absorbs ~50% of the incident ultraviolet radiation and it is the primary target of UV damage. However, in photosynthetic organisms chlorophyll and other pigments may contribute significantly to shielding DNA from UV radiation[8]. Therefore, besides damaging DNA a crucial part of the overall UV-B effect is

F. Ghetti et al. (eds.), Environmental UV Radiation: Impact on Ecosystems and Human Health and Predictive Models, 121–135.

related to the damage of pigments, lipids, amino acids as well as complex enzyme systems of the photosynthetic apparatus.

The reaction series of photosynthesis, which connects the light driven oxidation of water with the incorporation of carbon from atmospheric CO_2 into energy rich organic compounds takes place in the chloroplasts (Figure 1). The so-called light reactions are

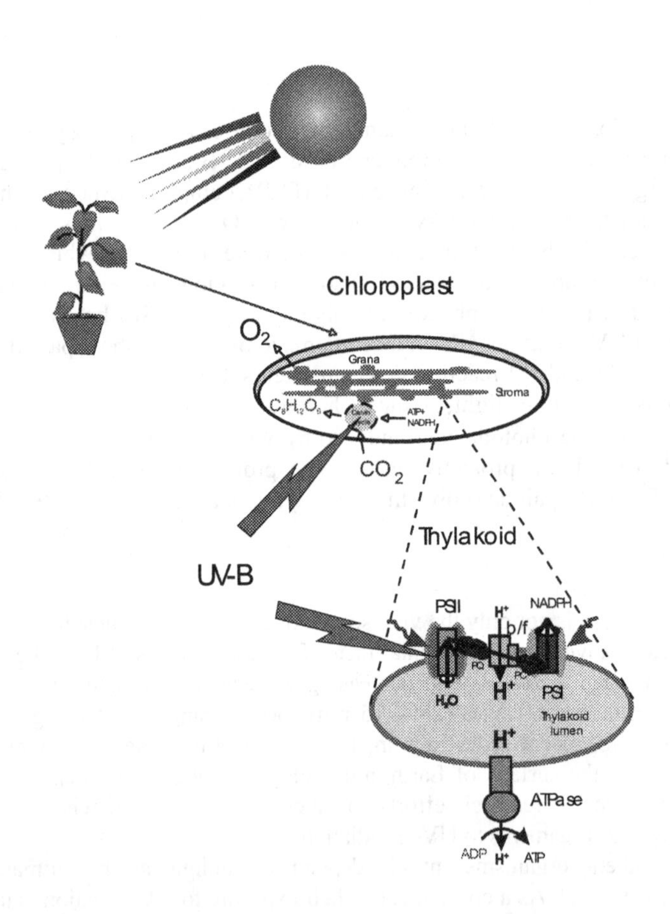

Figure 1. **The structure of the photosynthetic apparatus.** In plants photosynthesis takes place in the cell organ called chloroplast. The light reactions are mediated by electron/proton transport proteins embedded in the thylakoid membrane and result in the oxidation of water and the production of ATP and NADPH. These energy rich compounds are utilized in the process of CO_2 fixation mediated by the water soluble enzymes of the Calvin cycle. The main targets of UV-B radiation in plant cells are the nucleic acids, the Calvin cycle enzymes and the light-energy converting Photosystem II complex.

mediated by large pigment-protein complexes in the thylakoid membrane, which perform the oxidation of water as well as the production of ATP and NADPH. These compounds are utilized in the process of CO_2 fixation, performed by the soluble enzymes of the Calvin cycle. Previous literature data have demonstrated that the light reactions of photosynthesis, especially those mediated by the PSII complex[9-11], as well as the Calvin cycle[12-13] are susceptible to UV-induced damage. UV effects on DNA and on the Calvin cycle enzymes will be covered in other chapters of this book, therefore we will concentrate on the effects exerted by UV radiation on the light energy converting complex of PSII.

PSII is a water/plastoquinone oxido/reductase embedded in the thylakoid membrane. It contains over 25 subunits and performs the exclusive task of light-driven oxidation of water that serves as final electron source for all photosynthetic processes (Figure 2). The electrons liberated from water are transferred to membrane soluble plastoquinone molecules[14]. The redox cofactors of PSII electron transport are bound to or contained by the D1 and D2 protein subunits which form the reaction center of PSII.

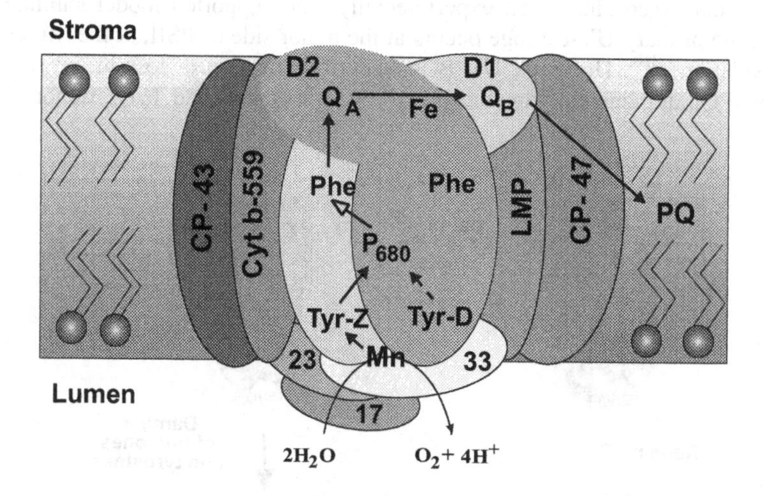

Figure 2. **The structure and function of the Photosystem II complex**. The reaction center of PSII consists of the D1 and D2 protein subunits, which bind the redox cofactors of light-induced electron transport: The Mn cluster of water oxidation, the redox-active tyrosine electron donors (Tyr-Z and Tyr-D), the reaction center chlorophyll (P_{680}), the primary electron acceptor phyophytin (PheO), and the first and second quinone electron acceptors Q_A and Q_B, respectively. The reaction center heterodimer is closely associated with cytochrome b-559, and surrounded by chlorophyll binding antenna (CP43 and CP47). The PSII complex also contains various low molecular mass polipeptides (LMP).

Water oxidation is catalyzed by a Mn cluster, and electrons liberated during this process are transferred to the reaction center chlorophyll, P680, via a redox active tyrosine residue, Tyr-Z, of the D1 protein. On the acceptor side of PSII, the electron produced by the light-induced charge separation event reduces a pheophytin molecule and then the first, Q_A, and second, Q_B, plastoquinone electron acceptor. Q_A is a firmly bound component of the reaction center complex, which undergoes one-electron reduction. In contrast, Q_B is a mobile electron carrier, which takes up two electrons sequentially from

Q_A before leaving its binding site formed by the D1 protein. The catalytic site of water oxidation is composed of four manganese ions bound by the D1 protein. During four subsequent photoacts the Mn cluster accumulates four oxidising equivalents, which are utilised in the oxidation of water into protons and molecular oxygen. This side-product of PSII function has been the source of atmospheric oxygen during the course of evolution, and hence is the origin of the UV-B screening ozone veil that protects life on Earth[15].

3. The mechanism of UV-induced damage of PSII

Damage by UV-B radiation

The mechanism of damage induced by UV-B light to the electron transport and protein structure of PSII has been addressed by a number of studies both in isolated and intact systems. According to an experimentally well supported model summarized in Figure 3, the primary UV-damage occurs at the donor side of PSII, at the Mn cluster of water oxidation[9-11,16]. However, UV-B induced modification or loss in the function of the Q_A and Q_B quinone electron acceptors[10,17-19], the Tyr-D and Tyr-Z donors[10,20] have

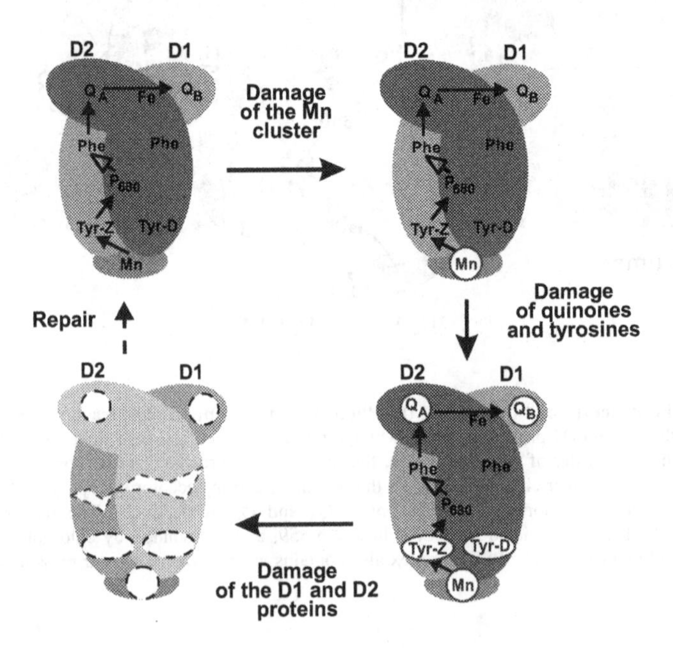

Figure 3. **The sequence of UV radiation induced damaging events in PSII.** The primary effect of UV radiation is the inactivation of the Mn cluster of the water-oxidizing complex. This is followed by the damage of quinone electron acceptors and tyrosine donors. Finally, both the D1 and D2 reaction center subunits are degraded. In intact cells the damage can be repaired via resynthesis of the damaged subunits.

also been observed. One manifestation of the UV-B effect on the acceptor side components is an apparent resistance against electron transport inhibitors, which act by replacing the mobile plastoquinone electron acceptor at the Q_B binding site. This effect, which is most likely related to the UV-B induced modification of the Q_B binding site, has been observed not only in oxygen evolving organisms[9,16,21], but also in isolated reaction centers of the purple bacterium *Rhodobacter sphaeroides*[22], that have analogous reaction center subunits and quinone acceptors as PSII. Besides the electron transport components, UV-B light also damages the D1 and D2 reaction center subunits of PSII, leading eventually to their degradation[17,23-25] (Figure 3), which can be repaired via *de novo* protein synthesis in intact cells[26-27].

In order to obtain more detailed information about the mechanism of UV-B induced damage of PSII we applied short light flashes to synchronize the water-oxidizing complex into various oxidation states, called S-states, in isolated thylakoid membranes. The synchronized samples were then illuminated with monochromatic UV-B laser flashes of 308 nm. Under the applied experimental conditions 80 repetitions of the flash protocol was sufficient to cause significant loss of oxygen evolution. The damage induced by the UV-B flashes shows a clear S-state dependence indicating that the water-oxidizing complex is most prone to UV damage in the S_2 and S_3 oxidation

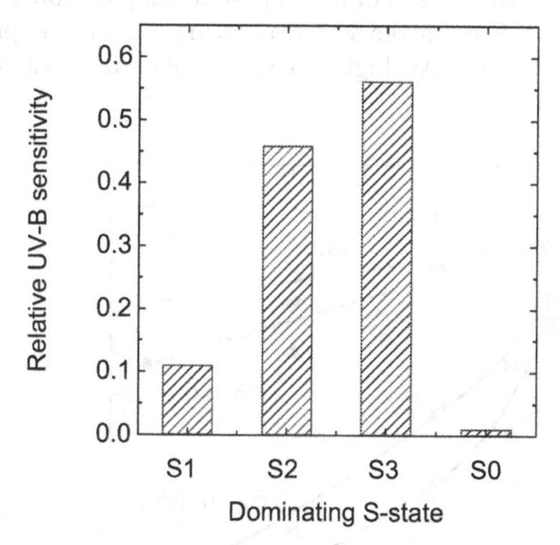

Figure 4. **The S-state dependence of the UV-B radiation induced damage**. The PSII centers were synchronized into different S-states in isolated spinach thylakoids by using Xe flashes, whose UV component was filtered out, and the synchronized samples were irradiated by 308 nm laser flashes. The fraction of damaged PSII centers was calculated from the loss of oxygen evolution using principal component analysis by taking into account the actual distribution of the different S-states after synchronization.

states (Figure 4). During the S-state transitions the catalytic Mn cluster of water oxidation is sequentially oxidized. Mn ions bound to organic ligands (such as amino acids) have pronounced absorption in the UV-B and UV-A regions in the Mn(III) and Mn(IV) oxidation states, which dominate the higher S-states, but not in the Mn(II) oxidation state, which occur in the lower S-states[28]. Thus the high UV sensitivity of

PSII in the S_2 and S_3 states indicates that UV absorption by the Mn(III) and Mn(IV) ions could be the primary sensitizer of UV-induced damage of the water-oxidizing machinery.

Damage by UV-A radiation

In contrast to the wealth of information available on the damaging mechanism of UV-B radiation our knowledge is more limited on the effects of UV-A (315-400 nm) radiation. The intensity of this spectral range in the natural sunlight is at least 10 times higher than that of UV-B, and its penetration to the Earth is not attenuated significantly by the ozone layer or other components of the atmosphere[29]. Thus, the damaging effects of UV-A radiation could be highly significant[30].

UV-A radiation has been shown to damage PSII to a considerably larger extent than PSI[31]. Within PSII, the slow rise of variable chlorophyll fluorescence together with the modification of the oscillatory pattern of flash-induced oxygen evolution indicates a damage of PSII donor side components[31]. This conclusion is supported by the appearance of a fast decaying phase of flash-induced chlorophyll fluorescence, which arises from the recombination of Q_A^- with intermediate electron donors, such as Tyr-Z^+, which are stabilized by the inactivation of the water-oxidizing complex. Further support for the primary effect of UV-A on the water-oxidizing complex is provided by low temperature EPR measurements. As Figure 5 shows, the so-called multiline signal,

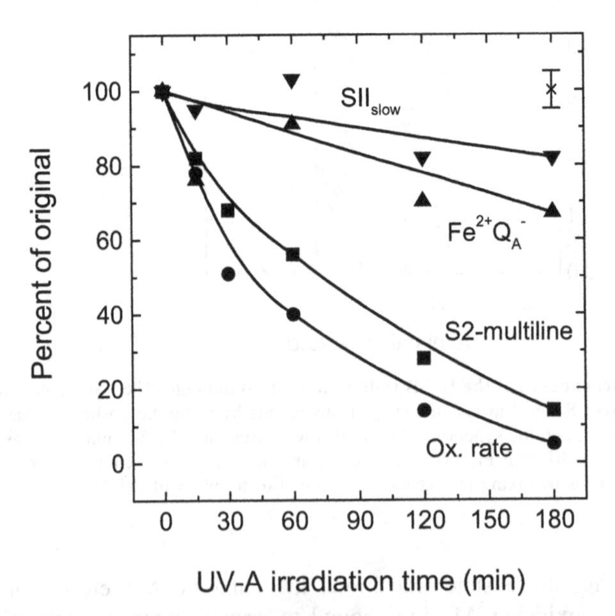

Figure 5. **The effect of UV-A radiation on the functioning of PSII redox components.** Isolated PSII membrane particles were irradiated with UV-A. The rate of oxygen evolution is compared with the amplitude of EPR signals arising from the S_2 state of the water-oxidizing complex (S_2-multiline), the reduced Q_A acceptor ($Q_A^-Fe^{2+}$) and the oxidized Tyrosine-D donor (Tyr-D^+).

which arises from the S_2 state of the water-oxidizing complex is lost much faster than the EPR signal which arises from the interaction of Q_A^- with the non-heme Fe^{2+} or from Tyr-D^+. Thus, the immediate cause for the loss of oxygen evolution is the inactivation of electron transport between the catalytic Mn cluster and the tyrosine electron donors. However, the loss of Q_A function points to additional UV-A induced alteration of PSII acceptor side components. This observation is in agreement with flash-induced thermoluminescence and chlorophyll fluorescence measurements which showed that the Q_B-binding pocket on the acceptor side is also modified[31]. Comparison of the characteristics of PSII damage induced by UV-A and UV-B radiation, shows that the two spectral ranges inhibit PSII by very similar or identical mechanisms, which target primarily the water-oxidizing complex. Although the damaging efficiency of UV-A is much smaller than that of UV-B, due to its higher intensity the UV-A component of sunlight appears to have the same overall potential to inactivate the light reactions of photosynthesis as UV-B.

The comparison of action spectra with the absorption of putative UV targets is often useful approach in identifying the critical target sites. The data shown in Figure 6 demonstrate that absorption by Mn (III) and Mn (IV) ions matches well the action spectrum of UV-induced inactivation of PSII in the whole UV-A and UV-B range, in contrast to PQ^- and Tyr-Z^+, which give only limited match. However, in the UV-C region additional targets, most likely PQ and/or Tyr-Z^+, are also involved.

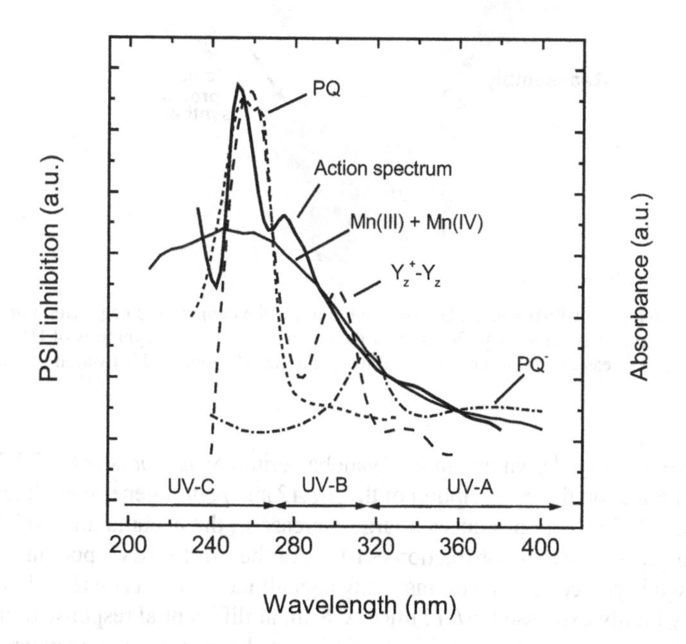

Figure 6. **The action spectrum of UV damage in comparison with the absorption spectra of the main UV targets in PSII**. Action spectrum of UV damage (thick solid line[2]). Absorption spectra: PQ (dotted line[32]) PQ^- (dashed-dotted line[32]), oxidized minus reduced Tyr-Z (dashed line[33]), Mn(III) + Mn(IV) bound to an organic ligand (thin solid line[28]).

4. The repair of UV-induced damage of PSII

Repair via de novo protein synthesis

One of the main defense responses of intact photosynthetic cells against UV-B damage is the repair of PSII via *de novo* synthesis of the D1 and D2 proteins[26] (Figure. 7).

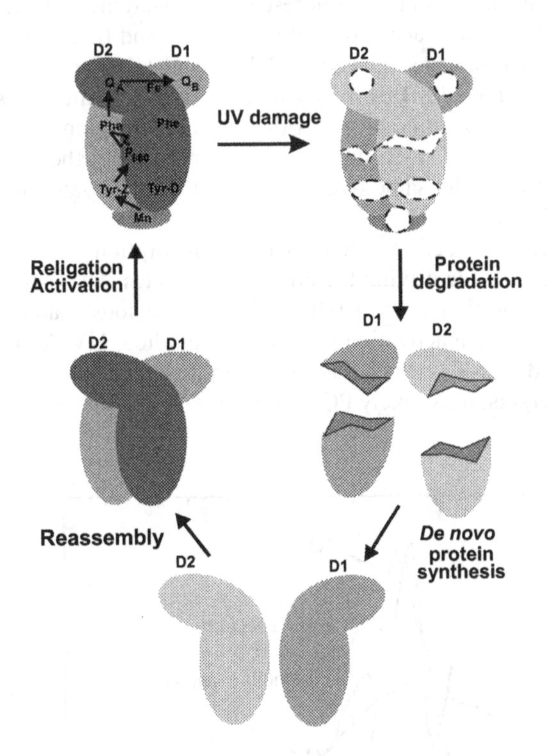

Figure 7. **The scheme of PSII repair**. UV radiation induces the damage and degradation of the D1 and D2 reaction center subunits. The first step of the repair is the *de novo* synthesis of D1 and D2. This is followed by reassembly of the reaction center, and the religation and activation of the redox cofactors.

We have recently shown that in the cyanobacterium *Synechocystis* 6803 UV-B light has differential effect on the transcription of the *psbA2* and *psbA3* genes encoding identical D1 proteins. UV-B irradiation induces a large increase in the accumulation of the *psbA3* mRNA, which gives only a minor fraction (<10 %) of the *psbA* mRNA pool under normal conditions in wild-type cells[27]. In contrast, only a small increase was observed in the level of the normally highly expressed *psbA2* mRNA. Similar differential response of the *psbD1* and *psbD2* genes, encoding identical D2 proteins, was also observed in response to UV-B light[34]. These results show that *psbA3* and *psbD2* act as UV-B stress genes which are needed for rapid repair of the damaged PSII reaction center subunits.

Since the process of protein synthesis is affected by a large number of factors, such as light, temperature and water content, changes in environmental conditions can significantly affect the protein repair capacity and in turn the UV-B sensitivity of photosynthetic organisms. An interesting example will be discussed below regarding the role of visible light in the regulation of protein repair.

The role of visible light in PSII repair

The visible component of sunlight light in itself can damage the photosynthetic apparatus, but it can also modify the redox state of UV targets as well as activate and regulate protein repair[26,35]. In order to clarify the role of visible light in UV-induced damage and repair of PSII, *Synechocystis* 6803 cells were exposed to UV-B and visible light in separate and simultaneous illumination protocols. The inactivation curves give straight lines on a semilogarithmic plot (Figure 8) indicating that damage of single molecular

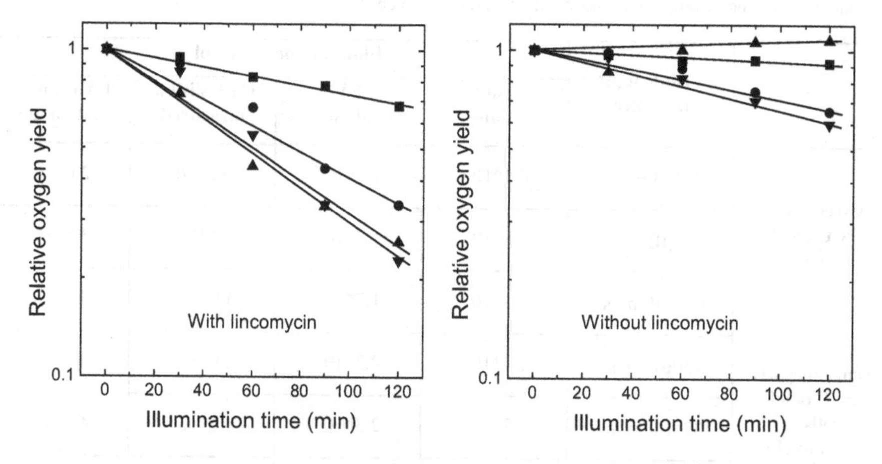

Figure 8. **The interaction of visible and UV-B light during PSII damage.** Cell suspension of the cyanobacterium *Synechocystis* 6803 was illuminated with visible light (squares), UV-B light (circles) and by the simultaneous application of the two spectral regions (up triangles). The experiments were performed without addition (right panel) and in the presence of the protein synthesis inhibitor lincomycin (left panel). The damage curves calculated by using the assumption that UV-B and visible light damages PSII via independent mechanisms are also shown (down triangles).

targets are responsible for the loss of PSII activity. When the repair of damaged PSII centers was prevented by protein synthesis inhibitor, the damage induced by UV radiation was enhanced by the presence of visible light. This enhancement could arise either from independent damage of PSII by the two spectral regions or from a synergistic interaction. In order to distinguish between these possibilities, we simulated the damage curves for the simultaneous illumination protocol using the assumption that UV-B and visible light inactivates PSII by independent mechanisms. The good correlation of the calculated and measured inactivation curves (Figure 8, left panel, Table I.) supports that UV-B and visible light damages PSII by independent mechanisms without synergistic interaction. However, the situation is quite different in intact cells, which are capable of *de novo* protein synthesis. On the first place, the damage induced by either visible or UV-B illumination applied separately is significantly smaller than that observed in the presence of protein synthesis inhibitor (Figure 8, right panel). This is in agreement with previous findings showing that repair of damaged PSII centers is a constitutive process during visible and UV-B illumination[26,35]. Even more important is the observation that the presence of low intensity visible light prevents the UV-induced loss of PSII activity.

At higher light intensities the UV-induced damage was not prevented, or even enhanced, and the difference between the slope of the measured and calculated inactivation curves became slower or disappeared as observed at 650 or 1300 $\mu Em^{-2}s^{-1}$ intensity, respectively (Table I).

Table I. **The slopes of the PSII inactivation curves under separate and simultaneous UV-B and visible illumination.** Experiments were performed by using *Synechocystis* 6803 cells as shown in Figure 8 and the slopes of the damage curves were calculated from the semi-logarithmic plots. UV-B intensity was 6 $\mu Em^{-2}s^{-1}$ in all experiments, but the intensity of the visible light varied between 130-1300 $\mu Em^{-2}s^{-1}$. In the last column the values show the predicted slopes of inactivation curves calculated by the assumption of independent damaging mechanisms.

	Visible light intensity	Illumination Protocol			
		Visible alone	UV alone	UV+ Visible measured	UV+ Visible calculated
Synechocystis PCC 6803 cells	130$\mu Em^{-2}s^{-1}$	$0.3*10^{-3}$	$1.6*10^{-3}$	$-0.5*10^{-3}$	$2.0*10^{-3}$
	650$\mu E\ m^{-2}s^{-1}$	$1.8*10^{-3}$	$1.7*10^{-3}$	$1.7*10^{-3}$	$3.5*10^{-3}$
	1300$\mu E\ m^{-2}s^{-1}$	$3.4*10^{-3}$	$1.7*10^{-3}$	$4.0*10^{-3}$	$4.4*10^{-3}$
Synechocystis PCC 6803 cells + lincomycin	130$\mu E\ m^{-2}s^{-1}$	$1.4*10^{-3}$	$4.2*10^{-3}$	$5.0*10^{-3}$	$5.6*10^{-3}$
	325$\mu E\ m^{-2}s^{-1}$	$2.3*10^{-3}$	$2.3*10^{-3}$	$4.5*10^{-3}$	$4.7*10^{-3}$
	1300$\mu E\ m^{-2}s^{-1}$	$3.5*10^{-3}$	$2.8*10^{-3}$	$5.3*10^{-3}$	$6.4*10^{-3}$

The role of DNA damage in PSII repair

DNA is a well known target of UV-B radiation not only in non-pigmented cells, but also in photosynthetic organisms. As discussed above the repair of UV-damaged PSII proceeds via *de novo* protein synthesis, which requires the transcription of DNA encoding the D1 and D2 reaction center subunits. Accumulation of damaged DNA could obviously hamper the transcription process and retard the protein-synthesis-dependent repair of PSII. In order to create a suitable experimental system to study this question we inactivated the *slr0854* gene encoding a type-I photolyase enzyme in *Synechocystis* 6803. This mutant was unable to grow under UV-B light complemented with visible illumination, condition which did not affect the wild-type cells (Figure 9). However, the photolyase deficient cells could grow as the wild-type under visible light, in the absence of UV-B radiation. This result shows that the photolyase encoded by the *slr0854* gene is indispensable for growth in the presence of UV-B radiation, but DNA damage which could occur under normal growth conditions is either not significant or can be repaired by systems other than photolyase.

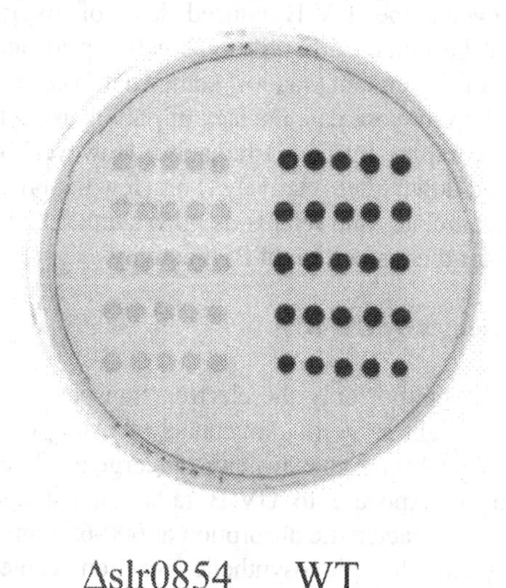

Figure 9. **The effect of photolyase (slr0854) inactivation on the growth of *Synechocystis* 6803 under UV-B.** Wild-type and Δslr0854 cells were inoculated on agar plate and grown under visible light for 2 days. The visible light was then complemented with UV-B, and the cells were grown for another 3 days.

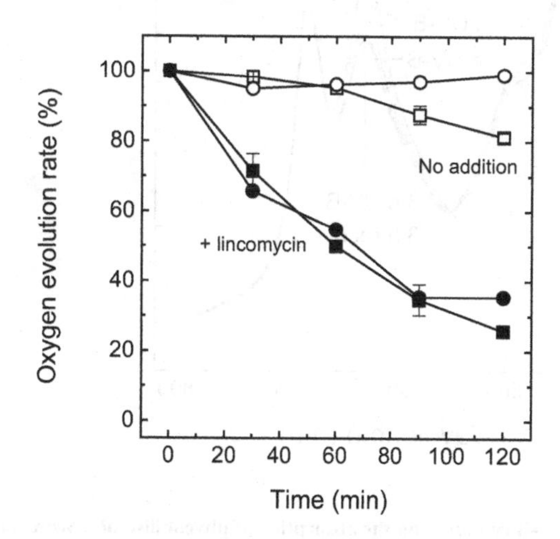

Figure 10. **The effect of photolyase (slr0854) inactivation on the UV-B induced loss of PSII activity.** Wild-type (circles) and Δslr0854 cells (squares) were cultured under visible light. The cultures were exposed to UV-B radiation complemented with visible light in the absence (open symbols) and presence of lincomycin (closed symbols).

To study the role of DNA damage in the protein-synthesis-dependent repair of PSII, the photolyase deficient cell cultures, which were grown under visible light, were exposed to UV-B radiation. When protein synthesis was blocked by lincomycin, the oxygen evolution activity was inhibited with the same rate in the mutant and the wild-type cells. In the absence of lincomycin, the UV-B-induced loss of oxygen evolution was not significantly different in the mutant and wild-type cells up to one hour, but after longer illumination the activity of the mutant was lost somewhat faster than that of the wild-type cells (Figure 10). According to these data the lack of photolyase activity does not influence the damage of the PSII complex by UV-B radiation. However, the efficiency of protein repair is somewhat slowed down after relatively short (few hours) exposure times, showing that unrepaired DNA accumulates and retards *de novo* synthesis of the D1 and D2 reaction center subunits required for the restoration of PSII activity.

5. Damage of phycobilisomes by UV-B

UV-B radiation affects not only the electron transport processes mediated by the PSII complex, but also the light harvesting antenna systems. In cyanobacteria light capture is performed mainly by phycobiliproteins which form large rod-like structures, the so called phycobilisomes. Prolonged exposure to UV-B radiation damages the phycobilisomes, leading to the loss of their characteristic absorption at 600-650 nm (Figure 11) and inhibits the transfer of energy to the photosynthetic reaction centers[8,36-37]. However, the UV-sensitivity of phycobilisomes relative to PSII has not been studied previously. In order

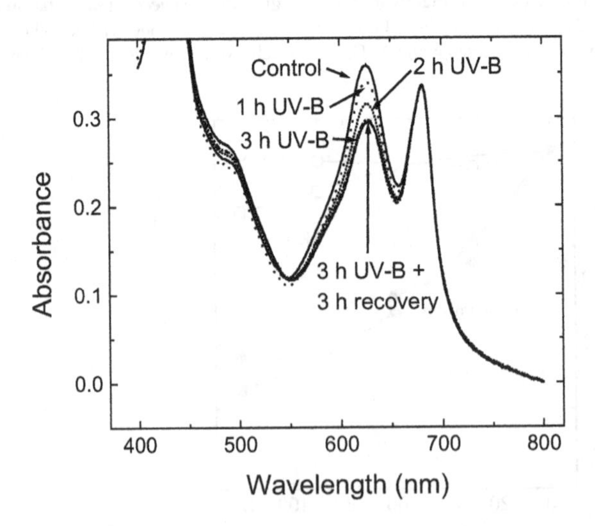

Figure 11. **The effect of UV-B radiation on the absorption of phycobilisomes.** *Synechocystis* 6803 cell were irradiated with 6 μEm^{-2}s^{-1} UV-B complemented with 250 μEm^{-2}s^{-1} visible light and the absorption spectrum of the cell suspension was measured after the indicated exposure times. After the UV-B treatment the samples were transferred to visible light for a 3 h recovery period.

to clarify this point we monitored the activity of PSII in parallel with the integrity of phycobilisomes, by measuring their absorption peak at 626 nm in *Synechocystis* 6803 cells. Under illumination with 6 $\mu Em^{-2}s^{-1}$ UV-B complemented with 250 $\mu Em^{-2}s^{-1}$ visible light oxygen evolution was stable during the course of the experiment (Figure 12 lower panel). However, the 626 nm peak of phycobilisomes was gradually lost, which was not restored when cells were transferred to visible light. A drastically different picture was observed when protein synthesis was inhibited by the addition of lincomycin (Figure 12 upper panel). Under these conditions oxygen evolution was completely inactivated within 1 hour, but the time course of the phycobilisome damage was the same as in the absence of lincomycin.

These results demonstrate that in the absence of protein repair UV-B induced damage of phycobilisomes occurs much slower than that of PSII. However, in cells which are capable of *de novo* protein synthesis, PSII is efficiently repaired when the UV-B radiation is removed from the illumination protocol. In contrast, restoration of phycobilisomes was observed only after 24 h (Figure 12 shows only the first 3 h of the recovery).

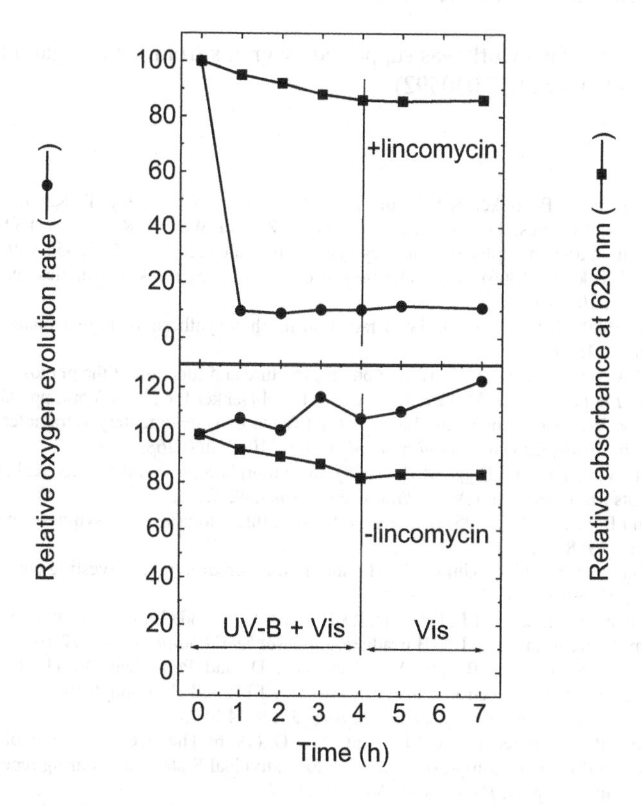

Figure 12. **UV-B-induced damage of phycobilisomes and oxygen evolution.** *Synechocystis* 6803 cells were irradiated as in Figure 10 in the presence (upper panel) and absence of lincomycin (lower panel). The integrity of phycobilisomes was quantified by measuring the absorbance at 626 nm (squares), and PSII activity was followed by measuring the rate of oxygen evolution.

Since under our conditions the doubling time of the cells is 10-12 h, it is quite likely that recovery of phycobilisome function requires the development of new cells, via cell division. Comparison of PSII and phycobilisomes provides an interesting example for a highly UV-sensitive, but well repaired component in contrast to a low sensitivity, but inefficiently repaired component.

6. Concluding remarks

Our results clearly demonstrate that PSII is a highly UV-sensitive component of the photosynthetic apparatus. However, due to its very efficient repair system, PSII damage becomes limiting for gross photosynthesis only when the repair capacity of the cells is slowed down by other environmental factors, such as low temperatures, high intensity visible light, drought, etc. Therefore, in plants exposed to UV-B radiation under additional environmental stress conditions, which retard the repair of PSII, inhibited electron flow from PSII can be limiting for the production of ATP and NADPH required for the Calvin cycle, and consequently for the overall photosynthetic process.

Acknowledgement. This work was supported by grants from the Hungarian Granting Agency OTKA (T 034321, T 030592)

References

1. Smith, R.C., Prézelin, B.B., Baker, K.S., Bidigare, R.R., Bouche,r N.P., Coley, T., Karentz, D., MacIntyre, S., Matlick, H.A., Menzies, D., Ondrusek, M., Wan, Z. and Waters, K J. (1995) Ozone depletion: Ultraviolet radiation and phytoplankton biology in antarctic waters. *Science*, 255: 952-959.
2. Jones, L.W. and Kok, B. (1966) Photoinhibition of chloroplast reactions. I. Kinetics and action spectra, *Plant Physiol.*, 41: 1037-1043.
3. Bornman, J.F. (1989) Target sites of UV-B radiation in photosynthesis of higher plants. *J. Photochem. Photobiol. B:Biol.*, 4: 145-158.
4. Vass, I. (1996) Adverse effects of UV-B light on the structure and function of the photosynthetic apparatus In *Handbook of Photosynthesis* (M. Pessarakli, ed.), Marcel Dekker Inc., New York, pp. 931-950.
5. Strid, A., Chow, W.S. and Anderson, J.M. (1990) Effects of supplementary ultraviolet-B radiation on photosynthesis in *Pisum Sativum*, *Biochim. Biophys. Acta*, 1020: 260-268.
6. Basiouny, F.M., Van, T.K. and Biggs, R.H. (1978) Some morphological and biochemical characteristics of C3 and C4 plants irradiated with UV-B, *Physiol. Plantarum*, 42: 29-32.
7. Fiscus, E.L. and Booker, F.L. (1995) Is increased UV-B a threat to crop photosynthesis and productivity?, *Photosynth. Res.*, 43: 81-92.
8. Lao, K. and Glazer, A.N. (1996) Ultraviolet-B photodestruction of a light-harvesting complex, *Proc. Natl. Acad. Sci. USA*, 93: 5258-5263.
9. Renger, G., Völker, M., Eckert, H.J., Fromme, R., Hohm-Veit, S. and Graber, P. (1989) On the mechanism of photosystem II deterioration by UV-B irradiation, *Photochem. Photobiol.*, 49: 97-105.
10. Vass, I., Sass, L., Spetea, C., Bakou, A., Ghanotakis, D. and Petrouleas, V. (1996) UV-B induced inhibition of photosystem II electron transport studied by EPR and chlorophyll fluorescence. Impairment of donor and acceptor side components, *Biochemistry*, 35: 8964-8973.
11. Post, A., Lukins, P.B., Walker, P.J. and Larkum, A.W.D. (1996) The effects of ultraviolet irradiation on $P680^+$ reduction in PS II core complexes measured for individual S-states and during repetitive cycling of the oxygen-evolving complex, *Photosynth. Res*,. 49: 21-27.
12. Strid, A., Chow, W.S. and Anderson, J.M. (1994) UV-B damage and protection at the molecular level in plants, *Photosynth. Res.*, 39: 475-489.
13. Nogués, S. and Baker, N.R. (1995) Evaluation of the role of damage to photosystem II in the inhibition of CO_2 assimilation in pea leaves on exposure to UV- B radiation, *Plant Cell Environ.*, 18: 781-787.
14. Andersson, B. and Styring, S. (1991) Photosystem II: Molecular organization, function, and acclimation, *Curr. Top. Bioenerg.*, 16: 1-81.

15. Kasting, J.F. (1993) Earth's Early Atmosphere, *Science*, 259: 920-926.
16. Hideg, É., Sass, L., Barbato, R. and Vass, I. (1993) Inactivation of oxygen evolution by UV-B irradiation. A thermoluminescence study, *Photosynth. Res.*, 38: 455-462.
17. Greenberg, B.M., Gaba, V., Canaani, O., Malkin, S., Mattoo, A.K. and Edelman, M. (1989) Separate photosensitizers mediate degradation of the 32-kDa photosystem II reaction centre protein in visible and UV spectral regions, *Proc. Natl. Acad. Sci. USA*, 86: 6617-6620.
18. Melis, A., Nemson, J.A. and Harrison, M.A. (1992) Damage to functional components and partial degradation of photosystem II reaction center proteins upon chloroplast exposure to ultraviolet-B radiation, *Biochim. Biophys. Acta*, 1100: 312-320.
19. Jansen, M.A.K., Gaba, V. and Greenberg, B.M. (1998) Higher plants and UV-B radiation: balancing damage, repair and acclimation, *Trends Plant Sci.*, 3: 131-135.
20. Yerkes, C.T., Kramer, D.M., Fenton, J.M. and Crofts, A.R. (1990) UV-photoinhibition: studies in vitro and in intact plants. In *Current Research in Photosynthesis, Vol. II.* (M. Baltscheffsky, ed.), Kluwer Academic Publisher, The Netherlands, pp. II.6.381-II.6.384.
21. Vass, I., Kirilovsky, D. and Etienne, A.-L. (1999) UV-B radiation-induced donor- and acceptor-side modifications of Photosystem II in the cyanobacterium *Synechocystis* sp. PCC 6803, *Biochemistry*, 38: 12786-12794.
22. Tandori, J., Máté, Z., Vass, I. and Maróti, P. (1996) The reaction centre of the purple bacterium *Rhodopseudomonas sphaeroides* R-26 is highly resistant against UV-B radiation, *Photosynth. Res.*, 50: 171-179.
23. Trebst, A. and Depka, B. (1990) Degradation of the D-1 protein subunit of photosystem II in isolated thylakoids by UV light, *Z. Naturforsch.*, 45c: 765-771.
24. Friso, G., Spetea, C., Giacometti, G.M., Vass, I. and Barbato, R. (1993) Degradation of photosystem II reaction center D1-protein induced by UVB irradiation in isolated thylakoids. Identification and characterization of C- and N-terminal breakdown products, *Biochim. Biophys. Acta*, 1184: 78-84.
25. Barbato, R., Frizzo, A., Friso, G., Rigoni, F. and Giacometti, G.M. (1995) Degradation of the D1 protein of photosysem-II reaction centre by ultraviolet-B radiation requires the presence of functional manganese on the donor side, *Eur. J. Biochem.*, 227: 723-729.
26. Sass, L., Spetea, C., Máté, Z., Nagy, F. and Vass, I. (1997) Repair of UV-B induced damage of Photosystem II via *de novo* synthesis of the D1 and D2 reaction centre subunits in *Synechocystis* sp. PCC 6803, *Photosynth. Res.*, 54: 55-62.
27. Máté, Z., Sass, L., Szekeres, M., Vass, I. and Nagy, F. (1998) UV-B induced differential transcription of *psbA* genes encoding the D1 protein of Photosystem II in the cyanobacterium *Synechocystis* 6803, *J. Biol. Chem.*, 273: 17439-17444.
28. Bodini, M.E., Willis, L.A., Riechel, T.L. and Sawyer, D.T. (1976) Electrochemical and Spectroscopic Studies of Manganese(II), -(III), and -(IV) Gluconate Complexes. 1. Formulas and Oxidation-Reduction Stoichiometry, *Inorg. Chem.*, 15: 1538-1543.
29. Holm-Hansen, O., Lubin, D. and Helbling, E.W. (1993) Ultraviolet radiation and its effects on organisms in aquatic environments. In *Environmental UV Photobiology* (A.R. Young, L.O. Björn, J. Moan and W. Nultsch, eds.), Plenum Press, New York, pp. 379-426.
30. Cullen, J.J., Neale, P. and Lesser, M.P. (1992) Biological weighting function for the inhibition of phytoplankton photosynthesis by ultraviolet radiation, *Science*, 258: 646-650.
31. Turcsányi, E. and Vass, I. (2000) Inhibition of Photosynthetic electron transport by UV-A radiation targets the Photosystem II complex, *Photochem. Photobiol.*, 72: 513-520.
32. Amesz, J. (1977) Plastoquinone. In *Encyclopedia of Plant Physiology, Vol. 5* (A. Trebst and M. Avron, eds.), Springer-Verlag, Berlin, pp. 238-246.
33. Dekker, J.P., van Gorkom, H.J., Brok, M. and Ouwehand, L. (1984) Optical characterization of photosystem II electron donors, *Biochim. Biophys. Acta*, 764: 301-309.
34. Viczián, A., Máté, Z., Nagy, F. and Vass, I. (2000) UV-B induced differential transcription of *psbD* genes encoding the D2 protein of Photosystem II in the cyanobacterium *Synechocystis* 6803, *Photosynth. Res.*, 64: 257-266.
35. Aro, E.-M., Virgin, I. and Andersson, B. (1993) Photoinhibition of photosystem II. Inactivation, protein damage and turnover, *Biochim. Biophys. Acta*, 1143: 113-134.
36. Sinha, R.P., Lebert, M., Kumar, A., Kumar, H.D. and Hader, D.-P. (1995) Disintegration of phycobilisomes in a rice field cyanobacterium *Nostoc* sp. following UV irradiation, *Biochem. Mol. Biol. Int.*, 37: 697-706.
37. Pandey, R., Chauhan, S. and Singhal, G.S. (1997) UVB-induced photodamage to phycobilisomes of *Synechococcus* sp. PCC 7942, *J. Photochem. Photobiol. B:Biol.*, 40: 228-232.

POTENTIAL EFFECTS OF UV-B ON PHOTOSYNTHESIS AND PHOTOSYNTHETIC PRODUCTIVITY OF HIGHER PLANTS

SALVADOR NOGUÉS[1], DAMIAN J. ALLEN[2] AND NEIL R. BAKER[3]
[1]*Departament de Biologia Vegetal, Universitat de Barcelona, Barcelona, Spain*
[2]*Agricultural Research Service, Photosynthesis Research Unit, Urbana (IL), USA*
[3]*Department of Biological Sciences, University of Essex, Colchester, UK*

1. Abstract

The effects of UV-B radiation on photosynthesis and photosynthetic productivity of higher plants are reviewed. When plants were exposed to large UV-B doses in a glasshouse in order to study the mechanistic basis of UV-B-induced inhibition of photosynthesis, direct effects on stomata and on Calvin cycle enzymes (i.e. large decreases in ribulose-1,5-bisphosphate carboxylase/oxygenase and in sedoheptulose-1,7-bisphosphate) without any significant effect on photosystem II were observed. When plants were continuously grown and developed under UV-B in a glasshouse, exposure to UV-B only decreased adaxial stomatal conductance (g_s) of the leaves and consequently increased the stomatal limitation of CO_2 uptake. Furthermore, field studies using realistic UV-B levels (i.e. a predicted 30% increase in this radiation) demonstrate a lack of UV-B effects on photosynthesis or on plant biomass. It is concluded that any predicted future increase in UV-B irradiation is unlikely to have a significant impact on the photosynthetic characteristics and productivity of higher plants.

2. Introduction

It has been predicted frequently that increasing UV-B radiation due to future stratospheric ozone depletion might affect the photosynthesis and photosynthetic productivity of higher plants. The photosynthetic productivity of a plant is determined by the quantity of photosynthetically active radiation intercepted and the efficiency with which this intercepted radiation is utilised for dry matter production. It is evident that UV-B can potentially impair the performance of the three main processes of photosynthesis, the thylakoid electron transport and photophosphorylation, the carbon reduction cycle, and the stomatal control of the CO_2 supply[1]. The activities of many components of photosynthesis are regulated by the rates of other photosynthetic reactions and therefore examination of a single process or component, such as the carboxylation velocity of ribulose-1,5-bisphophate carboxylase/oxygenase, the quantum efficiency of photosystem II photochemistry or stomatal conductance, does not allow identification of the primary UV-B-induced limitation.

F. Ghetti et al. (eds.), Environmental UV Radiation: Impact on Ecosystems and Human Health and Predictive Models, 137–146.

In order for UV-B to have any effects on photosynthetic productivity, UV-B must penetrate the leaf and be absorbed by chromophores associated with the photosynthetic apparatus or associated genes and gene products. Cellular components that absorb UV-B directly include nucleic acids, proteins, lipids and quinones[2]. Leaves contain water-soluble phenolic pigments, such as flavonoids, which strongly absorb UV-B radiation whilst not absorbing photosynthetically active radiation. These UV-B absorbing pigments are present throughout the leaf, but accumulate in adaxial epidermal cells[3-4]. UV-B radiation stimulates the expression of the genes encoding phenylalanine ammonia-lyase (PAL), the first stage of the phenylpropanoid pathway, and chalcone synthase (CHS), the key stage that commits the pathway to flavonoid synthesis (for reviews see Bornman and Teramura[5], Beggs and Wellmann[6], Jenkins *et al.*[7]).

In considering how UV-B can potentially affect the photosynthetic productivity of higher plants a clear distinction has to be made between studies made under glasshouse conditions at high UV-B doses and field studies using realistic UV-B levels. In the former, a clear distinction has also to be made between plants grown without UV-B and suddenly exposed to it, and plants continuously grown and developed under UV-B.

3. Glasshouse studies

Electron transport and photophosphorylation

Photoinhibitory damage of PSII could be primarily responsible for UV-B-induced inhibition of photosynthesis in higher plants. In many reviews, PSII damage has often been implicated as the major potential limitation to photosynthesis in UV-B irradiated leaves[8-10], as is the case in the photoinhibition of photosynthesis by excess photosynthetically-active radiation (400-700 nm)[11], although it has been suggested that the mechanism of UV-B induced damage may be different[12-13].

When pea (*Pisum sativum* L. cv. Meteor) plants were grown in a greenhouse and suddenly irradiated with a high UV-B dose (40 kJ m^{-2} d^{-1} weighted with a generalised plant action spectrum[16]) during 14 h decreases in CO_2 assimilation occurred (Table I), which were not accompanied by decreases in F_v/F_m, the maximum quantum efficiency of PSII photochemistry[17]. Increased exposure to very high UV-B doses resulted in a further loss of CO_2 assimilation and decreases in F_v/F_m, which were accompanied by a loss of the capacity of thylakoids isolated from the leaves to bind atrazine, but clearly PSII was not the primary target site involved in the onset of inhibition of photosynthesis in UV-B-irradiated plants. This results has been recently confirmed in other vascular plants[18].

In contrast, when these pea plants were grown throughout their development under a high irradiance of UV-B in a glasshouse, there was not a significant effect of UV-B on F_v/F_m or on relative quantum efficiency of PSII electron transport (ϕ_{PSII}, Table I) and the only effect was a decrease of adaxial stomatal conductance of the leaves[14].

It is now widely accepted that PSII is not directly affected by UV-B radiation and decreases in PSII efficiency are usually attributable to down regulation of electron transport, due to large reductions in carbon metabolism induced by UV-B radiation. Decreases in F_v/F_m usually appeared to be a secondary effect of UV-B radiation.

Table 1. Analysis of the responses of chlorophyll fluorescence, carbon assimilation to intercellular CO_2 concentration and stomatal conductance parameters in pea plants suddenly exposed for 14h or developed during 24d to high UV-B doses in a greenhouse. Parameters estimated were as follows: maximum (F_v/F_m) and relative (ϕ_{PSII}) quantum efficiency of PSII photochemistry in the dark measured at Photosynthethically active Photon Flux Density (PPFD) = 450 μmol m^{-2} s^{-1}; light saturated CO_2 assimilation rate (A_{sat}), maximum carboxylation velocity of Rubisco ($V_{c,max}$), maximum potential rate of electron transport contributing to RuBP regeneration (J_{max}) and stomatal limitation (l) to A_{sat} measured at PPFD = 1200 μmol m^{-2} s^{-1}, leaf temperature = 25 ± 0.5 °C and VPD = 1.5 kPa; and adaxial, abaxial and total (adaxial plus abaxial) g_s measured at PPFD = 450 μmol m^{-2} s^{-1}. Means of at least three replicates ± 1 SE. Data modified from Nogués et al.[14-15]

Parameter	Non-UV-B	Suddenly exposed for 14h	Developed during 24d
F_v/F_m	0.81±0.01	0.80±0.01	0.79±0.02
ϕ_{PSII}	0.54±0.03	0.45±0.02	0.47±0.03
A_{sat} (μmol m^{-2} s^{-1})	23.3±1.0	7.7±1.4	18.0±2.1
$V_{c,max}$ (μmol m^{-2} s^{-1})	99.5±7.1	42.2±11.1	91.3±15.2
J_{max} (μmol m^{-2} s^{-1})	238±11	99±28	332.9±50.9
Stomatal limitation (%)	11.1±1.4	19.8±1.9	34.0±1.6
Ada. g_s (mol m^{-2} s^{-1})	0.42±0.05	0.24±0.07	0.13±0.06
Aba. g_s (mol m^{-2} s^{-1})	0.50±0.07	0.20±0.03	0.44±0.11
Total g_s (mol m^{-2} s^{-1})	0.92±0.08	0.44±0.07	0.57±0.12

Calvin cycle

Decreases in photosynthesis on exposure of plants grown in the absence of UV-B to high UV-B doses have been attributed to deactivation or a loss of Rubisco activity[19]. Wilson *et al.*[20] reported the appearance of a 66 kDa protein during UV-B exposure in oilseed rape, pea, tomato and tobacco which they attributed to a photomodified form of the large subunit of Rubisco. Levels of mRNA coding for both the large (LSU, 55 kDa) and small (SSU, 14.5 kDa) submits of Rubisco have been also reported to decline[2]. Furthermore, decline in the maximum rate of electron transport contributing to RuBP regeneration also occurred in UV-B-irradiated leaves (Table I). Since PSII was excluded as a primary factor limiting CO_2 assimilation during UV-B exposure, other Calvin cycle enzymes must be involved in the photosynthetic decreases by UV-B[21]. Allen *et al.*[1] were the first to show large reductions in the content of sedoheptulose-1,7-bisphosphatase (SBPase), a key regulatory enzyme in the Calvin cycle.

The process by which UV-B radiation induces the loss of some Calvin cycle enzymes is unknown. The selective loss of specific enzymes suggest that this mechanism is not a non-specific oxidation of leaf proteins resulting from UV-B absorption by these proteins, or a UV-B-induced production of reactive oxygen species, unless these particular proteins are not easily repaired or are relatively unprotected by antioxidants. An alternative mechanism may be the induction by UV-B of proteases specific to the proteins that are degraded. It is possible that UV-B is inducing such a senescence-like response, as has been implicated in plant responses to tropospheric ozone where specific proteins are degraded that are spatially separated from the site of primary oxidative damage[22].

In contrast, when plants were developed from seed under high UV-B doses, there were no effects on photosynthesis (Table I[14]). A likely protective mechanism is an increase in the content of UV-B-absorbing compounds (i.e. flavonoids) found in these irradiated plants. The induction of UV-B-absorbing compounds in leaves is one of the best-described and most widespread responses of plant to UV-B.

Stomata

Changes in stomatal function play a major role in the UV-B-induced inhibition of CO_2 assimilation. Sudden exposure of pea plants to high UV-B doses largely decreases both adaxial and abaxial g_s (Table I[15]). However, when plants were grown and developed under UV-B for 24 d in a greenhouse, adaxial g_s was substantial decreased, with no major effects on abaxial g_s[15]. The effects on adaxial g_s were mediated by changes in aperture, as there was no reduction in stomatal density in these pea leaves. The adaxial stomata are the cells that are most exposed to UV-B radiation, because they are not screened by flavonoids in the epidermal layer.

It is clear that there is a direct effect of UV-B on stomata in addition to that caused by changes in the mesophyll photosynthesis of the leaves:

(i) the effects on adaxial g_s on plants grown and developed under UV-B for 24 d in a greenhouse were accompanied by no changes in any photosynthetic parameter measured

(ii) the effects of UV-B was largest on the exposed adaxial leaf surface (if UV-B was affecting mesophyll photosynthesis it presumably would have affected both leaf surfaces equally)

(iii) when leaves were inverted the light level on the different epidermis is changed by 10-50 fold, but photosynthesis should not be affected as the same total photon flux density is incident on the mesophyll. Therefore, we expected that the so-called "direct" response of guard cells to light, which acts independently of the response to c_i or to some mesophyll photosynthesis-related signal, would result in abaxial stomata opening and adaxial stomata closing. This was exactly what was observed in the control plants. In the UV-B treatments the opening response of abaxial stomata on inversion was either reduced or eliminated at highest dose, while the adaxial stomata (with a lower sensitivity to light) closed.

The mechanism for the UV-B effect on stomata is not clear. One potential mechanism is an inhibition of ATP synthesis by electron transport in guard cells thylakoids. A second mechanism may involve a direct inhibition by UV-B of the plasmalemma ATPase proton pump[1]. Alternatively, UV-B may not directly affect the generation of the guard cell turgor pressure, but rather the effect of this turgor on pore size through UV-B-induced changes in the elasticity of the cell walls or cytoskeleton of guard cells and the neighbouring epidermal cell. This possibility warrants further investigation.

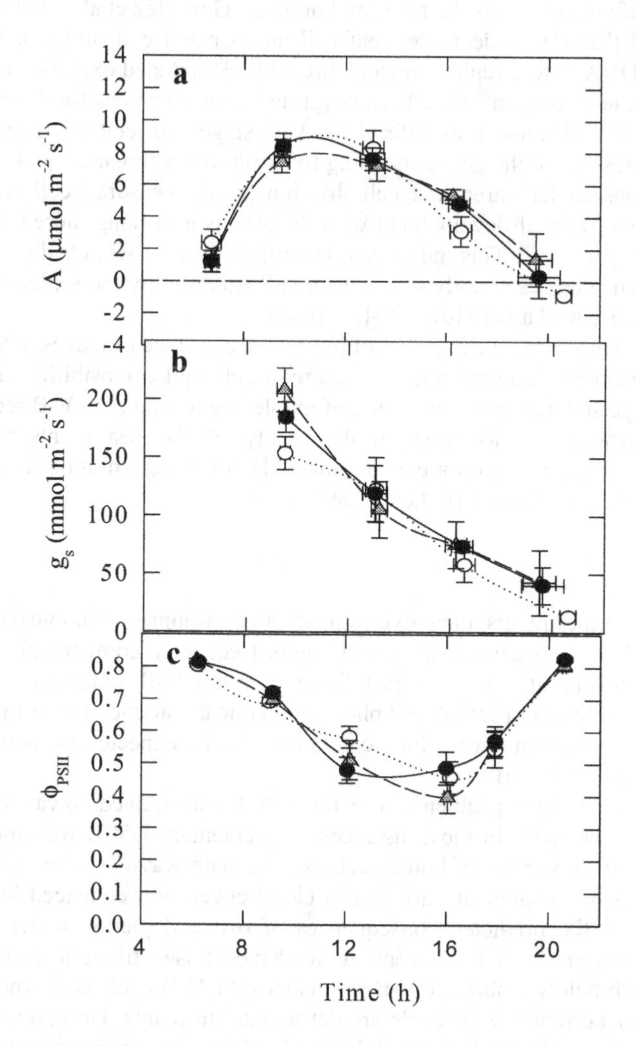

Figure 1. In situ diurnal trends in net CO_2 assimilation rate (A) (a), stomatal conductance (g_s) (b) and quantum efficiency of PSII photochemistry (ϕ_{PSII}) (c) in the field throughout a sunny day of summer 1997. Values given are the means and standard errors of four replicate blocks for AMBIENT UV (white circles), UV-A (grey triangles) and UV-B (black circles) treatments. Measurements were made on the youngest fully expanded leaf. Data partially from Allen et al.[23]

Photosynthetic productivity

Sudden exposure of plants to high UV-B irradiance for a few hours has no effect on plant productivity. However, pea plants grown under high UV-B exposure for 24 d had reduced plant height, leaf area, and total, leaf and root dry weights[14], but the number of leaves was not significantly affected. This is important since two factors determine total leaf area of plants, leaf size and number. Gonzalez et al.[24] and Nogués et al.[14] demonstrated that UV-B decreases leaf cell number and cell division. Repair of UV-B damage to DNA before replication and direct UV-B-induced oxidation of tubulin, delaying microtubule formation[25] have been suggested as mechanisms for direct slowing of cell division. UV-B may also affect the key stages of cell division through transcriptional repression of the genes encoding for a mitotic cyclin and a $p34^{cdc2}$ protein kinase[26]. Reductions in leaf area and cell division in all measured cell types were observed at all stages of leaf development with the effects becoming more pronounced approaching full expansion[14]. This indicates it is unlikely that UV-B acts directly on the dividing cells because leaves at early stages were still enclosed by the folded bracts and so would have experienced a very low UV-B exposure.

UV-B also inhibited cell expansion of the exposed surface of leaves. UV-B could reduce cell expansion by changing turgor pressure or cell wall extensibility, and Tevini and Iwanzik[27] suggested that direct oxidation of indole-acetic acid by UV-B reduces cell wall expansion. Therefore, decreases in the ability of the pea plants to capture photosynthetically active radiation are also a major factor in determining the reduction in the photosynthetic productivity of these plants.

4. Field studies

Glasshouse experiments may overestimate the response of photosynthesis to UV-B radiation. This is because such experiments frequently compare plants grown under very high levels of UV-B against those at non-UV-B radiation, and/or are conducted under low levels of UV-A and photosynthethically active photon flux density (PPFD) compared to field studies[1]. Such conditions can be expected to increase plant sensitivity to damage by UV-B.

To overcome these problems research effort was moved towards applying enhanced UV-B in the field. In most instances supplementary UV-B was applied as a fixed levels for a fixed number of hours each day ("square wave" systems). However, this approach takes no account of variation in cloud cover, and enhanced levels were frequently based on the predicted consequences of ozone depletion under clear sky conditions at the summer[28]. A few recent studies have utilised filters to compare near ambient UV-B with below ambient UV-B irradiances ("UV-B exclusion experiments") to identify whether current UV-B levels are detrimental to plants. However, exclusion experiments do not enable predictions to be made about the impact of future UV-B increases, in response to continued O_3 depletion. These problems can be overcome by the use of modulated systems, which provide a proportional UV-B supplement by modulating UV-B lamp output in proportional to ambient UV-B levels ("modulated supplementation systems"[23,29]).

When pea plants were grown in the field and subjected to a modulated 30% increase in ambient summer UV-B radiation, no significant effects on photosynthesis were found (Figure 1[23,29]). Moreover, UV-B exposure had no significant effects on plant biomass, leaf area or partitioning assimilates from shoots to roots (Table II[23]). The only UV-B effect in the field was an increase in leaf flavonoid content. This increase will give protection against the deleterious effect of UV-B irradiation.

Table 2. Growth and pigment parameters measured six weeks after planting for AMBIENT UV, UV-A, UV-B, well watered (WW) and drought (D) treated pea plants in the field. Values are the means of four replicate blocks. Leaf number is the number of true leaf pairs and leaf area includes both true leaf pairs and associated stipules. DW, Dry weight. Flavonoid and anthocyanin pigment relative contents were estimated from the absorbances at 300 (A_{300}) and 530 nm (A_{530}) respectively, of acidified methanol extracts of the youngest fully expanded leaf and expressed on a leaf area basis. Data partially from Allen et al.[23]

UV	Water	Plant DW (g)	Height (cm)	Leaf Number	Leaf Area (cm^2)
AMB					
	WW	2.99	38.0	10.8	339
	D	2.25	36.0	7.7	324
UV-A					
	WW	2.73	36.2	11.4	367
	D	2.78	35.8	8.0	310
UV-B					
	WW	3.04	39.8	10.0	366
	D	2.43	36.1	7.2	363

UV	Water	Root/shoot ($g\ g^{-1}$)	Shoot Number	Flavonoids ($A_{300}\ cm^{-2}$)	Anthocyanins ($A_{530}\ cm^{-2}$)
AMB					
	WW	0.155	1.75	4.78	0.527
	D	0.170	1.08	4.66	0.576
UV-A					
	WW	0.157	2.08	4.50	0.491
	D	0.185	1.50	5.17	0.612
UV-B					
	WW	0.158	1.75	5.10	0.504
	D	0.176	1.67	5.25	0.642

This absence of UV-B effects on photosynthesis and biomass contradicts the conclusion of many "square-wave systems". However, the data from Mepsted *et al.*[29] and Allen *et al.*[23] support the findings of other modulated UV-B field experiments which found no effects either on biomass accumulation in pea[30-31], rice, barley[31], oak[32], wheat and wild oat[33], or on photosynthetic parameters e.g. light-saturated CO_2 assimilation[34], stomatal conductance[34], maximum or steady-state ϕ_{PSII}[31,35]. The weight of evidence coming from these modulated UV-B field experiments suggests that current predictions for O_3 depletion will not adversely affect plant productivity and photosynthesis in crops directly. The discrepancy between the results from "square-wave" and modulated UV-B field supplementation systems does support the contention that "square-wave" systems may exaggerate plant responses to UV-B supplementation, as suggested by Fiscus *et al.*[36]. Caldwell *et al.*[35] demonstrated in the field with soybean exposed to modulated UV-B that a reduction in biomass was seen only when PPFD and UV-A were reduced to below half of the full sunlight levels. The higher ratio of UV-B to PPFD and UV-A irradiation is the likely cause of the much larger UV-B effects observed when plants are exposed to "square-wave" UV-B supplementation in controlled environments, glasshouses or in the field, compared to modulated UV-B field experiments.

The absence of any significant supplemental UV-A effect on photosynthesis (Fig. 1) or biomass (Table II) suggests that under these conditions the small increase in ambient UV-A produced by the lamps (< 2 %) did not have any appreciable effect. It is clear that a realistic increase in UV-B radiation will not have any impact on photosynthesis or on photosynthesis productivity of field-grown plants.

5. Conclusions

Although glasshouse studies are useful in identifying the mechanism by which UV-B radiation affects the photosynthetic capacity of leaves, they do not predict the impact of increase UV-B radiation on plants. In glasshouse studies, high UV-B irradiances affected two main processes of photosynthesis (stomata and Calvin cycle enzymes) and decreased leaf area and productivity. When the plants were acclimated to high UV-B radiation, UV-B absorbing compounds increased and UV-B effects were only found in adaxial stomata (because these cells are not screened by flavonoids). Interestedly, field studies using UV-B levels that may be realistically expected in the future found no effects on plants. It is concluded that an increase in UV-B radiation, associated with depletion in the stratospheric ozone layer, is unlikely to have any significant impact on photosynthesis or on the photosynthetic productivity of higher plants.

Acknowledgement. We are grateful to J.I.L. Morison for discussions on A/ci relationship and stomata.

References

1. Allen, D.J., Nogués, S. and Baker, N.R. (1998) Ozone depletion and increased UV-B radiation: is there a real threat to photosynthesis?, *J. Exp. Bot.*, 49: 1775-1788.
2. Jordan, B.R. (1996) The Effects of Ultraviolet-B Radiation on Plants: A Molecular Perspective, *Adv. Bot. Res.*, 22: 97-162.
3. Cen, Y.-P. and Bornman, J.F. (1993) The effect of exposure to enhanced UV-B radiation on the penetration of monochromatic and polychromatic UV-B radiation in leaves of *Brassica napus*, *Physiologia Plantarum*, 87: 249-255.
4. Ålenius, C.M., Vogelmann, T.C. and Bornman, J.F. (1995) A three-dimensional representation of the relationship between penetration of UV-B radiation and UV-screening pigments in leaves of *Brassica napus*, *New Phytologist*, 131: 297-302.
5. Bornman, J.F. and Teramura, A.H. (1993) Effects of Ultraviolet-B Radiation on Terrestrial Plants. In *Environmental UV Photobiology* (A.R. Young, L.O. Björn, J. Moan and W. Nultsch, eds), Plenum Press, New York, pp 427-472.
6. Beggs, C.J. and Wellmann, E. (1994) Photocontrol of flavonoid biosynthesis. In *Photomorphogenesis in Plants* (R.E. Kendrick and G.H.M. Kronenberg, eds), Kluwer Academic Publishers, The Netherlands, pp 733-751.

7. Jenkins, G.I., Fuglevand, G. and Christie, J.M. (1997) UV-B perception and signal transduction. In *Plants and UV-B: Responses to Environmental Change* (P.J. Lumsden, ed), Cambridge University Press, pp 95-111.
8. Stapleton, A.E. (1992) Ultraviolet radiation and plants: burning questions. *Plant Cell*, 4: 1353-1358.
9. Teramura, A.H. and Sullivan, J.H. (1994) Effects of UV-B radiation on photosynthesis and growth of terrestrial plants, *Photosynthesis Research*, 39: 463-473.
10. Fiscus, E.L. and Booker, F.L. (1995) Is increased UV-B a threat to crop photosynthesis and productivity?, *Photosynthesis Research*, 43: 81-92.
11. Baker, N.R. and Bowyer, J.R. (1994) *Photoinhibition of Photosynthesis: from Molecular Mechanisms to the Field*, βios Scientific Publishers, Oxford.
12. Friso, G., Spetea, C., Giacometti, G.M., Vass, I. and Barbato, R. (1994) Degradation of Photosystem II reaction centre D1-protein induced by UV-B radiation in isolated thylakoids. Identification and characterisation of C- and N- terminal breakdown products, *Biochim Biophys Acta*, 1184: 78-84.
13. Jansen, M.A.K., Gaba, V., Greenburg, B.M., Mattoo, A.K. and Edelman, M. (1996) Low threshold levels of ultraviolet-B in a background of photosynthetically active radiation trigger rapid degradation of the D2 protein of photosystem II, *Plant Journal*, 9: 693-699.
14. Nogués, S., Allen, D.J., Morison, J.I.L. and Baker, N.R. (1998) Ultraviolet-B radiation effects on water relations, leaf development and photosynthesis in droughted pea plants, *Plant Physiol.*, 117: 173-181.
15. Nogués, S., Allen, D.J., Morison, J.I.L. and Baker, N.R. (1999) Characterization of stomatal closure by ultraviolet-B radiation, *Plant Physiol.*, 121: 489-496.
16. Caldwell, M.M. (1971) Solar UV irradiation and the growth and development of higher plants. In *Photophysiology Vol. 6* (A.C. Giese, ed), Academic Press, New York, pp 131-177.
17. Nogués, S. and Baker, N.R. (1995) Evaluation of the role of damage to photosystem II in the inhibition of CO_2 assimilation in pea leaves on exposure to UV-B. *Plant Cell Environ.*, 18: 781-787.
18. Xiong, F.S. and Day, T.A. (2001) Effect of solar ultraviolet-B radiation during springtime ozone depletion on photosynthesis and biomass production of Antarctic vascular plants. *Plant Physiol.*, 125: 738-751.
19. Allen, D.J., McKee, I.F., Farage, P.K. and Baker, N.R. (1997) Analysis of the limitation to CO_2 assimilation on exposure of leaves of two *Brassica napus* cultivars to UV-B. Plant, *Cell Environ.*, 20: 633-640.
20. Wilson, M.I., Ghosh, S., Gerhardt, K.E., Holland, N., Babu, T.S., Edelman, M., Dumbroff, E.B. and Greenburg, B.M. (1995) In-vivo photomodification of Ribulose 1,5- Bisphosphate Carboxylase Oxygenase holoenzyme by ultraviolet-B radiation - formation of a 66-kiloDalton variant of the large subunit, *Plant Physiol.*, 109: 221-229.
21. Baker, N.R., Nogués, S. and Allen, D.J. (1997) Photosynthesis and Photoinhibition. In *Plants and UV-B: Responses to Environmental Change* (Lumsden PJ, ed), Cambridge University Press, pp 95-111.
22. Nie, G.-Y., Tomasevic, M. and Baker, N.R. (1993) Effects of ozone on the photosynthetic apparatus and leaf proteins during leaf development in wheat, *Plant Cell Environ.*, 16: 643-651.
23. Allen, D.J., Nogués, S., Morison, J.I.L., Greenslade. P.D., Mcleod, A.R. and Baker, N.R. (1999) A thirty percent increase in UV-B has no impact on photosynthesis in well-watered and droughted pea plants in the field, *Global Change Biology*, 5: 213-222.
24. Gonzalez, R., Mepsted, R., Wellburn, A.R. and Paul, N.D. (1998) Non-photosynthetic mechanisms of growth reduction in pea (*Pisum sativum*) exposed to UV-B radiation, *Plant Cell Environ.*, 21: 23-32.
25. Staxén, L., Bergounioux, C. and Bornman, J.F. (1993) Effect of ultraviolet radiation on cell division and microtubule organization in *Petunia hybrida* protoplasts, *Protoplas.*, 173: 70-76.
26. Logemann, E., Wu, S.-C., Schröder, J., Schmelzer, E., Somssich, I.E. and Hahlbrock, K. (1995) Gene activation by UV light, fungal elicitor or fungal infection in *Petroselinum crispum* is correlated with repression of cell cycle-related genes, *Plant Journal*, 8: 865-876.
27. Tevini, M. and Iwanzik, W. (1986) Effects of UV-B radiation on growth and development of cucumber seedlings. In *Stratospheric ozone reduction, solar ultraviolet radiation and plant life*, Vol G8 (R.C. Worrest and M.M. Caldwell, eds), Springer-Verlag, Berlin, pp 271-285.
28. Sullivan, J.H., Teramura, A.H. and Dillenburg, L.R. (1994) Growth and photosynthesis responses of field-grown sweetgum (*Liquidambar styracifolia*, Hamammelidaceae) seedlings to UV-B radiation, *Am. J. Bot.*, 81: 826-832.
29. Mepsted, R., Paul, N.D., Stephen, J., Corlett, J.E., Nogués, S., Baker, N.R., Jones, H.G. and Ayres, P.G. (1996) Effects of enhanced UV-B radiation on pea (*Pisum sativum*) grown under field conditions in the UK, *Global Change Biology*, 2: 325-334.
30. Day, T.A., Howells, B.W. and Ruhland, C.T. (1996) Changes in growth and pigment concentrations with leaf age in pea under modulated UV-B radiation field treatments, *Plant Cell Environ.*, 19: 101-108.

31. Stephen, J., Woodfin, R., Corlett, J.E., Paul, N.D., Jones, H.G. and Ayres, P.G. (1998) Is increased UV-B a threat to crops in the United Kingdom? A case study of barley and pea, *J. Agricult. Sci.*.
32. Newsham, K.K., McLeod, A.R., Greenslade, P.D. and Emmett, B.A. (1996) Appropriate controls in outdoor UV-B supplementation experiments, *Global Change Biology*, 2: 319-324.
33. Barnes, P.W., Jordan, P.W., Gold, W.G., Flint, S.D. and Caldwell, M.M. (1988) Competition, morphology and canopy structure in wheat (*Triticum aestivum* L.) and wild oat (*Avena fatua* L.) exposed to enhanced Ultraviolet-B radiation, *Functional Ecology*, 2: 319-330.
34. Flint, S.D., Jordan, P.W. and Caldwell, M.M. (1985) Plant protective responses to enhanced UV-B radiation under field conditions: leaf optical properties and photosynthesis, *Photochem. Photobiol.*, 41: 95-99.
35. Caldwell, M.M., Flint, S.D.and Searles, P.S. (1994) Spectral balance and UV-B sensitivity of soybean: a field experiment. *Plant Cell Environ.*, 17: 267-276.
36. Fiscus, E.L., Miller, J.E. and Booker, F.L. (1994) Is UV-B a hazard to soybean photosynthesis and yield? Results of an ozone-UV-B interaction study and model predictions. In *Stratospheric Ozone Depletion/UV-B Radiation in the Biosphere* (R.H. Biggs and M.E.B. Joyner, eds), Springer-Verlag, Berlin, pp 135-147.

DETECTING STRESS-INDUCED REACTIVE OXYGEN SPECIES IN PLANTS UNDER UV STRESS

ÉVA HIDEG
Institute of Plant Biology, Biological Research Center
P.O.Box 521, Szeged, H-6701 Hungary

1. Abstract

Sunlight is the energy source for photosynthesis in all land plants and in many aquatic organisms. On the other hand, solar irradiation may also appear as a stress, specially when it is combined with other factors, such as deviations from optimal growth conditions (temperature, water status) or with environmental pollutants (heavy metals, air pollutants, acidic rain). The adaptation and acclimation of photosynthetic organisms to changing environmental conditions is important, not only for the plant's survival but also for manking utilising them as food, raw material and energy source. Reactive oxygen species (ROS, also known as active oxygen species, AOS) are associated with stress and stress response in many ways: They may appear as primary elicitors, as propagators of oxidative damage, or as by-products. More recently, ROS are also considered as messenger molecules involved in signalling pathways, thus potential inducers of defence or adaptation. Until recently, the involvement of ROS in stress was usually presumed from products of oxidative damage. These techniques are useful and important, but - unlike more direct methods-they provide little information about the primary reactions. Also, due to increasing recognition of ROS in signal transduction - when they are present at low levels and do not cause much oxidative damage - methods for direct ROS detection *in vivo* are of special importance. After giving an overview on basic ROS chemistry, this chapter will illustrate direct and indirect methods of ROS detection with special emphasis on plants and their response to UV-B (280-320 nm) irradiation.

2. ROS and other free radicals in plants

2.1. PRODUCTION

ROS are produced from ground state (triplet, $^3O_2$) molecular oxygen, in two main pathways. One is via energy transfer from an other triplet molecule forming singlet oxygen ($^1O_2$), the other is via reduction (Fig. 1). The complete reduction of oxygen yields water, the incompletely reduced intermediates are free radicals and a non-radical form of ROS, H_2O_2 (for reviews see Refs. 1,2) (Fig. 2). The production of hydroxyl radicals is enhanced by the presence catalytic amounts of transition metals, specially Fe(II) or Cu(I) in the so called Fenton reaction. Hydroxyl radicals may attack

F. Ghetti et al. (eds.), Environmental UV Radiation: Impact on Ecosystems and Human Health and Predictive Models, 147–157.

polyunsaturated fatty acids to form carbon centred (lipid) radicals by hydrogen abstraction. In the presence of oxygen these are converted into lipid peroxy radicals, which are capable of hydrogen abstraction and thus of propagating oxidative stress in the membrane. Besides the chemical nature of the molecule, the life-time of ROS depends the environment where they are produced (e.g. lipid membrane or cytosol), temperature and several other factors.

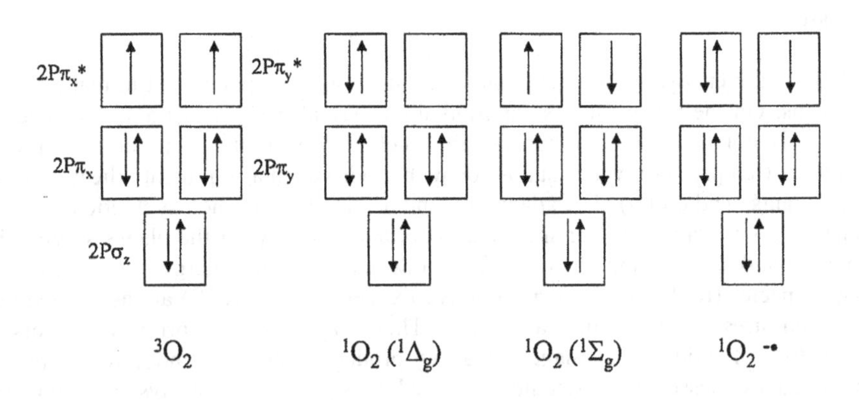

Fig. 1. Electron configuration on molecular orbits in ground state (triplet) and singlet oxygen and in the superoxide anion radical. In $^1\Delta_g$ the electron pair can be localised either on $2P\pi_x^*$ or on $2P\pi_y^*$. The unpaired electron of the superoxide radical can occupy either $2P\pi_x^*$ or $2P\pi_y^*$.

As the example of carbon centred radicals illustrates, free radicals other than oxygen centred ones are also important in oxidative damage. olyunsaturated lipids in biological membranes may be converted into carbon-centred lipid radicals upon hydrogen abstraction by $^{\bullet}OH$, hydroperoxide ($HO_2^{\bullet-}$) or other peroxide radicals. Superoxide does not participate in this process directly, only through its protonated form, $HO_2^{\bullet-}$. The unpaired electron is usually formed on bis-allyl methyl groups. Two carbon-centred radical groups may form cross links and stabilise as conjugated double bonds. These lower membrane fluidity and alter hydrophobicity. Carbon-centred radicals may also combine with oxygen, forming ROS capable of hydrogen abstraction and thus propagating oxidative damage in the membrane (for reviews see Refs. 2,3). Singlet oxygen may also yield hydroperoxides from lipids directly [4]. As lipids in thylakoid membranes are ca. 90% polyunsaturated, photosynthetic organisms are prone to oxidative damage [5-7].

ROS may also damage proteins yielding various carbon-centred (alkoxyl or $CO_2^{\bullet-}$ anion) free radicals and bound amino acid hydroperoxides [8]. By reacting with neighbouring amino acids, these latter may spread oxidative damage along the protein. This promotes protein fragmentation or may also produce free radicals [9].

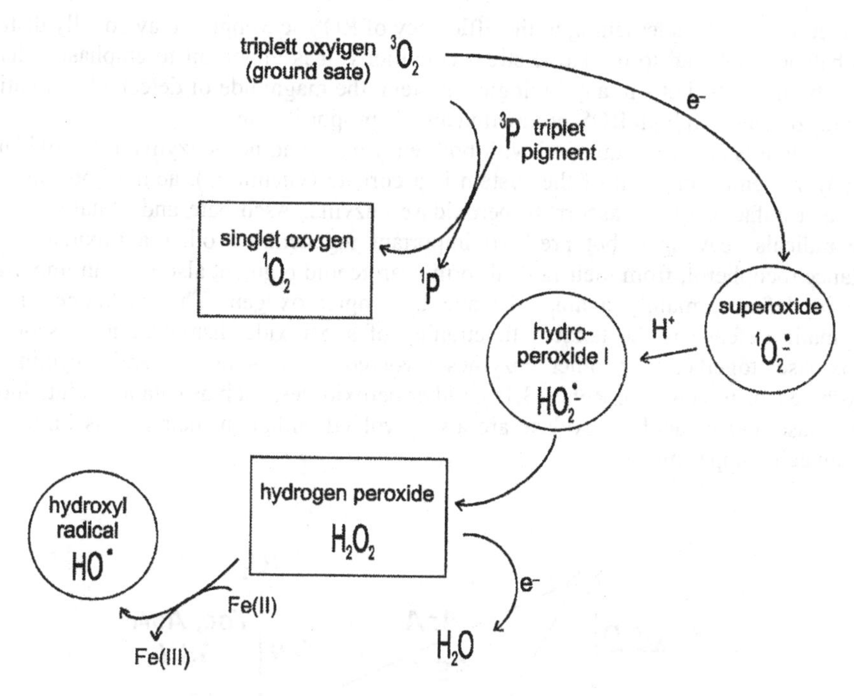

Fig. 2. ROS production from molecular oxygen. Free radical products are encircled, non-radical ROS products are framed in boxes.

Phenolic compounds may also act as antioxidants. Similarly to ascorbate, they react with ROS while being converted into a free radical. Radicals formed from antioxidants are less harmful than ROS and are usually recycled, reduced back to their original protective form by other antioxidants or enzymes. Their recycling may be delayed when ROS production is intense, or when enzymes are damaged. In this way, the accumulation of antioxidant radicals, such as ascorbate [10,11] or phenolic radicals [12] may serve as a stress marker.

It is important to note that longer life-time (lower reactivity) does not necessary mean that the ROS is "less damaging" than an other with shorter life-time (higher reactivity). Short lived ROS, such as $^1O_2$ and $^{\bullet}OH$, react and cause oxidative damage at the site of their production. Longer lived ones may travel in the membrane and then react relatively far from their production site where they may initiate reactions yielding more reactive ROS and thus propagate damage, or may initiate signalling [1].

2.2. DEFENSE

Under physiological conditions ROS production is controlled in plants, these potentially hazardous chemicals are processed directly at or close to the production site by the antioxidant systems and thus their concentration is kept low. Enhanced ROS production as well as lessening in the efficiency of ROS scavenging may equally disturb this balance and lead to oxidative stress conditions. It is important to emphasise that - not only in plants, but - in any biological system, the magnitude of detectable oxidative damage depends on both ROS production and disproportioning.

Plants are well equipped with both enzymatic and non-enzymatic antioxidants (Fig.3). A central element of the system is ascorbate (vitamin C), acting both directly and as a cofactor of the ascorbate peroxidase enzyme. Ascorbate and glutathione are free radicals scavenger, but are also important regenerating other antioxidants, for instance tocopherol, from their radical form. Carotenoid pigment also play an important role in defence, mainly against damage by singlet oxygen. The backbone of the enzymatic defence is the tandem functioning of superoxide dismutase and ascorbate peroxidase, together with other enzymes involved in the synthesis and recycling of ascorbate (for reviews see Refs. 13,14). Other peroxidases, such as catalase, glutathione peroxidase and guiacol peroxidase are also involved, although their role is limited to certain cell compartments.

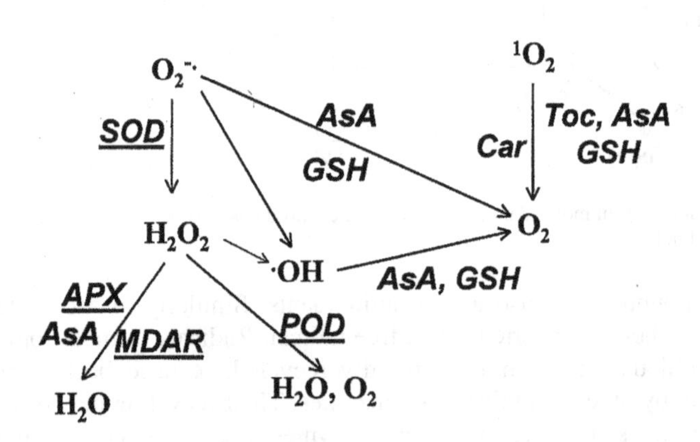

Fig. 3. Schematic pathways of disproportioning ROS in plants. Enzymes are underlined, non-enzymatic antioxidants are in italics. APX: ascorbate peroxidase, AsA: ascorbate, Car: β-carotin, GSH, reduced glutathione, MDAR: monodehydro-ascorbate reductase, POD: peroxidase including catalase, SOD: superoxide dismutase, Toc: α-tocopherol

3. The oxidative nature of UV-B (280-320 nm) stress

Changes in the biotic environment, such as increasing solar UV irradiation-or the appearance of abiotic stress factors may result in the oxidative damage of structures (membrane lipid, protein) or functions (redox component, enzyme) causing smaller bioproduction and – under extreme conditions – even the irreversible damage. Regarding its central role in the damaging process and potential involvement as activator of the defence system, ROS detection is of special importance.

3.1 INDIRECT INDICATIONS

The occurrence of oxidative stress is often presumed from detecting damaged products. In sugar beet leaves, UV-B irradiation was found to initiate lipid peroxidation [15,16]. In isolated, photosynthetically active membrane preparations, UV-B results in loss of electron transport activity and in fragmentation of reaction centre proteins D1 and D2 [17]. This is also observed in cyanobacteria, although a repair cycle based on de novo protein synthesis may counterbalance damage *in vivo* (for review Ref. 18).

UV-B induced changes in the activity of antioxidant enzymes strongly suggests the involvement of ROS. Although reports on the effect of UV-B irradiation on antioxidants may vary with plant species, developmental stage and experimental conditions, the general trend is activation of both enzymatic and non-enzymatic defence (for review see Ref. 19). UV-B irradiation upregulated catalase and gluthatione peroxidase in *Nicotiana plumbaginifolia* [20] induced most peroxidase-related enzymes *Arabidopsis* [21], increased glutathione reductase transcript levels in *Pisum sativum* [22] and *Arabidopsis* [23]. Reports on SOD are more controversial, ranging from little or no [20,22] to marked activation [16,23]. UV-B was also found to induce pathogenesis-related proteins in *Arabidopsis* [24]. The effect of UV-B is – in some aspects – similar to other oxidative stresses. Plants respond to UV-B stress and pathogen infection concerning both the free radical related ultraweak light emission [25] and the induction of a stress response protein [26].

UV-B exposure increases the synthesis of UV absorbing pigments. [27]. Besides UV screening, flavonoid compounds may also act as ROS scavenging antioxidants [28-30]. Their central role is well illustrated by the increased susceptibility of barley [31] and *Arabidopsis* [32,33] mutants with deficient flavonoid biosynthesis to UV-B

Increased levels of the free radical related ultraweak light emission upon UV-B exposure of Hibiscus [34], sugar beet [15] and Brassica leaves [35] also indicated oxidative stress.

3.2. DIRECT EVIDENCE

Although the paramagnetic nature of free radicals offers the possibility of detection by electron paramagnetic resonance (EPR) spectroscopy, their high reactivity and small concentration obstructs the application of this method. However, the use of spin traps helps to overcome this difficulty making it possible to detect not only radical natured ROS, but also singlet oxygen. As illustrated in Fig.4, spin traps are diamagnetic (i.e. have no EPR signal), but form relatively stable, paramagnetic spin adducts when reacting with ROS (for reviews see Refs. 36,37). Various spin traps are available commercially.

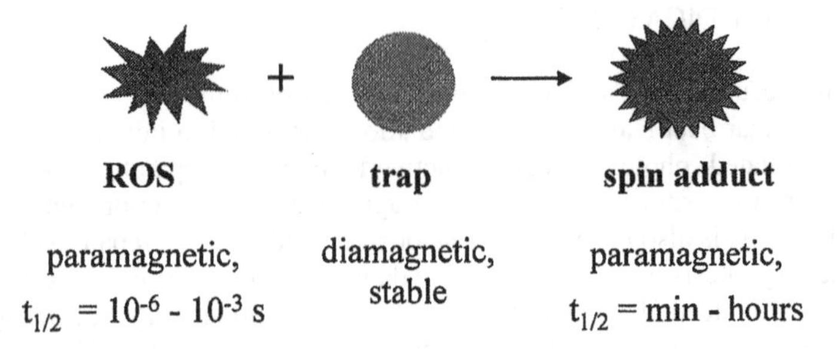

Fig. 4. The principle of ROS detection by spin trapping EPR spectroscopy

One of these, DMPO [38] is of special importance, because DMPO-adducts have characteristic EPR spectra, reflecting the chemical nature of the trapped free radical. Experiments with the DMPO showed that various free radicals are produced upon UV-B irradiation in photosynthetic organisms both *in vitro* and *in vivo* [39]. As more purified preparations containing smaller functional subunits of photosystem II showed mainly hydroxyl radical production (Hideg and Vass, unpublished), these ROS appear as primary products. Beside hydroxyl radicals, more composite preparations feature carbon centred free radicals, which are probably secondary ROS products. The central role of hydroxyl radicals is supported by the observation, that the presence of hydroxyl radical trap, POAN [40] at high concentration retarded UV-B induced protein damage in isolated thylakoid membranes [41].

In leaves, EPR based ROS traps are difficult to apply directly, because the high water content of the sample may hamper EPR measurements. Also, ROS probes successfully applied in animal physiology may be metabolised in plants into inactive forms [42]. Double (fluorescent and spin) sensors provide better possibilities. These are fluorescent probes, containing a fluorophore and an ROS trap. Upon reacting with ROS, the latter is converted into a nitroxide radical (leaving the possibility of EPR detection) and causing structural changes which partly quench the probe's fluorescence [42,43]. The principle of double sensors is illustrated in Fig.5.

Using a double sensor, we have shown that although the two different aspects of light stress, damage by excess photosynthetically active irradiation (photoinhibition) and UV-B irradiation both result in damaging the function and structure of special part (the reaction centre of photosystem II) of the photosynthetic apparatus, the oxidative nature of the two processes is different [44].

Fluorescence from the double sensor can also be measured using laser scanning microscopy. In this way, the non-invasive imaging technique makes spatial information available on the production site of ROS (Hideg et al. 2001). In Fig.6, a comparison of short exposure to photoinhibition by excess photosynthetically active radiation (PI) and UV-B irradiation illustrates the difference. Stress conditions were chosen to result in equal loss of photosynthetic function in 30 min PI and 20 min UV-B, the fluorescence of the infiltrated singlet oxygen trap (DanePy, [43]) decreased only during PI but not in UV-B stress in chloroplasts (Fig.6). This experiment confirmed that singlet oxygen is not the dominant ROS form in UV-B induced oxidative damage. It is also interesting to note UV induced morphological changes in cell structure.

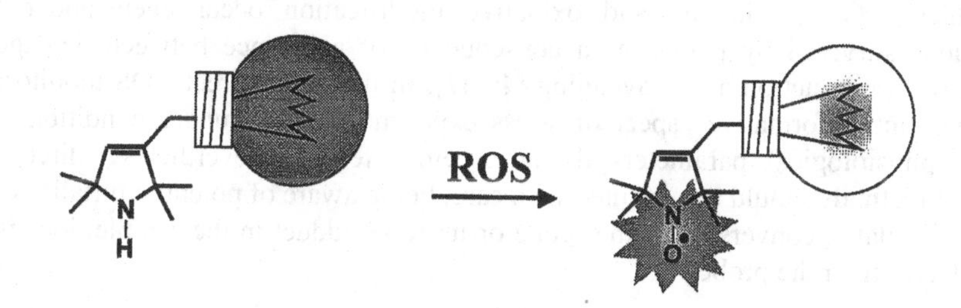

Fig. 5. The principle of ROS detection by a double (fluorescent and spin) sensor. Upon reacting with ROS, the sensor is converted into a nitroxide radical and its fluorescence is partly quenched.

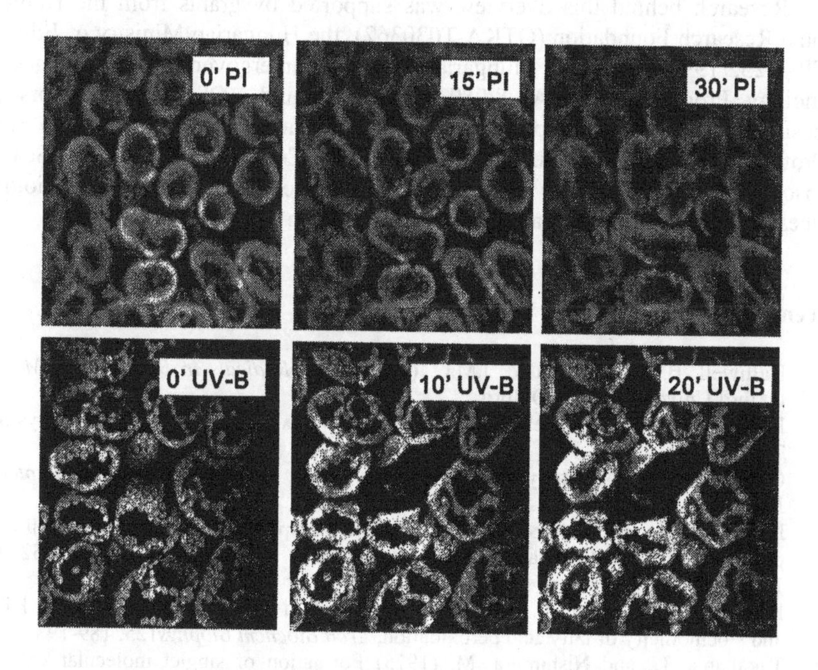

Fig. 6. Singlet oxygen detection in Arabidopsis leaves under stress by excess photosynthetically active radiation (upper panel) and by UV-B irradiation (lower panel). Leaves were infiltrated with the singlet oxygen sensor, which penetrates into the chloroplasts. Both types of light stress conditions (30 min PI and 20 min UV-B) resulted in 45% loss of photosynthetic activity.

4. Concluding Remarks

ROS production and thus the potential danger of oxidative damage is an inevitable consequence of life in oxygenic atmosphere. Plants have developed a multifold system of protection against stress by UV-B (280-320 nm) irradiation, ranging from morphological and metabolic changes to lessen radiation intensity inside the plant through a system of ROS neutralising antioxidants to the repair of oxidatively damaged molecules. ROS production and oxidative modification occur even under stress conditions survived by plants, as a consequence of a balance between damage and repair, ROS production and scavenging [46,47]. In this way, direct ROS monitoring is an important, informative aspect of stress experiments, even under conditions when certain physiological parameters do not seem affected. Nevertheless, direct ROS trapping methods should be executed with care, being aware of potential pitfalls, such as non ROS related conversion of the probe or its ROS adduct in the sample, localisation and selectivity of the probe.

Acknowledgements

Research behind this overview was supported by grants from the Hungarian National Research Foundation (OTKA T030362), the Hungarian Ministry of Education (FKFP 0252/1999) and the Hungarian-Japanese Intergovernmental Science and Technology Program (TéT JAP-8/989. Spin traps and ROS double sensors were synthesised at the Department of Organic and Medicinal Chemistry, University of Pécs by Prof. Kálmán Hideg, Drs. Tamás Kálai and Cecília P. Sár. Laser Scanning Microscopy experiments were carried out at the Research Institute for Biological Sciences (RIBS) Okayama (Japan) in collaboration with Dr. Ken'ichi Ogawa.

References

[1] Halliwell, B. and Gutteridge, J.M.C. (1999) *Free Radicals in Biology and Medicine.* Oxford University Press, Oxford

[2] Elstner, E.F. (1982) Oxygen acivation and oxygen toxicity, Annu. Rev. Plant Physiol. 33, 73-96

[3] Girotti, A. (1990) Photodynamic lipid peroxidation in biological systems, *Photochem. Photobiol.* 51, 497-509

[4] Pederson, T.C. and Aust, S.D. (1973) The role of superoxide and singlet oxygen in lipid peroxidation formed by xanthine oxidase, *Biochem. Biophys. Res. Commun.* 52, 1071-1078

[5] Heath, R.L. and Packer, L. (1968) Photoperoxidation of isolated chloroplasts. I. Kinetics and stochiometry of fatty acid peroxidation, *Arch Biochem Biophys*125, 189-198

[6] Takahama, U. and Nishimura, M. (1975) Formation of singlet molecular oxygen in illuminated chloroplasts, *Plant Cell Physiol* 16, 737-748

[7] Knox, J.P. and Dodge, A.D. (1985) Singlet oxygen and plants, *Phytochemistry* 24, 889-896.
[8] Davies, M.J., Fu, Sh. and Dean, R.T. (1995) Protein hydroperoxides can rise to reactive free radicals, *Biochem J* 305, 643-649
[9] Hawkins, C.L. and Davies, M.J. (2001) Generation and propagation of radical reactions in proteins, *Biochim Biophys Acta* 1504, 196-219
[10] Heber, U., Miyake, C., Mano, J., Ohno, C. and Asada, K. (1996) Monodehydroascorbate radical as a sensitive probe of oxidative stress in intact leaves as detected by electron paramagnetic resonance spectrometry, *Plant Cell Physiol* 37, 1066-1072
[11] Hideg, É., Mano, J., Ohno, Ch. and Asada, K. (1997) Increased levels of monodehydroascorbate radical in UV-B irradiated broad bean leaves, *Plant Cell Physiol* 38, 684-690.
[12] Sakihama, Y., Mano, J., Sano, S., Asada, K. and Yamasaki, H. (2000) Reduction of phenoxyl radicals mediated by monodehydroascorbate reductase, *Biochem. Biophys. Res. Commun.* 279, 949-954
[13] Foyer, Ch. H. (1993) Ascorbic acid, in Alscher, R.G. and Hess, J.L., (eds.) *Antioxidants in Higher Plants*, CRC Press, Boca Raton, pp. 31-58
[14] Asada, K. (1999) The water-water cycle in chloroplasts: scavenging of active oxygen and dissipation of excess photons, *Annu. Rev. Plant Physiol. Plant Mol. Biol.* 50, 601-639
[15] Panagopoulos, I., Bornman, J.F. and Björn, L.O. (1990) Effects of ultraviolet radiation and visible light on growth, fluorescence induction, ultraweak luminescence and peroxidase activity in sugar beet plants, *J. Photochem. Photobiol. B. Biol.* 8, 73-87
[16] Malanga, G. and Puntarulo, S. (1995) Oxidative stress and antioxidant content in *Chlorella vulgaris* after exposure to ultraviolet-B radiation, *Physiol. Plant.* 94, 672-679
[17] Salter, H.A., Koivuniemi, A. and Strid, A. (1997) UV-B and UV-C irradiation of spinach PSII preparations *in vitro*. Identification of different fragments of the D1 protein depending upon irradiation wavelength, *Plant Physiol. Biochem.* 35, 809-817
[18] Vass, I. (1997) Adverse effects of UV-B light on the structure and function of the photosynthetic apparatus, in: M. Pessarakli (ed.), *Handbook of Photosynthesis*, Marcel Dekker Inc, Ch 60, pp. 911-930
[19] Jansen, M.A.K., Gaba, V. and Greenberg, B.M. (1998) Higher plants and UV-B radiation: balancing damage, repair and acclimation, TIPS 3, 131-135
[20] Willekens, H., van Camp, W., van Montague, M. and Inze, D. (1994) Ozone, sulfur dioxide and ultraviolet B have similar effects on mRNA accumulation of antioxidant genes in *Nicotiana plumbaginifolia* (L.), *Plant Physiol.* 106, 1007-1014
[21] Rao, M.V., Paliyath, G. and Ormrod, D.P. (1996) Ultraviolet-B-and ozone-induced biochemical changes in antioxidant enzymes of *Arabidopsis thaliana, Plant Physiol.* 110, 125-136
[22] Strid, A. (1993) Alteration in expression of defence genes in *Pisum sativum* after exposure to supplementary ultraviolet-B radiation, *Plant Cell Physiol.* 34, 949-953
[23] Rao, M.V. and Ormrod D.P. (1995) Impact of UVB and O_3 on the oxygen free radical scavenging system in *Arabidopsis thaliana* genotypes differing in flavonoid biosynthesis, *Photochem. Photobiol.* 62, 719-726
[24] Surplus, S.L., Jordan, B.R., Murphy, A.M., Carr, J.P., Thomas, B. and Mackerness, S. A.-H. (1998) Ultraviolet-B-induced responses in *Arabidopsis thaliana*: role of salicylic acid and reactive oxygen species in the regulation of transcripts encoding photosynthetic and acidic pathogen-related proteins, *Plant Cell Environ.* 21, 685-694

[25] Panagopoulos, I., Bornman, J.F. and Björn, L.O. (1992) Response of sugarbeet plants to ultraviolet-B (280-320 nm) radiation and Cercospora leaf spot disease, *Physiol. Plant.* 84, 140-145
[26] Green, R. and Fluhr, R. (1995) UV-B induced PR-1 accumulation is mediated by active oxygen species, *The Plant Cell* 7, 203-212
[27] Tevini, M., Braun, J. and Frieser G. (1991) The protective function of the epidermal layer of rye seedlings against ultraviolet-B radiation. *Photochem. Photobiol.* 53, 329-333
[28] Takahama, U., Youngman, R.J. and Elstner, E.F. (1984) Transformation of quercetin by singlet oxygen generated by a photosensitized reaction, *Photobiochem. Photobiophys.* 7, 175-181
[29] Husain, S.R., Cilard, J. and Cilard, P. (1987) Hydroxyl radical scavenging activity of flavonoids, *Phytochem.* 28, 2489-2491
[30] Torel, J., Cillard, J. and Cillard, P. (1988) Antioxidant activity of flavinoids and reactivity with peroxy radical, *Phytochem.* 25, 383-385
[31] Reuber, S., Bornman, J.F. and Weissenböck, G. (1996) A flavonoid mutant of barley (Hordeum vulgare L.) exhibits increased sensitivity to UV-B radiation in the primary leaf, *Plant Cell Environ.* 19, 593-601
[32] Lois R. and Buchanan B.B. (1994) Severe sensitivity to ultraviolet radiation in an Arabidopsis mutant deficient in flavonoid accumulation, *Planta* 194, 504-509
[33] Landry, L.G., Chapple C.C.S. and Last R.L. (1995) Arabidopsis mutants lacking phenolic sunscreens wxhibit enhanced ultraviolet-B injury and oxidative damage, *Plant Physiol.* 109, 1159-1166
[34] Panagopoulos, I., Bornman, J.F. and Björn, L.O. (1989) The effect of UV-B and UV-C radiation on Hibiscus leaves determined by ultraweak luminescence and fluorescence induction, *Physiol. Plant.* 78, 461-465.
[35] Cen, Y.-P. and Björn, L.O. (1994) Action spectra from enhancement of ultraweak light emission by ultraviolet radiation (270-340 nm) in leaves of Brassica napus, *J. Photochem. Photobiol. B. Biol.* 22, 125-129
[36] Evans CA (1979) Spin trapping, *Aldrich Acta* 12, 182-188
[37] Janzen EG (1971) Spin trapping, *Acc Chem Res* 4, 31-39
[38] Janzen EG, Liu J I-P (1973) Radical addition reactions of 5,5-dimethyl-1-pyrroline-1-oxide. ESR spin trapping with a cyclic nitrone, *J Magn Reson* 9, 510-512
[39] Hideg, É. and Vass, I. (1996) UV-B induced free radical production in plant leaves and isolated thylakoid membranes, *Plant Sci* 115, 251-260
[40] Sár., P.C., Hideg, É., Vass, I., Hideg, K. (1998) Synthesis of α-aryl *N*-adamant-1-yl nitrones and using them for spin trapping of hydroxyl radicals, *Bioorg Med Chem Lett* 8, 379-384
[41] Hideg, É., Takátsy, A., Sár, P.C., Vass, I., Hideg, K. (1999) Utilizing new adamantyl spin traps in studying UV-B-induced oxidative damage of photosystem II, *J Photochem Photobiol B: Biol* 48, 174-179
[42] Hideg, É., Vass, I., Kálai, T. and Hideg, K. (2000) Singlet oxygen detection with sterically hindered amine derivatives in plants under light stress, *Meth Enzymol* 319, 77-85
[43] Kálai, T., Hideg, É., Vass, I. and Hideg, K. (1998) Double (fluorescent and spin) sensors for detection of reactive oxygen species in the thylakoid membrane, *Free Rad Biol Med* 24, 649-652
[44] Hideg, É., Kálai, T., Hideg, K. and Vass, I. (2000) Do oxidative stress conditions impairing photosynthesis in the light manifest as photoinhibition? *Philos Trans Roy Soc London B* 355, 1511-1516
[45] Hideg, É, Ogawa, K., Kálai, T., Hideg, K. (2001) Singlet oxygen imaging in Arabidopsis thaliana leaves under photoinhibition by excess photosynthetically active radiation, *Physiol Plant* 112, 10-13

[46] Dat, J., Vandenabeele, S., Vranová, E., Van Montagou, M., Inzé, D. and Van Breusegem, F. (2000) Dual action of the active oxygen species during plant stress responses, *Cell. Mol. Life Sci.* 57, 779-795.

[47] Asada, K. (1994) Production and action of active oxygen species in photosynthetic tissues, in Foyer, Ch. and Mullineaux, P.M. (eds.) *Causes of Photooxidative Stress and Amelioration of Defence Systems in Plants*, CRC Press, Boca Raton FL, pp. 77-104

NON-DAMAGING AND POSITIVE EFFECTS OF UV RADIATION ON HIGHER PLANTS

M.G. HOLMES
Department of Plant Sciences, University of Cambridge, Cambridge, CB2 3EA, U.K.

1. Introduction

Excess ultraviolet-B (UV-B; 280-320nm) radiation is detrimental to the growth and development of higher plants and other organisms, primarily because it damages DNA. At high levels, UV-B radiation can also damage the photosynthetic system and membrane lipids. DNA damage resulting from excessive UV-B radiation can occur by two main ways. The main cause is the direct absorption of photons by DNA. The lesser cause is by the absorption of photons by other molecules, resulting in the production of active oxygen species or free radicals which then damage DNA in subsequent reactions. By either means, the DNA is damaged. Unlike motile organisms which can make avoidance responses, most plants have to endure ambient UV-B radiation levels and their only options are to repair damage as efficiently as possible and, preferably, to minimise that damage by protective mechanisms. Damage effects and their interpretation are described in detail elsewhere in this volume.

The majority of studies on the responses of plants to UV-B radiation have concentrated on the potential damage caused by excessive UV-B radiation as a result of stratospheric ozone depletion. These have reported on the specific damage caused, the susceptible targets, the mechanisms of damage, and the methods of damage repair. By contrast, UV radiation also has many non-damaging effects on plants. Separating non-damaging from damaging effects of UV radiation on plants is not always straightforward. Morphological and biochemical responses are often used to assess possible UV-induced damage. There are many responses which have been reported in the absence of obvious damaging effects; examples include de-etiolation, flavonoid biosynthesis, dimensions of hypocotyls and leaves, leaf number, branching and, possibly, cotyledon curling.

The purpose of this chapter is to outline our current understanding of both non-damaging and positive effects of UV radiation on higher plants. Topics covered include antioxidants, photoprotection (e.g. flavonoids and phenylpropanoid derivatives). Indirect effects such as fungal pathogens, phytophagous insects and litter decomposition are also discussed. Photomorphogenic responses to UV radiation are becoming better understood. The putative UV-B photoreceptor(s) is discussed. These responses are also considered in terms of co-action between UV-B and other wavelengths, using phytochrome and cryptochrome as examples.

F. Ghetti et al. (eds.), Environmental UV Radiation: Impact on Ecosystems and Human Health and Predictive Models, 159–177.

2. Antioxidants and pathogenesis-related products

Photooxidative stress results from the light-dependent production of active oxygen species[1]. The reader is also referred to the more detailed chapter in this volume. Active oxygen species can be produced during photosynthesis by donation of energy or electrons directly to oxygen and also by UV radiation. Active oxygen species are usually destroyed rapidly by antioxidative systems; at the same time, they act to indicate adverse conditions to the plant such as low temperatures, pollution, pathogens and excessive UV-B radiation. It is now well established in mammalian systems that UV radiation, apart from causing damage, can induce transcription in certain genes which result in biological responses which have a protective function. The equivalent of this mammalian UV response also appears to exist in plants[2,3]. As A.-H.-Mackerness[4] points out, specific sets of genes are activated and repressed by UV-B radiation and these signalling pathways are shared with other oxidative stresses such as pathogen infection, ozone exposure and wounding[5].

It is essential that a plant has optimised its antioxidative metabolism if it is to withstand excessive levels of UV-B radiation. Most studies have naturally centred on antioxidant enzymes induced by UV-B radiation. Although their importance relative to each other is not known because studies have been carried out in different species, there is evidence for the increase in the expression and activity of the antioxidant enzymes ascorbate peroxidase, superoxide dismutase and glutathione reductase in response to UV-B radiation[6-8].

There is also evidence for the induction of pathogenesis-related[9,10] and defencin proteins[11] by UV-B radiation. UV-B radiation may control the production of the pathogen-induced protective factors salicylic acid, jasmonic acid and ethylene[4,12]. In the limited space available here, jasmonic acid is used as an example. Jasmonic acid production is induced by UV-B radiation in both tomato[12] and Arabidopsis[5]. It is considered to activate defence responses in plants which have been attacked by herbivores and pathogens[13-16]. Endogenous jasmonic acid levels rise dramatically when plants are attacked by pathogens[14] and it also induces the induction of systemic defence responses[14,17-19]. Jasmonic acid synthesis results in the induction of other pathways involved in UV-B protection including pathogenesis related proteins and enzymes involved in flavonoid biosynthesis[4]. The information available is indicative that these may be other UV-B-induced non-damaging or even positive effects, but further research is required on this subject area before any firm conclusions can be reached as to whether their production is part of a true defence mechanism.

3. Screening pigments

The potential of higher plants for absorbing UV-B radiation has been studied for many years[20,21]. Along with the activation of antioxidants and the processes of photo- and dark repair, the production of screening pigments provides a third main defence against excessive UV-B radiation. Flavonoids and phenylpropanoid derivatives such as sinapate esters are powerful absorbers of UV-B radiation. These chemicals, which largely accumulate in the vacuoles of epidermal cells in higher plants, are very transparent to potentially useful photosynthetically active radiation (PAR; 400-700 nm). It is established that mutants in the synthesis of phenylpropanoids exhibit enhanced sensitivity to UV-B radiation[21-25]. Mazza et al[26] demonstrated the importance of phenylpropanoids in reducing UV penetration into leaf tissue using *Arabidopsis* transparent testa mutants in the field. The potential antioxidant role of these pigments is discussed in this volume in the chapter by Bornman.

Although there are nine major sub-groups produced in the various branches of the flavonoid biosynthesis pathways, it is common practice in the literature to use the term flavonoid to refer to all flavonoids with the exception of anthocyanins[27] and this approach will be followed here. Anthocyanins are 3- or 3,5-glycosides of anthocyanidins and are a major subgroup of flavonoid biosynthesis [see Winkel-Shirley[28] for a recent update]. The main function of anthocyanins appears to be to provide the red (R), purple and blue (B) colouration of flowers and other organs, but there is also evidence for them having a protective function against UV-B radiation[29], drought and cold[30]. Specific biological reponses are often controlled by more than one photoreceptor, and UV-B appears to be involved in the control of several of these reponses as the following examples for flavonoid and anthocyanin synthesis demonstrate.

Beggs et al[31] have demonstrated that there is a specific UV-B induction of anthocyanin and/or flavonoid production in at least 11 species. Using parsley cell suspension cultures and seedlings, Wellmann (in Beggs and Wellmann[27]) defined a specific photomorphogenetic UV-B effect on flavonoid synthesis. He observed that phytochrome and the BAP were only effective if the tissue had received UV-B radiation. A detailed action spectrum for flavonoid synthesis in parsley cell suspension cultures (Fig. 1) showed a clear action maximum near 295 nm with rapidly declining effectiveness at both shorter and longer wavelengths; had this been a damage response, increasing effectiveness at shorter wavelengths would have been expected. It has also been shown that the synthesis of chalcone synthase in *Arabidopsis*, which plays a key early role in the formation of flavonoids, is regulated by UV-B, UV-A/B and R/FR photoreceptors[32]. Wade et al[33] have shown that phytochrome B regulates the cryptochrome and UV-B signalling pathways.

A different spectral response is seen in the action spectrum for isoflavonoid synthesis in bean (*Phaseolus vulgaris* L.) is clearly different from that for flavonoid synthesis in parsley (Fig. 1[34]). Not only is the shape different with no decrease in effectiveness at lower wavelengths, but the response can be photorepaired by subsequent irradiation with UV-A or B radiation[34] thereby indicating that the radiation is perceived by DNA rather than the UV-B photoreceptor. Nevertheless, the effect of UV radiation acting through DNA is positive rather than negative because it results in the production of protective pigments

Photocontrol of anthocyanin production can involve several photoreceptors and the photoreceptors involved differ between species and even between organs, as the following examples show. UV-B radiation has long been known to induce anthocyanin synthesis[38]. Anthocyanin synthesis in mustard (*Sinapis alba* L.) cotyledons, for example, is controlled by phytochrome alone and is R/FR reversible (data of Schmidt in Mohr[39]). In the cotyledons of mustard, however, a specific promotory B effect on anthocyanin production exists (data of Drumm-Herrel in Mohr[39]) which indicates that photoreceptor responses can be organ-specific within the same species. The role of UV-B radiation varies between organs and species.

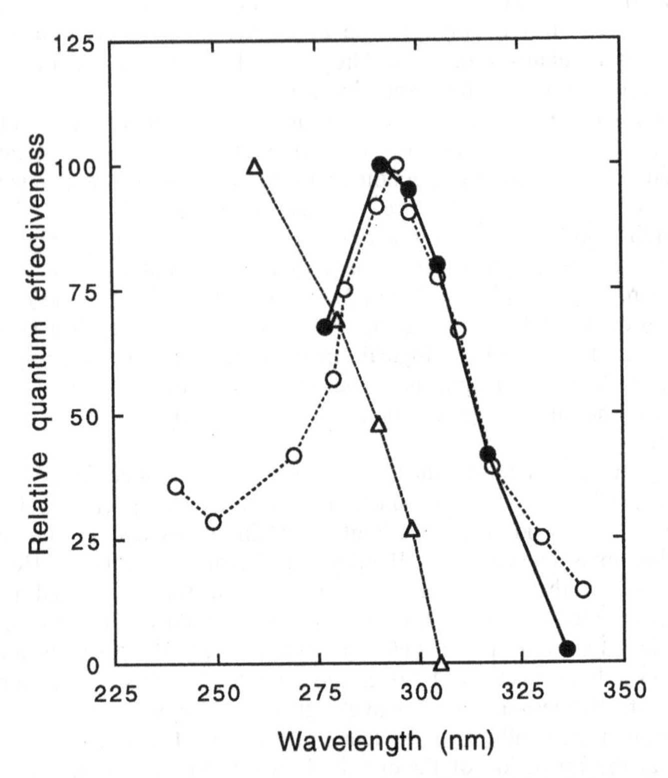

Figure 1. UV action and response spectra for flavonoid synthesis. Solid circles = in parsley (*Petroselinum hortense* D.) cell cultures[36,37]; open circles = same as previous one (Wellmann cited in ref. 27); triangles = action spectrum for isoflavonoid (coumestrol) synthesis in bean (*Phaseolus vulgaris* L.) primary leaves[35]. To facilitate comparison, the data are normalised to 100.

In tomato (*Lycopersicon esculentum* L.), anthocyanin synthesis is also controlled by phytochrome, but the response (as with mustard hypocotyls) is amplified by UV-B radiation. Interestingly, a tomato mutant deficient in phytochrome was induced to form anthocyanins by UV-B radiation[40].

A specific UV-B photoreceptor is involved in the induction of anthocyanin synthesis in the mesocotyls of milo (*Sorghum vulgare* Pers.) seedlings. However, synthesis is modulated by phytochrome and Pfr (the active form) must be present. R or FR radiation alone do not result in pigmentation because there is an absolute requirement for B or UV light[44]. It was later shown that not only the B/UV-A photoreceptor and phytochrome controlled this response, but also that a specific UV-B photoreceptor was involved[45,46]. Further details are given by Mohr[39].

Action spectra for the regulation of anthocyanin synthesis in maize (*Zea mays* L.) are shown in Fig. 2[35,37]. As far as the authors could ascertain, these were photomorphogenic rather than damage responses and therefore represent excitation of the UV-B photoreceptor.

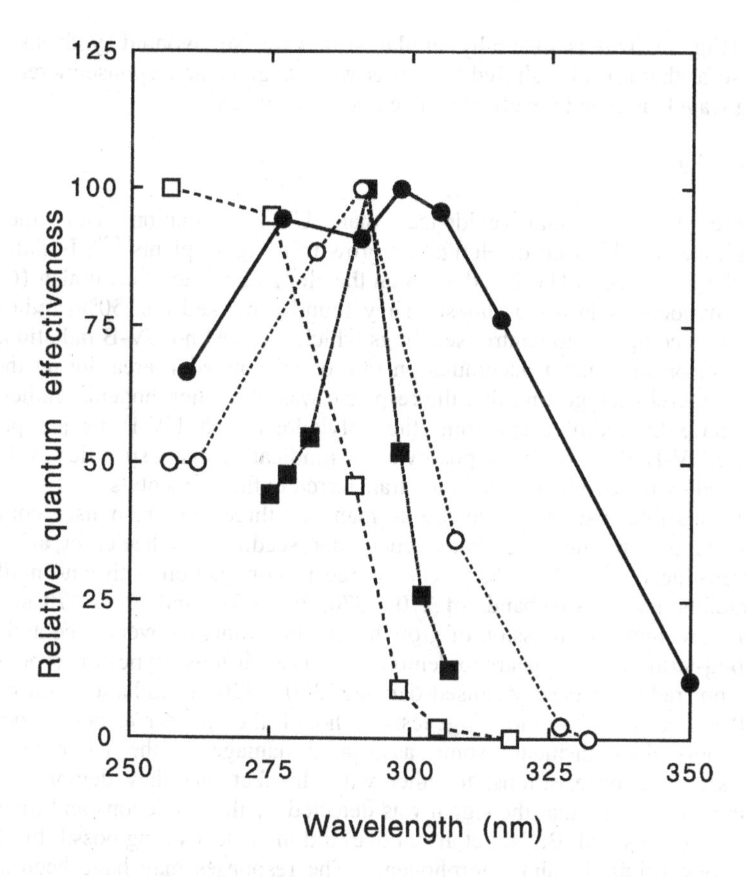

Figure 2. UV action and response spectra for control of anthocyanin production. Solid circles = maize (*Zea mays* L.) coleoptiles[35]; open circles = broom sorghum (*Sorghum bicolor* L.) against a background of R radiation[41]; solid squares = broom sorghum (*Sorghum bicolor* L.) against a background of R radiation[42]; open squares = inhibition of phytochrome-induced anthocyanin synthesis in mustard (*Sinapis alba* L.) cotyledons[43]. To facilitate comparison, the data are normalised to 100.

Although action spectroscopy is a useful tool in aiding the identification of photoreceptors, caution is often needed in their interpretation. In broom sorghum (*Sorghum bicolor*), the complex photoreceptor interaction in controlling anthocyanin synthesis is different again. In this species, UV-B, UV-A, B and R wavelengths induce anthocyanin synthesis[41,42]. Subsequent irradiation with FR radiation nullifies all but the UV-B response, thereby suggesting the use of alternative photoreceptors to control the induction of the response. Action spectra for the induction of anthocyanin synthesis in sorghum are shown in Fig. 2. The UV irradiation treatments, which themselves produce Pfr, were carried out under a background of R light. However, a damaging response could also be induced by UV radiation with the response increasing as the wavelength was decreased[41]. An example of an action spectrum for a damage response in the context of anthocyanin synthesis is that for the inhibition of P-induced anthocyanin synthesis in mustard[43] (Fig. 2). This is spectrally similar to that for isoflavonoid synthesis in bean (Fig. 1) insofar that it is also shifted to shorter wavelengths, the response increases with decreasing wavelength and the effect can be photoreactivated.

4. Elongation growth

There is substantial evidence that UV-B radiation can induce a photomorphogenic inhibition of elongation growth in higher plants[47-50]. Ballaré et al[51] measured the influence of UV-B radiation on the elongation rate of cucumber (*Cucumis sativus L.*) hypocotyls in a glasshouse. They found a marked (ca. 50%) reduction in elongation rate compared to control seedlings which received no UV-B radiation. There was no effect on dry matter accumulation rate or on cotyledon area during the same time period, thereby suggesting that the response was photomorphogenic, rather than a damage effect. In addition, covering the cotyledons with UV-B-opaque polyester nullified the UV-B effect on the hypocotyls, thus indicating that a specific UV-B signal was perceived by the cotyledons and then transferred to the hypocotyls.

It is possible that there may be as many as three photoresponses controlling hypocotyl elongation rate in etiolated cucumber seedlings. Wheeler et al[52] used a position transducer and a broadband UV source in conjunction with cut-on filters to produce radiation with wavebands of 270 - 370, 300 - 370 and 320 - 370 nm. Their analysis of percentage suppression of growth, response time, recovery time and the lag time for long-term effects appears to demonstrate three different types of response. The 270 - 300 nm radiation clearly caused damage; 300 - 320 nm radiation and radiation above 320 nm caused photomorphogenesis, although the "further response" described by the authors does insinuate some associated damage in the latter two cases. Nevertheless, these observations, together with the fact that they demonstrated (by masking the cotyledons) that the signal was detected by the cotyledons and transmitted to the hypocotyls (as with Ballaré et al[51] above) did indicate a strong possibility that the responses were primarily photomorphogenic. The responses may have been species-specific because the authors later reported that similar responses had been found in pea, but not in sunflower[52].

Studies by Ballaré et al[53] of hypocotyl elongation in etiolated tomato seedlings revealed the spectral dependency of the UV-B inhibitory response. They measured a response spectrum between about 265 and 335 nm (4h exposure to 1.0 $\mu mol\ m^{-2}\ s^{-1}$ monochromatic radiation against a background of 63 $\mu mol\ m^{-2}\ s^{-1}$ PAR) and found an action maximum at 297 nm for the inhibition of hypocotyl elongation (Fig. 3a). There was no evidence in this report that a damage response was being studied. Chemicals which interfere with flavin and probably pterin photochemistry eliminated the photoresponse to UV-B radiation (see references to pterins below).

Studies by Ballaré et al[53] of hypocotyl elongation in etiolated tomato seedlings revealed the spectral dependency of the UV-B inhibitory response. They measured a response spectrum between about 265 and 335 nm (4h exposure to 1.0 $\mu mol\ m^{-2}\ s^{-1}$ monochromatic radiation against a background of 63 $\mu mol\ m^{-2}\ s^{-1}$ PAR) and found an action maximum at 297 nm for the inhibition of hypocotyl elongation (Fig. 3a). There was no evidence in this report that a damage response was being studied. Chemicals which interfere with flavin and probably pterin photochemistry eliminated the photoresponse to UV-B radiation (see references to pterins below).

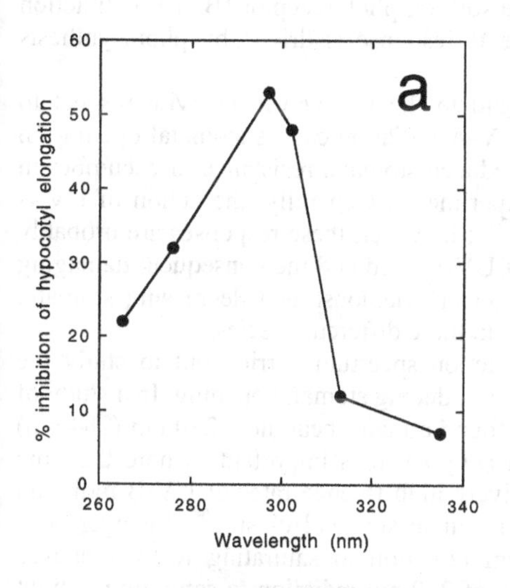

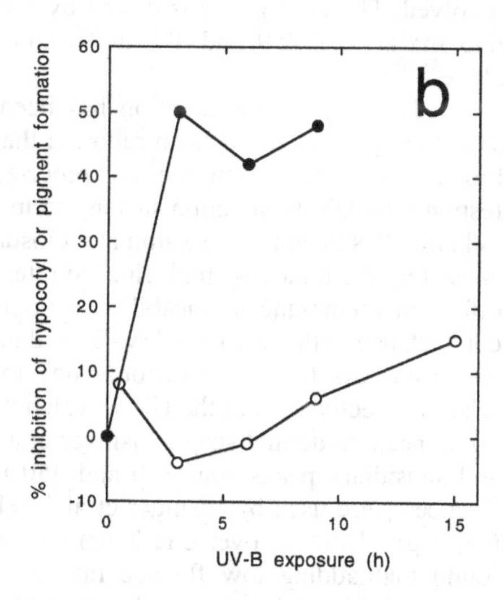

Figure 3. (a) Response spectrum for the inhibition of the elongation of etiolated tomato hypocotyls by UV radiation (4h exposure to 1.0 μmol m^{-2} s^{-1} monochromatic radiation against a background of 63 μmol m^{-2} s^{-1} PAR). Redrawn after Ballaré[53]. (b) Time course for the inhibition of elongation and the formation of UV absorbing pigments in etiolated tomato hypocotyls in response to 3.1 μmol m^{-2} s^{-1} UV-B radiation. Control plants received no UV-B radiation. Redrawn after Ballaré et al[54].

The possible adaptive significance of a UV-B photomorphogenic response is unclear. It has been suggested[37] that the UV-B inhibitory effect may be a mechanism by which plants delay cell division and thereby reduce the potentially damaging effects of UV-B radiation on their DNA before protective measures such as protective pigment formation have been fully induced. This hypothesis may be supported by the observations of Ballaré et al[54] on tomato which showed that the inhibition of hypocotyl elongation rate in response to UV-B radiation had a lag time which was much shorter than that for the formation of bulk UV-B absorbing pigments (Fig. 3b).

Faster production of UV-B absorbing compounds than those found in tomato have been observed, for example in etiolated rye[55], but comparative growth measurements were not made. Ballaré et al[54] have suggested that the rapid inhibition of growth response may delay the emergence of the seedling from the soil until photoprotective mechanisms are established. However, rapid inhibition of growth can also be induced by the B, R and FR wavebands and much lower fluence rates of these wavelengths are required.

5. Stomatal opening

Stomata play an essential role in gas exchange and the regulation of their aperture is a critical factor in optimising CO2 uptake and water loss. Light plays a predominant role in controlling stomatal opening and action spectra have demonstrated peaks of activity in both the B and R wavebands[56]. At least two photoreceptors are involved. The B response is driven by a B absorbing photoreceptor (BAP) with action maxima at 420, 450 and 470 nm[57], and the R response is driven by photosynthesis alone[58-60].

Relatively little attention has been paid to the UV waveband with respect to stomatal opening. It has been reported that UV-A radiation causes stomatal opening in broad bean[61] and Teramura et al[62] reported reduced stomatal resistance in cucumber in response to UV-B radiation in long-term experiments. Generally, the action of UV-B radiation has been to cause stomatal closure[63-66]; however, these responses are probably caused by the relatively high fluence rates of UV-B used and the consequent damaging effect on membrane permeability[67]. Nogués et al[68] demonstrated decreasing stomatal conductance with increasing UV-B irradiance in three different species.
There appears to have been only one true action spectrum carried out to study the relative effectiveness of the UV wavelengths in inducing stomatal opening. In a study of broad bean epidermal strips, Eisinger et al[69] found a major peak near 280 nm (284 nm) and subsidiary peaks near 350 and 450 nm (Fig. 4). It is important to note that low fluences were used by Eisinger et al[69]; relatively high fluence rates of UV-B radiation (ca. 2 μmol m^{-2} s^{-1} over one hour) tend to result in stomatal closure[64]. Eisinger et al found that adding low fluence rate 280 nm radiation to saturating R light caused additional stomatal opening whereas addition of 280 nm radiation to saturating B light did not. From this observation, and evidence in the literature, the authors conclude that rather than a specific UV-B photoreceptor being involved, the UV-B and B photoreceptors are synonymous and that the 280 nm response results from energy being transferred to the chromophore of the BAP from the protein moiety of the protein-pigment complex.

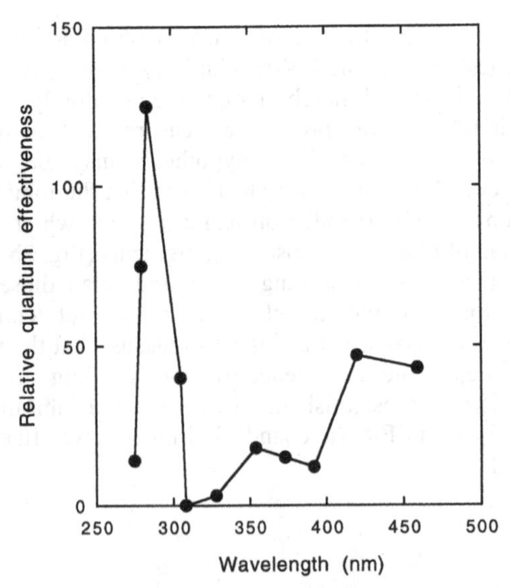

Figure 4. Action spectrum for the induction of stomatal opening in broad bean epidermal strips. Samples were exposed to monochromatic radiation for one hour. Redrawn after Eisinger et al[69].

6. Cotyledon opening

When a dicotyledonous plant emerges from the soil, it is essential that the cotyledons open quickly to maximise the absorption of PAR. Although phytochrome is known to be essential for this de-etiolation process[70], UV-B radiation also appears to play a substantial role. Boccalandro et al[71] found that daily irradiation of etiolated *Arabidopsis* seedlings with UV-B radiation for 2.5 h followed by a pulse of R light resulted in substantial cotyledon opening. By contrast, a R pulse, or 2.5h of R light, or 2.5h of UV-B light followed by a far-red (FR) light pulse resulted in only slight cotyledon opening. This implies a photoresponse controlled by a UV-B absorbing photoreceptor, the effectiveness of which is determined by phytochrome. The response to UV-B radiation increased with fluence rate up to 7.58 μmol m^{-2} s^{-1} in Boccalandro et al's[71] study but decreased with higher fluence rates, thereby suggesting an initial photomorphogenic effect followed by a damage effect at high fluence rates. 7.58 μmol m^{-2} s^{-1} of UV-B radiation is a very high fluence rate for maximal response. However from their experiments on phytochrome, cryptochrome and DNA repair mutants, the authors convincingly conclude that cotyledon opening is greatly amplified by a specific UV-B photoreceptor which is modulated by phytochrome B.

7. Cotyledon curling

Cotyledon curling is an alleged photomorphogenic effect of UV-B radiation. Wilson and Greenberg[72] have provided evidence that cotyledon curling in *Brassica napus* seedlings in response to UV-B radiation is not always a damage response.

8. Tendril curling

UV-B radiation induces coiling in broom sorghum first internodes[73]. The response correlates well with pyrimidine dimer formation[74] in this species, thereby indicating that this may be a DNA damage response. Application of jasmonic acid or methyl jasmonate, which are associated with UV-B radiation effects[4,12], to *Bryonia dioica* plants results in tendril coiling[75]. Brosché and Strid[76] found that relatively low levels of UV-B radiation (but not UV-A) added to broad-band white light (WL) caused coiling in pea (*Pisum sativum* L.) tendrils. The coiling response, which required continuous irradiation, occurred within less than 8h in both attached and detached tendrils and was reversible if the tendrils were returned to WL for 5d. In contrast to the observations with *B. dioica*, methyljasmonate applied in the absence of UV-B did not induce coiling. Although the evidence is not conclusive, it is unlikely that the reaction of pea was a purely damage response at the low level of UV-B radiation used; it is also possible, as with anthocyanin and flavonoid production, that different species use different photoreceptors.

9. Fungal and viral diseases

Laboratory and glasshouse studies have been inconclusive. Four out of ten studies recorded that added UV-B radiation resulted in lower fungal and viral infection; however, six of these recorded that additional UV-B radiation resulted in higher infection by fungi and viruses[77]. Orth et al[78] found that cucumber (*Cucumis sativus* L.) plants pre-irradiated with UV-B radiation were more susceptible to subsequent infection by fungal pathogens; interestingly, UV-B irradiation after infection had no effect on the fungal pathogen load.

Outdoor studies may provide a potentially more realistic analysis of the role of UV-B radiation in influencing fungal and viral diseases. UV-B exclusion studies on tea (*Camellia sinensis* L.) have shown that removal of UV-B radiation results in increased blister blight[79]. Newsham et al[80] studied the abundance and distribution of phylloplane fungi on oak (*Quercus robur* L.) leaves and compared plants with a 30% elevation above the ambient level of erythemally-weighted UV-B radiation with control (added UV-A only) and plants receiving only ambient radiation. Over a four month period they found that the abundance of these non-phytopathenogenic fungi decreased with time on the adaxial leaf surface (but not on the abaxial surface) in the plants with added UV-B radiation compared to either the control or ambient radiation plants. In addition, the isolation frequency of the fungi decreased with increasing fluence rate and the authors were able to demonstrate that ambient temperature and rainfall had no influence on the response. Our own outdoor data (Holmes, unpublished) confirm that general phylloplane fungi are almost eliminated on the adaxial leaf surfaces of *Q. robur* using similar radiation treatments; furthermore, this tendency continued over 7 years and the plants themselves had a slightly higher dry weight accumulation rate UV-B+A radiation than under UV-A or ambient treatments. Isoflavonoids, which can be induced by both stress and UV radiation, have antifungal properties (see above).

Overall, current research on this subject area is inconclusive. However, outdoor studies, which are more representative than those carried out indoors, tend to indicate that UV-B radiation has a positive (non-damaging) effect on the plants insofar that fungal infections are reduced by the UV-B radiation.

10. Insect herbivory

Insect herbivory of plants which have received different amounts of UV-B radiation has received substantially more attention than the subject of viral and fungal pathogens. McCloud and Beranbaum[81] reported that caterpillars of the moth *Trichoplusia ni* have lower mortality and better growth when fed on leaves of glasshouse-grown (i.e. a UV-B deficient radiation environment) lemon plant (*Citrus jambhiri*) leaves than leaves which had received UV-B. The authors ascribed their observations to the furanocoumarins produced in response to UV-B radiation.

Using growth chamber experiments, Hatcher and Paul[82] found that larvae of the moth *Autographa gamma* L. consumed less leaf tissue of pea (*Pisum sativum* L.) when the plants had previously received UV-B radiation. They suggested that the reduced consumption was caused by the increased nitrogen content of the leaves. A study of moth larvae (*Operophtera brumata* L.) by Lavola et al[83] resulted in a different response. In this case the larvae preferred to graze on leaves of silver birch (*Betula pendula* Roth.) which had received UV-B radiation rather than on those leaves which had not.

Outdoor exclusion or attenuation experiments in which various wavebands of natural global UV radiation are attenuated by placing filters above plants have been used for several decades. UV-B exclusion experiments have generally, but not exclusively, demonstrated that the removal of UV-B radiation from ambient daylight using filters above plants increases insect herbivory. Ballaré et al[84] found that reducing ambient UV-B radiation by covering *Datura ferox* L. plants with UV-B opaque polyester filters resulted in increased insect herbivory compared to plants which had grown under UV-B transmitting cellulose diacetate filters.

Mazza et al[85] used similar filters to study thrips (*Caliothrips phaseoli*) predation on soybean grown both in controlled environment and field conditions (Fig. 5). They observed that not only did the thrips prefer plants which had not been exposed to UV-B radiation, but that they also appeared to detect UV-B radiation and deliberately avoid it. In addition, they also reported that the soybean worm preferred leaves which had not been attacked previously by the thrips. Clearly, the non-damaging effects of UV-B radiation on plants are not straightforward: complex interactions between radiation, plant and predator exist.

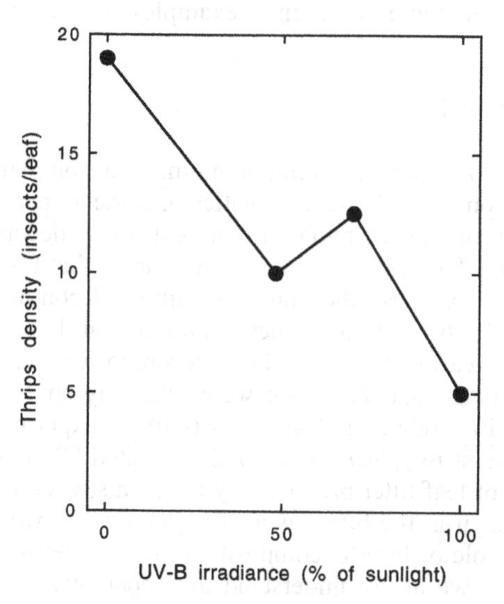

Figure 5. Effects of solar UV-B radiation on thrips density. Redrawn after Mazza et al[85].

Following an earlier study[86,87] examined the effects of excluding natural UV-B radiation from an ecosystem of plants growing in Tierra del Fuego in Argentina which is an area which receives enhanced UV-B as a result of ozone depletion over Antarctica. The authors were therefore studying the growth and herbivory of plants which either received enhanced UV-B (10 - 13% ozone depletion) with plants receiving only 15-20% of ambient UV-B levels. Over three complete growing seasons they found negligible differences in the growth of the dominant evergreen shrub *Chiliotrichum diffusum*. Marginal negative effects of ambient UV-B radiation on vegetative growth (about 12%) were found on the interspersed hebaceous species *Gunnera magellanica* and *Blechnum penna-marina*. The most striking observation was that insect herbivory (measured as leaf area consumed) was increased by between 25 and 75% in *G. magellanica* when it received reduced UV-B radiation.

Gwynn-Jones et al[88] studied the effects of supplementary UV-B radiation simulating a 15% reduction in the ozone column on a sub-arctic forest heath ecosystem in northern Sweden. They observed contrasting responses in which UV-B supplementation resulted in increased insect herbivory in the dwarf shrub *Vaccinium myrtillus* L. but reduced herbivory in *Vaccinium uliginosum* L. Contrasting responses were observed by Lindroth et al[89].

Salt et al[90] also used supplementary UV-B radiation to simulate a 15% ozone depletion (seasonally adjusted) to study psyllid (*Strophingia ericae* (Curtis)) populations on heather (*Calluna vulgaris* L.). They found that enhanced UV-B caused a gradual reduction in *S. ericae* populations compared with controls over 27 months. They tentatively proposed that reduced levels of the amino acid isoleucine in the plants exposed to supplementary UV-B radiation may have caused the lower psyllid population.

From the examples reported here, it can be concluded that exposure of plants to UV-B radiation, or to more UV-B radiation than the control plants, reduces insect herbivory in 8 cases, but increases it in 2 examples. However, the interactions are notably complex[91].

11. Leaf litter decomposition

Leaf litter decomposition has an important influence on plant growth[92] because it results in the production of soil organic matter and the mineralization of elements. Studies on the effect of UV-B radiation on leaf litter decomposition have been somewhat contradictory. Two outdoor studies have shown that UV-B radiation added to backgound daylight decreases the rate of litter decomposition (*Vaccinium*[93]; *Calamagrostis*[94]); by contrast, two studies reported that UV-B radiation added to backgound daylight increases the rate of litter decomposition (*Triticum*[95]; *Quercus*[96]). Newsham et al[97] point out that the above were relatively short-tem studies (between 0.15 and 1.33 years). In a relatively long-term (4.08 y) experiment they have clearly demonstrated that, at least for *Quercus robur* L., elevated UV-B levels accelerated the rate of decomposition of leaf litter produced by the plants when the litter was placed in a fully randomised design in the litter layer of a *Quercus/Fraxinus* woodland. Bearing in mind the important role of litter decomposition in ecosystems, it is clear that further studies are necessary if we are to understand the apparently important role of UV-B radiation in modulating the decomposition of leaf litter.

Further recent and relevant review reading in the general subject area of UV-B radiation effects on ecosystems at various trophic levels including insects, pathogens, symbionts and decomposers can be found in Ballaré et al[98], Paul et al[99,100] and van de Staaij et al[101].

12. The UV-B photoreceptor

Evidence for a specific UV-B photoreceptor has been mentioned in passing on several occasions in this article. Action spectroscopy is a useful tool for indicating the absorptive properties of functional photoreceptors. Under ideal conditions, the action spectrum would show close spectral similarity to the absorption spectrum of the target photoreceptor. In practice, the similarity is often distorted because the target photoreceptor is usually surrounded by tissue with selective absorptive and scattering properties so the measured action spectrum can be distorted. This can be a particular problem in the UV because many molecules exhibit changing absorbance across this waveband.

Most of the action spectra presented here (Figs. 1, 2, 3a, 4) provide strong implications of a photoreceptor with an absorption maximum between about 290 and 300nm. Only higher plants are considered in detail here, but it is worth mentioning that similar action maxima have been found for carotenogenesis in the fungus *Verticillium agaricum*[102] and for the induction of sporulation in the ascomycetes *Alternaria tomato* and *Helminthosporium oryzae*[103]. A similar UV peak was shown for anthocyanin synthesis in the aquatic plant *Spirodela oligorhiza*[104], but the existence of another action maximum in the R waveband and lack of further relevant experiments precludes substantive conclusions.

There appears to be no spectral evidence for either phytochrome (see also Mohr[39]) or flavins (see also Björn[105]) being specific UV-B photoreceptors. Although phytochrome and the flavins absorb in the UV-B region, they do not have specific absorption peaks similar to those shown in Figs. 1, 2, 3a and 4 above. It is also unlikely that phytochrome or the BAP can act as specific UV-B photoreceptors when natural daylight contains relatively high levels of R, FR and B radiation (compared to UV-B) which these photoreceptors absorb strongly compared to UV-B radiation. There is no current indication that phototropin plays a role. There is some evidence that pterins are a specific UV-B photoreceptor and in terms of spectral evidence pterins appear to be likely candidates[105]. Hsaio and Björn[106] suggested this possibility for the photoreceptor responsible for carotenoid formation in *V. agaricum.* Furthermore, studies with *Phycomyces* mutants indicates an interaction of pterins in photoreception in the near UV and B spectral regions[106,107]; however, such detailed studies have not been carried out on higher plants.

Although virtually nothing is known about the molecular structure of the UV-B photoreceptor(s), it is likely that current genetic and biochemical approaches will rapidly advance our understanding.

13. Ozone depletion and the UV-B photoreceptor

Björn[105] has shown that stratospheric ozone depletion with its concomitant increase in UV-B radiation levels at the Earth's surface will probably have negligible effect on the estimated perfomance of the UV-B photoreceptor compared to its effects on DNA. He used the well-established data of Wellmann[36] and Beggs and Wellmann [27] for the induction of flavonoids by UV radiation. Björn calculated radiation amplification factors (RAF) of 0.58 and 0.42, respectively. The data were calculated for clear skies in mid-summer at 55.7^{o} N and compared 300 Dobson units of stratospheric ozone with 270 units (10% depletion). These RAF values are strikingly low compared to comparable calculations for damage effects on DNA which are typically in the range of 1 to 2. Björn convincingly concluded that the effects of stratospheric ozone depletion on the performance of the UV-B photoreceptor are low compared to the damaging effects of UV-B radiation on DNA.

14. Conclusions

In conclusion, it can be stated that UV-B radiation has several effects on higher plants which are not only non-damaging, but can be positive for plant development and survival. Some of these effects are indirect, such as the possible reduction in phylloplane fungi and the reduction in attacks by phytophagous insects. Others are photomorphogenic and act directly on the plant. The induction of photoprotective screening pigments and antioxidants by UV-B radiation is of obvious adaptive advantage.

As Briggs and Olney[109] point out, it is now known that plants have at least nine photoreceptors: five phytochromes (responding primarily to R and FR light), two B/UV-A light photoreceptors (cryptochromes), a phototropism photoreceptor (phototropin), and a photoreceptor with properties akin to both phytochrome and phototropin ('superchrome'[110]. Identification of the UV-B photoreceptor(s) and the roles played by it/them will add to our understanding of the adaptation of plants to the shortest wavelengths of ambient solar radiation.

References

1. Foyer, C.H., Lelandais, M. and Kunert, K.G. (1994) Photooxidative stress in plants. *Physiol. Plant.* 92: 696-717.
2. Bender, K., Blattner, C., Knebel, A., Iordanov, M., Herrlich, P. and Rahmsdorf, H.J. (1997) UV-induced signal transduction. *J. Photochem. Photobiol. B:Biol.*, 37: 1-17
3. Herrlich, P., Blattner, C., Knebel, A., Bender, K. and Rahmsdorf, H.J. (1997) Nuclear and non-nuclear targets of genotoxic agents in the induction of gene expression. Shared principles in yeast, rodents, man and plants. *Biological Chemistry*, 378: 1217-1229.
4. A.-H.-Mackerness, S. (2000) Plant responses to ultraviolet-B (UV-B: 280-320nm) stress: What are the key regulators?. *Plant Growth Regulation*, 32: 27-39.
5. A.-H.-Mackerness, S. and Thomas, B. (1999) Effects of UV-B radiation on plants: Gene expression and signal transduction pathways. In *Plant responses to environmental stress* (M.F. Smallwood, C.M. Calvert and D.J. Bowles, eds.), Bios Scientific, Oxford, pp. 17-24.
6. A.-H.-Mackerness, S., Surplus, S.L., Jordan, B.R. and Thomas, B. (1998) Effects of supplementary UV-B radiation on photosynthetic transcripts at different stages of development and light levels in pea: Role of ROS and antioxidant enzymes, *Photochem. Photobiol.*, 68: 88-96.
7. Rao, M.V., Paliyath, C., Ormrod, D.P. (1996) Ultraviolet-B and ozone-induced biochemical changes in antioxidant enzymes of *Arabidopsis thaliana. Plant Physiology*, 110: 125-136.
8. Strid, A., Chow, W.S., Anderson, J.M. (1994) UV-B damage and protection at the molecular level in plants. *Photosynthesis Research*, 39: 475-489.
9. Green, R. and Fluhr, R. (1995) UV-B induced PR1 accumulation is mediated by active oxygen species. *Plant Cell*, 7: 203-212.
10. Surplus, S.L., Jordan, B.R., Murphy, A.M., Carr, J.P., Thomas, B. and A.-H.-Mackerness, S. (1998) Ultraviolet-B-induced responses in *Arabidopsis thaliana*: role of salicylic acid and reactive oxygen species in the regulation of transcripts encoding photosynthetic and acidic pathogenesis-related proteins. *Plant Cell Environ.*, 21: 685-694.
11. A.-H.-Mackerness, S., Surplus, S.L., Blake, P., John, C.F., Buchanan-Wollaston, V., Jordan, B.R. and Thomas, B. (1999) UV-B induced stress and changes in gene expression in *Arabidopsis thaliana*: Role of signaling pathways controlled by jasmonic acid, ethylene and reactive oxygen species. *Plant Cell Environ.*, 22: 1413-1423.
12. Conconi, A., Smerdon, M.J., Howe, G.A. and Ryan, C.A. (1996) The octadecanoid signalling pathway in plants mediates a response to ultraviolet radiation. *Nature*, 383: 826–829.
13. Creelman, R.A. and Mullet, J.E. (1995) Jasmonic acid distribution and action in plants: Regulation during development and response to biotic and abiotic stress. *Proc. Natl. Acad. Sci. USA*, 92: 4114-4119.

14. Penninckx, I.A.M.A., Eggermont, K., Terras, F.R.G., Thomma, B.P.H.J., De Samblaux, G.W., Buchala, A., Metraux, J. Manners, J.M. and Broekart, W.F. (1996) Pathogen-induced systemic activation of a plant defensin gene in *Arabidopsis* follows a salycylic acid-independent pathway. *Plant Cell*, 8: 2309-2323.
15. Ryan, C.A. (1992) The search for the proteinase inhibitor-inducing factor, PIIF, *Plant. Mol. Biol.*, 19: 123-133.
16. Sembdner, G. and Parthier, B. (1993) The biochemistry, physiological and molecular actions of jasmonates. *Annu. Rev. Plant Physiol. Plant Mol. Biol.*, 44: 569-589.
17. Epple, P., Apel, K. and Bohlman, H. (1995) An *Arabidopsis thaliana* thionin gene is inducible via a signal transduction pathway different from that for PR proteins. *Plant. Phys.*, 109: 813-820.
18. Penninckx, I.A.M.A., Thomma, B.P.H.J., Buchala, A., Metraux, J.M. and Broekart, W.F. (1998) Concomitant activation of jasmonate and ethylene response pathways is required for induction of a plant defensin gene in *Arabidopsis*. *Plant Cell*, 10: 2103-2113.
19. Vijayan, P., Shockley, J., Levesque, Q.A., Cook, R.J. and Browse, J. (1998) a role for jasmonate in pathogen defence of Arabidopsis. *Proc. Natl. Acad. Sci. USA*, 95: 7209-7214.
20. Robberecht, R., Caldwell, M. M. (1986) Leaf UV optical properties of *Rumex patientia* L. and *Rumex obtusifolius* L. in regard to protective mechanism against solar UV-B radiation injury. In *Stratospheric Ozone Reduction, Solar Ultraviolet Radiation and Plant Life* (R.C. Worrest and M.M. Caldwell, eds.), NATO ASI Series, Vol. G8, Springer-Verlag, Berlin, pp. 251-259.
21. Landry, L. G., Chapple, C. C. S., Last, R. L. (1995) *Arabidopsis* mutants lacking phenolic sunscreens exhibit enhanced ultraviolet-B injury and oxidative damage. *Plant Physiology*, 109: 1159-1166.
22. Li, J.Y., Oulee, T.M., Raba, R., Amundson, R.G. and Last, R.L. (1993) *Arbidopsis* flavonoid mutants are hypersensitive to UV-B irradiation. *Plant Cell*, 5: 171-179.
23. Lois, R. and Buchanan, B. B. (1994) Severe sensitivity to ultraviolet radiation in an *Arabidopsis* mutant deficient in flavonoid accumulation. 2. Mechanisms of UV resistance in *Arabidopsis*. *Planta*, 194: 504-509.
24. Stapleton, A.E. and Walbot, V. (1994) Flavonoids can protect maize DNA from the induction of ultraviolet radiation damage. *Plant Physiol.*, 105: 881-889.
25. Reuber, S., Bornman, J.F. and Weissenbock, G. (1996) Phenylpropanoid compounds in primary leaves of rye (*Secale cereale*) light regulation of their biosynthesis and the possible role in UV-B protection. *Physiol. Plant.*, 97: 160-168.
26. Mazza, C.A., Boccalandro, H.E., Giordano, C.V., Battista, D., Scopel, A.L. and Ballaré, C.L. (2000) Functional significance and induction by solar radiation of ultraviolet-absorbing sunscreens in field-grown soybean crops. *Plant Physiology*, 122: 117–125.
27. Beggs, C.J. and Wellmann, E. (1994) Photocontrol of flavonoid biosynthesis. In *Photomorphogenesis in plants, Vol. 2* (R.E. Kendrick and G.H.M. Kronenberg, eds.), Kluwer Academic, Dordrecht, pp. 733-750.
28. Winkel-Shirley, B. (2001) Flavonoid biosynthesis. A colorful model for genetics, biochemistry, cell biology, and biotechnology. *Plant Physiol.*, 126: 485-493.
29. Takahashi, A., Takeda, K. and Ohnishi, T. (1991) Light-induced anthocyanin reduces the extent of damage to DNA in UV-irradiated *Centaurea cyanus* cells in culture. *Plant Cell Physiol.*, 32: 541-547.
30. Chalker-Scott, L. (1999) Environmental significance of anthocyanins in plant stress responses. *Photochem. Photobiol.*, 70: 1–9.
31. Beggs, C.J., Schneider-Ziebert, U. and Wellmann, E. (1986) UV-B radiation and adaptive mechanisms in plants. In *Stratospheric Ozone Reduction, Solar Ultraviolet Radiation and Plant Life* (R.C. Worrest and M.M. Caldwell, eds.), Springer-Verlag, Berlin, pp. 235-250.
32. Fuglevand, G., Jackson, J.A. and Jenkins, G. I. (1996) UV-B, UV-A, and blue light signal transduction pathways interact synergistically to regulate chalcone synthase gene expression in Arabidopsis. *Plant Cell*, 8: 2347-2357.
33. Wade, H.K., Bibikova, T.N., Valentine, W.J. and Jenkins, G.I. (2001) Interactions within a network of phytochrome, cryptochrome and UV-B phototransduction pathways regulate chalcone synthase gene expression in *Arabidopsis* leaf tissue. *The Plant Journal*, 25: 675-685.
34. Beggs, C.J., Stolzer-Jehle, A. and Wellmann, E. (1985) Isoflavonoid formation as an indicator of UV stress in bean (*Phaseolus vulgaris* L.) leaves, *Plant Physiol.*, 79: 630-634.
35. Beggs, C.J. and Wellmann, E. (1985) Analysis of light-controlled anthocyanin formation in coleoptiles of *Zea mays* L.: The role of UV-B, blue, red and far-red light, *Photochem. Photobiol.*, 41: 481-486.
36. Wellmann, E. (1975) UV Dose-dependent induction of enzymes related to flavonoid biosynthesis in cell suspension cultures of parsley. *FEBS Lett.*, 51: 105-107.
37. Wellmann, E. (1983) UV radiation in photomorphogenesis. In *Encyclopedia of plant physiology, vol. 16 B (new series), Photomorphogenesis* (W. Shropshire and H. Mohr, eds.), Springer-Verlag, Berlin, pp. 745-756.

38. Arthur, J.M. (1936) Radiation and anthocyanin pigments. In *Biological effects of radiation 2* (B.M. Duggan, ed.) McGraw-Hill, New York, pp. 109-118.
39. Mohr, H. (1994) Coaction between pigment systems. In *Photomorphogenesis in plants* (R.E. Kendrick and G.H.M. Kronenberg, eds.), Kluwer, Netherlands, pp. 353-373.
40. Brandt, K., Giannini, A. and Lercari, B. (1995) Photomorphogenic responses to UV radiation III: a comparative study of UVB effects on anthocyanin and flavonoid accumulation in wild-type and aurea mutant of tomato (*Lycopersicon esculentum* Mill.). *Photochem. Photobiol.*, 62: 1081–1087.
41. Yatsuhashi, H., Hashimoto, T., Shimizu, S. (1982) Ultraviolet action spectrum for anthocyanin formation in broom sorghum first internodes. *Plant Physiol.*, 70: 735-741.
42. Hashimoto, T. Shichijo, C. and Yatsuhashi, H. (1991) Ultraviolet action spectra for the induction and inhibition of anthocyanin synthesis in broom sorghum seedlings. *J. Photochem. Photobiol. B: Biol.*, 11: 353-363.
43. Wellmann, E., Schneider-Ziebert, U. and Beggs, C.J. (1984) UV-B inhibition of phytochrome-mediated anthocyanin formation in *Sinapis alba* L. cotyledons. Action spectra and the role of photoreactivation. *Plant Physiol.*, 75: 997-1000.
44. Drumm, H. and Mohr, H. (1978) The mode of interaction between blue (UV) light photoreceptor and phytochrome in anthocyanin formation of the *Sorghum* seedling. *Photochem. Photobiol.*, 27: 241-248.
45. Drumm-Herrel, H. and Mohr, H. (1981) A novel effect of UV-B in a higher plant (*Sorghum vulgare*). *Photochem. Photobiol.*, 33: 391-398.
46. Oelmüller, R and Mohr, H (1985) Mode of coaction between blue/UV light and light absorbed by phytochrome in light-mediated anthocyanin formation in the milo (*Sorghum vulgare* Pers.) seedling. *Proc. Natl. Acad. Sci. USA*, 82: 6124-6128.
47. Steinmetz, V. and Wellmann, E. (1986) The role of solar UV-B in growth regulation of cress (*Lepidium sativum* L.) seedlings. *Photochem. Photobiol.*, 43: 189-193.
48. Ensminger, P. A. (1993) Control of development in plants and fungi by far-UV radiation. *Physiol. Plant.*, 88: 501 - 508.
49. Goto, N, Yamamoto, K.T. and Watanabe, N. (1993) Action spectra for inhibition of hypocotyl growth of wild-type plants and of the *hy2* long-hypocotyl mutant of *Arabidopsis thaliana* L. *Photochem. Photobiol.*, 57: 867-871.
50. Barnes, P.W., Ballaré, C.L. and Caldwell, M.M. (1996) Photomorphogenic effects of UV-B radiation on plants: consequences for light competition, *J. Plant Physiol.*, 148: 15-20.
51. Ballaré, C.L., Barnes, P.W. and Kendrick, R.E. (1991) Photomorphogenic effects of UV-B radiation on hypocotyl elongation in wild-type and stable photochrome-deficient mutant seedlings of cucumber. *Physiol. Plant.*, 83: 652-658.
52. Wheeler, S.L., Barnes, P.W. and Shinkle, J.R. (1997) Separate short-term growth responses of dicot seedlings to different regions within the UV-B spectrum. Plant Biology '97 (ASPP), St. Louis, Missouri, USA, Abstract No. 424.
53. Ballaré, C.L., Barnes, P.W. and Flint, S.D. (1995) Inhibition of hypocotyl elongation by ultraviolet-B radiation in de-etiolating tomato seedlings. 1. The photoreceptor. *Physiol. Plant.*, 93: 584-592.
54. Ballaré, C.L., Barnes, P.W. and Flint, S.D. (1995) Inhibition of hypocotyl elongation by ultraviolet-B radiation in de-etiolating tomato seedlings. 2. Time-course, comparison with flavonoid responses and adaptive significance. *Physiol. Plant.*, 93: 593-601.
55. Braun, J. and Tevini, M. (1993) Regulation of UV-protective pigment synthesis in the epidermal layer of rye seedlings (*Secale cereale* L. cv. Kustro). *Photochem. Photobiol.*, 57: 318-323.
56. Kuiper, P.J.C. (1964) Dependence upon wavelength of stomatal movement in epidermal tissue of *Senecio odoris*. *Plant Physiol.*, 39: 952–955.
57. Karlson, P.E. (1986) Blue light regulation of stomata in wheat seedlings. II. Action spectrum and search for action dichroism. *Physiol. Plant.*, 66: 207–210.
58. Lurie, S. (1978) The effect of wavelength of light on stomatal opening. *Planta*, 140: 245–249
59. Ogawa, T. (1981) Blue light response with stomata with starch-containing (Vicia faba) and starch-deficient (*Allium cepa*) guard cells under background illumination with red light. *Plant Sci. Lett.*, 22: 103-108
60. Sharkey, T.D., Raschke, K. (1981) Effect of light quality on stomatal opening in leaves of *Xanthium strumarium* L. *Plant Physiol.*, 68: 1170–1174
61. Ogawa, T., Ishikawa, H., Shimada, K., Shibata, K. (1978) Synergistic action of red and blue light and action spectrum for malate formation in guard cells of *Vicia faba* L. *Planta*, 142: 61–65
62. Teramura, A.H., Tevini, M., Iwanzik, W. (1983) Effects of ultraviolet-B on plants during mild water stress. I. Effects on diurnal stomatal resistance. *Physiol. Plant.*, 57: 175–180

63. Mirecki, R.M. and Teramura, A.A. (1984) Effects of ultraviolet-B irradiance on soybean. V. The dependence of plant sensitivity on the photosynthetic photon flux density during and after leaf expansion. *Plant Physiology*, 74: 475-480.
64. Negash, L. and Bjorn, L.O. (1986) Stomatal closure by UV radiation. *Physiol Plant*, 66: 360–364.
65. Grammatikopoulos, G., Karabourniotis, G., Kyparissis, A., Petropoulou, Y., Manetas, Y. (1994) Leaf hairs of olive (*Olea europaea*) prevent stomatal closure by ultraviolet-B radiation. *Australian Journal of Plant Physiology*, 21: 293-301.
66. Olszyk, D., Dai, Q.J., Teng, P., Leung, H., Luo, Y., Peng, S.B.(1996) UV-B effects on crops: Response of the irrigated rice ecosystem. *Journal of Plant Physiology*, 148: 26-34.
67. Negash, L., Jensén, P. and Björn, L.O. (1987) Effect of ultraviolet radiation on accumulation and leakage of $^{86}Rb^{+}$ in guard cells of *Vicia faba*. *Physiol. Plant.*, 69: 200-204.
68. Nogués, S., Allen, D.J., Morison, J.I.L. and Baker, N.R. (1999) Characterization of stomatal closure caused by ultraviolet-B radiation. *Plant Physiol.*, 121: 489–496.
69. Eisinger, W., Swartz, T.E., Bogomolni, R.A. and Taiz, L. (2000) The ultraviolet action spectrum for stomatal opening in broad bean. *Plant Physiol.*, 122: 99–105.
70. Mohr, H. (1972) *Lectures on Photomorphogenesis*. Springer, Berlin.
71. Boccalandro, H.E., Mazza, C.A., Mazella, M.A., Casal, J.J. and Ballaré, C.L. (2001) Ultraviolet B radiation enhances a phytochrome-B-mediated photomorphogenic response in *Arabidopsis*. *Plant Physiol.*, 126: 780-788.
72. Wilson, M. I. and Greenberg, B. M. (1993) Specificity and photomorphogenic nature of ultraviolet-B induced cotyledon curling in *Brassica napus* L. *Plant Physiol.*, 102: 671-677.
73. Hashimoto, T., Ito, S. and Yatsuhashi, H. (1984) Ultraviolet light-induced coiling and curvature of broom sorghum first internodes. *Physiol. Plant.*, 61: 1–7.
74. Hada, M., Tsurumi, S., Suzuki, M., Wellman, E. and Hashimoto, T. (1996) Involvement and non-involvement of pyrimidine dimer formation in UV-B effects on *Sorghum bicolor* Moench seedlings. *J. Plant Physiol.*, 148: 92-99.
75. Falkenstein, E., Groth, B., Mithöfer, A. and Weiler, E.W. (1991) Methyljasmonate and alpha-linolenic acid are potent inducers of tendril coiling. *Planta*, 185: 316–322.
76. Brosché, M. and Strid, Å. (2000) Ultraviolet-B radiation causes tendril coiling in *Pisum sativum*. *Plant Cell Physiol.*, 41: 1077-1079.
77. Manning, W.J. and Tiedemann, A.V. (1995) Climate change: potential effects of increased atmospheric carbon dioxide (CO2), ozone (O3), and ultraviolet-B (UV-B) radiation on plant diseases. *Environmental Pollution*, 88: 219–245.
78. Orth, A.B., Teramura, A.H. and Sisler, H.D. (1990) Effects of ultraviolet-B radiation on fungal disease development in *Cucumis sativus*. *Amer. J. Botany*, 77: 1188–1192.
79. Gunasekera, T.S., Paul, N.D. and Ayres, P.G. (1997) The effects of ultraviolet-B (UV-B: 290–320 nm) radiation on blister blight disease of tea (*Camellia sinensis*). *Plant Pathology*, 46: 179–185.
80. Newsham, K.K., Low, M.N.R., McLeod, A.R., Greenslade, P.D. and Emmett, B.A. (1997) Ultraviolet-B radiation influences the abundance and distribution of phylloplane fungi on pedunculate oak (*Quercus robur* L.). *New Phytol.*, 136: 287-297.
81. McCloud, E.S. and Berenbaum, M.R. (1994) Stratospheric ozone depletion and plant-insect interactions: effects of UVB radiation on foliage quality of *Citrus jambhiri* for *Trichoplusia ni*. *J. Chem. Ecol.*, 20: 525–535.
82. Hatcher, P.E. and Paul, N.D. (1994) The effect of elevated UV-B radiation on herbivory of pea by *Autographa gamma*. *Entomol. Exp. Appl.*, 71: 227–233.
83. Lavola, A., Julkunen-Tiitto, R., Roininen, H. and Aphalo, P. (1998) Host plant preference of an insect herbivore mediated by UV-B and CO2 in relation to plant secondary metabolites. *Biochemical Systematics and Ecology*, 26: 1-12.
84. Ballaré, C.L., Scopel, A.L., Stapleton, A.E. and Yanovsky, M.J. (1996) Solar ultraviolet-B radiation affects seedling emergence, DNA integrity, plant morphology, growth rate, and attractiveness to herbivore insect in *Datura ferox*. *Plant Physiol.*, 112: 161-170.
85. Mazza, C.A., Zavala, J., Scopel, A.L. and Ballaré, C.L. (1999) Perception of solar ultraviolet-B radiation by phytophagous insects. Behavioral responses and ecosystem implications. *Proc. Natl. Acad. Sci. USA*, 96: 980-985.
86. Rousseaux, M.C., Ballaré, C.L., Scopel, A.L., Searles, P.S. and Caldwell, M.M. (1998) Solar ultraviolet-B radiation affects plant-insect interactions in a natural ecosystem of Tierra del Fuego (southern Argentina). *Oecologia*, 116: 528-535.

87. Rousseaux, M.C., Scopel, A.L., Searles, P.S., Caldwell, M.M., Sala, O.E. and Ballaré, C.L. (2001) Responses to solar ultraviolet-B radiation in a shrub-dominated natural ecosystem of Tierra del Fuego (southern Argentina). *Global Change Biology*, 7: 467-478.
88. Gwynn-Jones, D., Lee, J.A. and Callaghan, T.V. (1997) Effects of enhanced UV-B radiation and elevated carbon dioxide concentrations on a sub-arctic forest heath ecosystem. *Plant Ecology*, 128: 242-249.
89. Lindroth, R.L., Hofman, R.W., Campbell, D.B., McNabb, W.C. and Hunt, D.Y. (2000) Population differences in *Trifolium repens* L. response to ultraviolet-B radiation: foliar chemistry and consequences for two lepidopteran herbivores. *Oecologia*, 122: 20-28.
90. Salt, D.T., Moody, S.A., Whittaket, J.B. and Paul, N.G. (1998) Effects of enhanced UVB on populations of the phloem feeding insect *Strophingia ericae* (Homoptera: Psylloidea) on heather (*Calluna vulgaris*). *Global Change Biology*, 4: 91-96.
91. McLeod, A.R., Rey, A. Newsham, K.K., Lewis, G.C. and Wolferstan, P. (2001) Effects of elevated ultraviolet radiation and endophytic fungi on plant growth and insect feeding in *Lolium perenne, Festuca rubra, F. arundinacea and F. pratensis*. *J. Photochem. Photobiol. B:Biol.*, 62: 97-107.
92. Swift, M.J., Heal, O.W. and Anderson, J.M. (1979) *Decomposition in terrestrial ecosystems*. Blackwell, Oxford.
93. Gehrke, C., Johanson, U., Gwynn-Jones, D., Björn, L.O., Callaghan, T.V. and Lee, JA (1995) The impact of enhanced ultraviolet-B radiation on litter quality and decomposition processes in *Vaccinium* leaves from the sub-Arctic. *Oikos*, 72: 213-222.
94. Rozema, J., Tosserams, M., Nelissen, H.J.M., van Heerwaarden, L., Broekman, R.A. and Flierman, N. (1997) Stratospheric ozone reduction and ecosystem processes: enhanced UV-B radiation affects chemical quality and decomposition of leaves of the dune grassland species *Calamagrostis epigeios*. *Plant Ecology*, 128: 284-294.
95. Yue, M., Li, Y. and Wang, X. (1998) Effects of enhanced ultraviolet-B radiation on plant nutrients and decomposition of spring wheat under field conditions. *Environ. Exp. Bot.*, 40: 187-196.
96. Newsham, K.K., Greenslade, P.D., Kennedy, V.H. and McLeod, A.R. (1999) Elevated UV-B radiation incident on *Quercus robur* leaf canopies enhances the decomposition of resulting leaf litter in soil. *Global Change Biology*, 5: 40-409.
97. Newsham, K.K., Anderson, J.M., Sparks, T.H., Splatt, P., Woods, C. and McLeod, A.R. (2001) UV-B effect on *Quercus robur* leaf litter decomposition persist over four years. *Global Change Biology*, 7: 479-483.
98. Ballaré, C.L., Scopel, A.L. and Mazza, C.A. (1999) Effects of solar UV-B radiation on terrestrial ecosystems: case studies from southern South America. In *Stratospheric Ozone Depletion. The Effects of Enhanced UV-B Radiation on Terrestrial Ecosystems* (J. Rozema, ed.), Backhuys, Leiden, pp.292-311.
99. Paul, N.D., Rasanayagam, S., Moody, S.A., Hatcher, P.E. and Ayres, P.G. (1997) The role of interactions between trophic levels in determining the effects of UV-B on terrestrial ecosystems. *Plant Ecology*, 128: 296-308.
100. Paul, N., Callaghan, T., Moody, S., Gwynn-Jones, D., Johanson, U. and Gehrke, C. (1999) UV-B impacts on decomposition and biogeochemical cycling. In *Stratospheric Ozone Depletion. The Effects of Enhanced UV-B Radiation on Terrestrial Ecosystems* (J. Rozema, ed.), Backhuys, Leiden, pp. 101-114.
101. van de Staaij, X., Rozema, J. and Aerts R. (1999) The impact of solar UV-B radiation on plant-micro-organism interactions at the soil-root interface. In *Stratospheric Ozone Depletion. The Effects of Enhanced UV-B Radiation on Terrestrial Ecosystems* (J. Rozema, ed.), Backhuys, Leiden, pp. 117-133.
102. Hsaio, K.C. and Björn, L-O. (1982) Aspects of photoinduction of carotenogenesis in the fungus *Verticillium agaricinum*. *Physiol. Plant.*, 54: 235-238.
103. Kumagai, T. (1983) Action spectra for blue and near ultraviolet reversible photoreaction in the induction of fungal conidiation. *Physiol. Plant.*, 57: 468-471.
104. Ng, Y.L., Thimann, K.V. and Gordon, S.A. (1964) The biogenesis of anthocyanin. X. The action spectrum for anthocyanin formation in *Spirodela oligorhiza*. *Arch. Biochem. Biophys.*, 107: 550-558.
105. Björn, L.O. (1999) UV-B effects: Receptors and targets. In *Concepts in photobiology* (G.S. Singhal, G. Renger, K.-D. Sopory and Govindjee, eds.), Narosa Publishing House, New Delhi, pp. 821-832.
106. Hsaio, K.C. and Björn, L-O. (1984) A possible photoreceptor pigment for light-induced carotenogenesis in *Verticillium agaricinum*. *Photochem. Photobiol.*, 39 (suppl.): 16S.
107. Galland, P. and Senger, H. (1988) The role of pterins in the photoreception and metabolism of plants. *Photochem. Photobiol.*, 48: 811-820.
108. Galland, P. and Senger, H. (1991) Flavins as possible blue light photoreceptors. In *Photoreceptor Evolution and Function* (M.G. Holmes, ed.), Academic Press, London, pp. 65-124.

109. Briggs, W.R. and Olney, M.A. (2001) Photoreceptors in plant photomorphogenesis to date. Five phytochromes, two cryptochromes, one phototropin, and one superchrome. *Plant Physiol.*, 125: 85-88.
110. Nozue, K., Kanegae, T., Imaizumi, T., Fukuda, S., Okamoto, H., Yeh, K.-C., Lagarias, J. C. and Wada, M. (1998). A phytochrome from the fern *Adiantum* with features of the putative photoreceptor NPH1. *Proc. Natl. Acad. Sci. USA*, 95: 15826-15830.

IMPACT OF UV RADIATION ON THE AQUATIC ENVIRONMENT

DONAT-P. HÄDER
Friedrich-Alexander-Universität, Institut für Botanik und Pharmazeutische Biologie, Staudtstr. 5, D-91058 Erlangen, Germany

1. Abstract

Aquatic ecosystems produce about 50% of the global biomass and play an important role in atmospheric carbon dioxide cycling. Since the primary producers are confined to the euphotic zone for energetic reasons, they are simultaneously exposed to short wavelength radiation. Solar UV affects growth, reproduction, photosynthetic production and many other physiological processes. Cyanobacteria are important ubiquitous prokaryotes which populate terrestrial and aquatic habitats. They account for up to 40 % of the marine biomass production and are important components of wet land ecosystems such as rice paddy fields. These organisms are also highly impaired by solar UV, but they and other motile microorganisms have developed mitigating strategies to protect themselves from this stress. One protection strategy is based on vertical migrations within the water column or a microbial mats. However, both motility and orientation are impaired by UV radiation. Another means of protection is achieved by the production of screening pigments including mycosporine-like amino acids (MAA) or scytonemins. MAAs are also produced by phytoplankton and macroalgae. In several organisms action spectra were measured which indicate that MAA synthesis is induced by UV in most cases. These sunscreen pigments prevent short wavelength radiation from reaching the UV sensitive DNA where it induces thymine dimers. Remaining dimers are removed by photorepair which involves the enzyme photolyase. The photosynthetic apparatus is another main target in primary aquatic biomass producers. Inhibition of the photosynthetic electron transport chain can be determined by oxygen measurements or by pulse amplitude modulated (PAM) fluorescence. Plants reduce the potentially deleterious effects of solar UV by decreasing the photosynthetic electron transport in photosystem II, a process called photoinhibition. Despite the dramatic effects of even ambient solar UV on individual species and physiological responses, the effect of ozone depletion on whole ecosystems is surprisingly low and close to the noise level induced by all other environmental factors such as mixing layer depth, cloud cover and temperature.

2. Introduction

Marine and freshwater aquatic ecosystems account for about half of the primary biomass production on our planet. The aquatic primary producers incorporate about as

F. Ghetti et al. (eds.), Environmental UV Radiation: Impact on Ecosystems and Human Health and Predictive Models, 179–191.

much atmospheric carbon as all terrestrial ecosystems taken together, which amounts to 100 gigatons in terms of carbon dioxide[1]. Marine ecosystems play an overwhelming role in the productivity since they represent 99.5 % of the water surface. In these ecosystems phytoplankton are the dominant biomass producers, but macroalgae also have a significant share in the biomass productivity although they are restricted to coastal areas and the continental shelves[2]. Macroalgae form the basis of an intricate food web, serve as shelter for larval fish and crustaceans and have a significant economic importance since several hundred thousand tons of seaweeds are harvested every year for food production and technological exploitation.

During the last two decades dramatic stratospheric ozone depletion has been observed over the Antarctic continent[3]. Several years later, this was also found over the North pole[4]. Significant decreases in total ozone column are also recorded at high and mid latitudes with concomitant increases in solar UV-B radiation (280 – 315 nm, C.I.E. definition) at the Earth's surface[5]. These trends are predicted to continue well into the current century with a return to pre-80s levels by the year 2065.

The high energetic short wavelength radiation affects most forms of life on this planet. It is responsible for increased incidences of skin cancer in humans, higher rates of cataracts, immunosuppression as well as other diseases. UV-B has been reported to affect terrestrial and aquatic ecosystems and may have significant consequences for the chemistry of the troposphere (e.g. photochemical smog formation) and biogeochemical cycles[3]. Solar short wavelength radiation penetrates deep into the water column of many freshwater and marine ecosystems, where primary productivity occurs in the euphotic zone[6-7]. Growing evidence indicates that further ozone depletion will impair light-dependent responses of aquatic primary producers, such as photosynthesis, photoorientation and photoprotection[8].

Significant increases in solar UV may affect biomass productivity in aquatic ecosystems with negative consequences for all levels of the intricate food webs and food production for animals and humans[9-10]. Further consequence may be changes in species composition and ecosystem integrity.

Aquatic ecosystems are important global sinks for atmospheric carbon dioxide, and decreased biomass productivity causes reduced sink capacity for CO_2[11] which consequently has negative effects on global warming[12].

3. Cyanobacteria

Cyanobacteria are an evolutionary ancient group of prokaryotes which possess oxygenic photosynthesis that closely resembles that of higher plants. However, they lack chlorophyll *b* and plastid envelopes but contain phycobiliproteins as accessory pigments which are organized in high molecular-weight phycobilisomes located on the outer thylakoid surface like in red algae. Cyanobacteria are key players in both freshwater and marine aquatic ecosystems and can constitute up to 40 % of marine biomass. Furthermore, many cyanobacteria can fix atmospheric nitrogen either as individual organisms or in symbiosis with many other species including protists, animals and plants[13]. Only prokaryotic organisms are able to use atmospheric nitrogen; all other organisms take up nitrogen in the form of nitrate, nitrite, ammonium or organic compounds. In rice paddy fields cyanobacteria are of high importance as a biological fertilizer[14], but they are also indispensable for other agricultural and natural soils as well as for marine habitats. Any reduction in cyanobacterial biomass productivity, e.g. by enhanced solar UV-B, could be relevant on a global scale.

Solar UV-B affects growth, survival, pigmentation, motility, as well as the enzymes of nitrogen metabolism and CO_2 fixation in cyanobacteria[15]. In some species growth and survival decrease within a few hours of UV-B irradiation. These organisms are major constituents in wet lands forming extended microbial mats. In order to survive the extreme solar UV stress they must possess a high potential of adaptation to diverse environmental factors.

The photosynthetic apparatus is strongly impaired by solar UV radiation. The phycobiliproteins, which constitute a large share of the accessory pigments, are readily cleaved[16-17]. The chromophoric groups, the phycobilins, are bleached with a faster kinetics than chlorophyll *a* or the carotenoids. At lower doses, which do not bleach the pigments, the energy transfer to the reaction center of photosystem II is impaired[16]. In parallel to phycobiliprotein destruction, an increased synthesis of the same pigments is induced under mild UV-B stress. Since they strongly absorb in the UV-B range and the thylakoids, on which they are located, form a peripheral layer around the sensitive centroplasma containing the DNA, one of the functions of phycobilins might be that of an effective UV screen[17]. It has been calculated that these pigments can intercept more than 99% of UV-B radiation before it penetrates to the genetic material.

Inhibition of photosynthesis by solar UV radiation has been demonstrated in a number of marine and freshwater cyanobacteria. The key enzyme, RuBISCO (ribulose-1,5-bisphosphate carboxylase/oxygenase), which is responsible for the CO_2 incorporation into organic material, was severely affected by UV-B treatment[16]. Ammonium uptake was decreased by 10 % after exposure to solar radiation. The nitrogen fixing enzyme nitrogenase is strongly affected by UV-B as well as the ammonia-assimilating enzyme glutamine synthetase[18], while nitrate reductase was stimulated by UV-B.

During evolution cyanobacteria have developed a number of mitigating and adaptive strategies to protect themselves from the deleterious effects of excessive radiation, including the avoidance of high light areas, the synthesis of UV absorbing pigments, DNA repair and recovery of the photosynthetic apparatus as well as the production of chemical scavengers to detoxify the highly reactive oxidants produced photochemically[19]. UV-absorbing pigments include scytonemins and mycosporine-like amino acids (MAAs), as well as a number of chemically unidentified pigments[18]. The protective mechanisms developed by cyanobacteria are discussed in more detail by Sinha and Häder in this volume.

4. Phytoplankton

Solar UV-B is an important ecological stress factor that affects growth, survival and distribution of phytoplankton, which are, as indicated above, the most important biomass producer in aquatic ecosystems. Phytoplankton are a diverse group of normally unicellular microalgae from many different classes which inhabit the euphotic zone of the water column where they receive sufficient solar photosynthetic available radiation (PAR). The lower limit of the euphotic zone is defined as the depth where photosynthetic biomass production balances respiratory losses. Since this limit depends on the light availability, the term is defined physiologically rather than physically. Mutual shading, absorption and light scattering by other particles within the water column limit light penetration. The euphotic zone may extend from a few decimeters in turbid coastal waters to up to 200 m in clear oceanic waters. The transparency of a body of water strongly depends on the wavelength: short wavelength radiation is more strongly absorbed than blue-green light. UV-B, even though attenuated, penetrates well into the euphotic zone[6] and affects the phytoplankton organisms.

Many phytoplankton organisms have a limited capacity to either actively swim or use buoyancy and undergo vertical migration to move to and stay at depths which are optimal for their growth and reproduction. This typical vertical distribution pattern of phytoplankton is disturbed by passive mixing due to high wind and waves[20].

The primary consumers (zooplankton) include unicellular and multicellular organisms which feed on the primary producers. The following level in the food web are free swimming organisms (nekton) including krill, molluscs, fishes and crab larvae, feeding in turn small fishes, molluscs and crustaceans. The final consumers are large fishes, birds and mammals including humans. During each transition from one trophic level to the next the amount of biomass is reduced by a factor of about ten. In addition to the direct effects of solar UV-B radiation, the consumers are affected indirectly when the productivity of the primary producers decreases. To evaluate the UV-related stress in phytoplankton the following questions need to be answered:

- What is the spectrally weighted distribution of solar radiation in dependence of time and depth?
- What are the biological weighting functions (BWF) of the physiological processes in ecologically important phytoplankton species?
- What is the spatial and temporal pattern of phytoplankton distribution within the euphotic zone, affected by wind and waves?

We are far from quantitatively understanding the intricate processes, but some of the questions have been partially answered.

Dissolved organic carbon (DOC), derived from decaying organic material, limits the penetration of UV-B into the water column[21]. Bacteria take up DOC and recycle the nutrients. In turn, the bacteria are food to heterotrophic nanoflagellates. Solar UV affects the heterotrophic bacteria and flagellates which have only limited protection and mitigating strategies. Simultaneously, short wavelength radiation breaks down the high molecular weight DOC into smaller fragments which can then be taken up easier by bacteria. By this process the transparency of the water column for solar UV increases[22].

Biologically weighting functions (BWF) have been determined to quantify the spectrally weighted sensitivity of photosynthesis to UV and visible radiation in several phytoplankton species[23]. Most BWF have a maximum in the UV-B, but also show significant sensitivity in the UV-A. Most BWF have roughly the same shape but in detail vary by species and the physiological process studied[23].

A number of models have been developed to quantify the impact of increased solar radiation on phytoplankton productivity[23], using physical and physiological parameters, water column dynamics and species distribution. One major problem is that it is difficult to use short-term observations to predict longer-term (days to years) ecological responses[23-24].

The targets of solar UV in phytoplankton are manifold, and short wavelength radiation affects DNA and cellular proteins, impairs growth and reproduction, photosynthesis[25-27] as well as motility and orientation in phytoplankton[28]. Ammonium and nitrate uptake is also affected by solar radiation[29]. The organisms respond by the synthesis of heat-shock proteins and changes in the cellular amino acid pools. DNA is a central target in the cell. It has a high absorption in the UV, and solar short wavelength radiation both damages DNA and delays DNA synthesis in many organisms[30].

Many phytoplankton use protective and mitigating strategies to prevent excessive harm by solar radiation. In addition to vertical migration and DNA repair, many (but not all) phytoplankton organisms produce UV screening pigments. In the marine dinoflagellate *Gyrodinium dorsum* short wavelength radiation induces the synthesis of mycosporine-like amino acids (MAA). HPLC and spectroscopy revealed that *G. dorsum* contains a complex mixture of several MAAs. A polychromatic action spectrum for the induction of MAA has a major peak at 320 nm[31-32]. However, exposure to excessive short wavelength UV-B radiation resulted in decreased overall MAA production. In addition, there was a shift in the absorption of the MAA mixture towards shorter wavelengths indicating that short wavelength UV-B may change the MAA composition[31].

MAAs may serve several functions in the cell and operate as osmotic regulator, as antifreeze, in addition to UV absorption. The protective effect of MAAs against UV radiation was demonstrated in *G. dorsum.* In one batch of cells MAA synthesis was induced by UV-A + PAR, while in the control (irradiated with PAR only) no MAAs were induced. Both samples (with high and low MAA content, respectively) were subsequently subjected to high UV-B irradiation. Cells with low MAA content suffered a complete loss of motility within 3 h, while the cells with high MAAs content survived at least two times longer.

A library of photoprotective compounds in cyanobacteria, phytoplankton and macroalgae has been compiled and made available in a database accessible on the Internet (http://www.biologie.uni-erlangen.de/botanik1/index.html) to provide easy access to the various photoprotective compounds reported in the literature[33]. The data base includes absorption maxima, molecular structures and extinction coefficients of some of the important photoprotective compounds.

To identify a MAA, both the absorption peaks and the R_f values (HPLC) need to be determined using a standard procedure. MAAs are water-soluble substances characterized by a cyclohexenone or cyclohexenimine chromophore linked to the nitrogen substituent of an amino acid or its imino alcohol. Their absorption maxima range from 310 to 360 nm, and they have an average molecular weight of around 300[34].

5. Macroalgae

While phytoplankton is free to move in the water column[9], most macroalgae are sessile and restricted to their growth site with its specific environmental properties[35].

Macroalgae have to thrive with the prevailing irradiance regime of solar radiation governed by the daily and seasonal changes as well as the tidal rhythm. Most macroalgae are adapted to lower irradiances than terrestrial plants so that they face a serious light stress when exposed to high irradiances during low tide and high solar angles[9], but they have developed strategies to adapt their photosynthetic apparatus to the changing light conditions and to protect themselves against excessive radiation. Macroalgae use the same mechanism of photoinhibition as higher plants to decrease their photosynthetic activity during high light exposure by reversibly reducing the photosynthetic electron transport chain[36]. The excess energy is thermally dissipated[37]. However, even higher solar irradiances may cause photodamage which is not as readily reversible as photoinhibition.

Different algal species have a clearly different behaviour upon irradiation[38-42]; especially their ability to cope with enhanced UV radiation varies widely among species[43]. A number of ecologically important algae have been studied during the last few years, found in the Atlantic, Pacific, Mediterranean, North Sea and Baltic Sea.

The phenomenon of photoinhibition can be studied by oxygen exchange measurements. Oxygen sensors have been used in field incubators as well as in microsensor studies. Oxygen gradient analysis has also been carried out in cyanobacterial mats indicating that the photosynthetic rates were negligible near the surface and maximal deeper in the mat[44].

Alternatively, the photosynthetic efficiency can be determined by PAM (Pulse Amplitude Modulated) fluorescence measurements developed by Schreiber et al.[45]. PAM fluorescence determines the transient changes of chlorophyll fluorescence and can be used to calculate the photochemical and non-photochemical quenching of the photosynthetic apparatus[46]. This method reveals the regulatory processes and the physiological status of the photosynthetic apparatus *in vivo*[37,47]. Often there is a strong correlation between photosystem II chlorophyll fluorescence and oxygen evolution[48] but it may not be a good indicator for growth and biomass production[49].

There is a distinction between two different types of photoinhibition: dynamic photoinhibition and chronic photoinhibition[50]. Dynamic photoinhibition is associated with a decrease of the effective quantum yield and with an increase in non-photochemical quenching (qN), related to the conversion of violaxanthin to zeaxanthin which is a potent quencher for excess excitation energy both in algae and in higher plants[51]. Simultaneously, an increment of pH-dependent processes or membrane energization (qE) is found, which lowers the efficiency of PS II[50].

Shade plants often show chronic photoinhibition which is characterized by slow reversibility if at all. It is found in algae which are exposed to excessive irradiation. The extent of chronic photoinhibition can be calculated by 1-qP. It has been known that photosynthetic active radiation (PAR) induces photoinhibition, but recent measurements indicate a strong role for solar ultraviolet radiation both in terrestrial and aquatic plants. Especially the short wavelength range of UV-B affects photosynthesis in several species of marine macroalgae even though its relative energy contribution is disproportionally smaller than that of PAR in the solar spectrum, and many inhibitory effects of UV-B on photosynthesis and chlorophyll fluorescence of several species of marine benthic algae and phytoplankton have been documented[52-53].

Recent miniaturization and computer-control facilitate measurements of both fluorescence and oxygen exchange in the field or even in the water column[45,54-55]. Measurements at the growth site are advantageous, since the transport of the specimen to the laboratory may cause artefacts due to thermal stress and changes in irradiance and salinity. Recently an underwater PAM instrument has been developed (Diving PAM underwater fluorometer, Walz, Effeltrich, Germany) which allows to measure the quantum yield of fluorescence under water on site.

Macroalgae have a distinct and fixed pattern of vertical distribution in their habitat[56] ranging from the supralittoral (above high water mark) through the eulittoral (intertidal zone) to the sublittoral zone where they are never exposed to air. One important factor controlling the abundance and species distribution of algae is solar exposure which ranges from the bright solar radiation at the surface and in rock pools to shaded habitats in crevices or under overhanging rocks where light exposure is strongly attenuated.

Photosynthetic production strongly depends on depth and varies with the species: surface-adapted macroalgae such as the brown *Cystoseira, Padina* and *Fucus* or green *Ulva* and *Enteromorpha* have maximal oxygen production close to the surface[39,53] while deeper water algae thrive best deeper in the water column (the green algae *Cladophora, Caulerpa*, most red algae)[40-42]. This is even more obvious in algae adapted to shaded habitats in crevices, under overhanging rocks or in the understorey of kelps[39,57]. In most species studied, respiration is inhibited to a far smaller degree than photosynthesis.

Almost all macroalgae show a pronounced photoinhibition after various times of exposure to unfiltered solar radiation at the surface at least at high zenith angles[39,58-60]. Even algae growing in rock pools, where they are exposed to extreme solar irradiances, show photoinhibition during local noon[59-62]. Algae adapted to deep water or shaded conditions are inhibited even faster and stronger when exposed to direct solar radiation[40]. The next question is: how long does it take an organism to recover from photoinhibition after exposure? The recovery time depends on the species and its depth adaptation. Surface-adapted species recover much faster from photoinhibition than low light-adapted algae. Also the time required for recovery strongly depends on the degree of inhibition. Algae adapted to shaded conditions or deep water may not recover at all or only partially after massive inhibition indicating that they experienced chronic rather than dynamic photoinhibition.

From an ecological standpoint it is even more relevant to follow the photosynthetic quantum yield in specimens exposed at their natural growth site over a whole day[39-42,48]. Most of these studies show a strong decrease during local noon and high values in the morning and evening especially when the tidal changes are not very pronounced. In habitats with high tidal changes as in the Atlantic or Pacific the photosynthetic quantum yield strongly depends on the depth of the water column above the organisms. From this it is obvious that we have to reconsider our conception of the pattern of maximal photosynthesis in supralittoral and subtidal macroalgae. These algae show optimal photosynthetic quantum yield either early in the morning and evening hours during low tide or when high tides coincide with high solar angles, resulting in a complicated pattern of the photosynthetic yield. Future research will have to show if this behaviour is endogenously regulated.

The regulatory mechanisms to relieve light stress include modulation of antenna size, thermal dissipation of excess excitation energy, involvement of antioxidant enzymes and repair of photooxidative damage[63]. The photosynthetic apparatus is protected by decreasing the electron transport chain (photoinhibition) and involvement of the violaxanthin cycle[50]: at moderate irradiances zeaxanthin is converted to antheraxanthin and finally to violaxanthin, while the reverse sequence is found during exposure to excessive irradiation. Zeaxanthin is located in the vicinity of the excited chlorophyll to facilitate thermal dissipation of excess excitation energy before the chlorophyll can undergo intersystem crossing to the triplet state from which the energy may be transferred to oxygen to produce the highly photooxidative singlet oxygen[63].

The photoprotective role of the xanthophyll cycle has been investigated mostly in microalgae[64] and to less extent in macroalgae, e.g. the green algae *Ulva rotundata*[65] and *Ulva lactuca*[66] and the brown algae *Dictyota dichotoma*[67] and *Lobophora variegata*[38]. Red algae do not have the xanthophyll cycle.

Passive protection against excessive visible or UV radiation relies on the production of screening pigments such as carotenoids or UV-absorbing mycosporine-like amino acids (MAAs). MAAs have been found in green, red and brown algae from tropical, temperate and polar regions[68]. Wood[69] described that apical pieces of *Euchema striatum* produced high concentrations of UV absorbing pigments under UV exposure. The absorption maxima of MAA range throughout the UV-A and UV-B regions. Macroalgae may contain several MAAs with different absorbance maxima. The concentration of MAAs is correlated with depth distribution and UV exposure[70]. In tropical algae, enhanced levels of carotenoids and UV-absorbing compounds were found in tissues from the canopy compared to tissues from understory sites[71]. MAAs may also have other biological functions including osmotic and antifreeze mechanisms[68].

6. Ecosystems

The Antarctic aquatic ecosystem

The Southern Ocean is characterized by large scale spatial and temporal variability in productivity[72]. Therefore it is difficult to isolate UV-B specific effects from other environmental effects[23]. At high latitudes, variability in solar elevation, cloud cover, deep vertical mixing and the cover of ice and snow significantly complicate the analysis of UV-B effects on phytoplankton. Recent estimates of the effect of 50 % ozone reduction on total water column productivity agree with earlier findings showing reductions between <5 % and 6 %[6,73].

There is convincing evidence of UV-B damage to phytoplankton, but long-term predictions are difficult because of acclimation and adaptation phenomena[74,75], as well as other factors[23]. Several models have been developed[23] to estimate ecosystem productivity. Vertical mixing is a major factor modulating UV-B effects in phytoplankton which needs to be taken into account in these models[23,73]. Container studies clearly showed that photosynthesis of Antarctic phytoplankton is inhibited by ambient UV[6,76]. The difficulty comes in the generalization of these experimental results to Antarctic waters where mixing significantly modulates the exposure of phytoplankton to UV-B. Neale and coworkers[23] found that near-surface UV strongly inhibits photosynthesis under all modelled conditions and that inhibition of photosynthesis can be altered by vertical mixing. A sudden 50 % reduction in stratospheric ozone can decrease daily integrated water column photosynthesis by as much as 8.5 %. This result confirms the finding of Smith and coworkers[6] who specifically measured in the marginal ice zone (MIZ), where melt water provides stability and minimizes vertical mixing. The conclusions of numerous studies are that Antarctic ozone depletion can inhibit primary productivity, but that natural variability in exposure of phytoplankton to UV, due to vertical mixing and cloud cover, has a major role in modifying the impact on net photosynthesis. But it is clear that solar UV is a significant environmental stressor, and its effects are enhanced by ozone depletion.

The Arctic aquatic ecosystem

The Arctic differs in many respects from the Antarctic[77] being a nearly closed water mass with limited water exchange with the Atlantic and Pacific oceans. It is boarded by 25% of the global continental shelf and receives about 10% of the world river discharge which causes a pronounced stratification and is responsible for high concentrations of particulate and dissolved organic carbon (POC and DOC), which strongly attenuate the penetration of solar UV into the water column[77]. These substances are photochemically attacked by UV-B which increases the palatability and thus enhances the uptake by bacteria and finally increases the transparency of the water column. Macroalgae play a more pronounced role in the Arctic than in the Antarctic. The Arctic aquatic ecosystem is one of the most productive ecosystems on earth and is a source of fish and crustaceans for human consumption. Productivity in the Arctic ocean is higher and more heterogeneous than in the Antarctic ocean[78]. Because of the shallow water and the stable stratification of the water layer, the phytoplankton may be exposed to relatively high levels of solar UV-B. In addition, many economically important fish spawn in shallow waters, and many of the eggs and early larval stages are found at or near the surface, so that harvests of fish for human consumption may be affected. However, currently we cannot accurately estimate if ozone-related impacts will influence fishes and other important marine crops.

The Arctic waters are nitrogen and phosphorus limited. This problem is aggravated by the fact that nitrogen and phosphorus uptake is UV-B sensitive[79] which also has an effect on the biogeochemical cycles.

7. Potential Consequences

The most direct effect of UV radiation on productivity is loss of biomass which relays through the food web and affects food sources for human consumption. As different species are differently affected by solar UV, changes in species composition is considered. In general smaller organisms are more prone to UV damage than larger ones[21]. The third consequence of a significant reduction in aquatic ecosystem productivity is reduced uptake capacity for atmospheric carbon dioxide, resulting in the potential augmentation of global warming. While there is significant evidence that increased UV-B exposure is harmful to aquatic organisms, damage on the ecosystem level is still uncertain.

References

1. Siegenthaler, U. and Sarmiento, J.L. (1993) Atmospheric carbon dioxide and the ocean. *Nature*, 365: 119-125.
2. Niell, F.X., Fernández, C., Figueroa, F.L., Figueiras, F.G., Fuentes, J.M., Pérez-Lloréns, J.L., Garcia-Sánchez, M.J., Hernández, I., Fernández, J.A., Espejo, M., Buela, J., García-Jiménez, M.C., Clavero, V. and Jiménez, D. (1996) Spanish Atlantic coasts. In *Marine Benthic Vegetation Ecological Studies* (W. Schramm and P. Nienhuis, eds.), Springer-Verlag, Berlin, pp. 265-281.
3. Madronich, S., McKenzie, R.L., Björn, L.O. and Caldwell, M.M. (1998) Changes in biologically active ultraviolet radiation reaching the Earth's surface. *J. Photochem. Photobiol. B: Biol.*, 46: 5-19.
4. Stolarski, R. (1997) A bad winter for Arctic ozone. *Nature*, 389: 788-789.
5. Kerr, J.B. and McElroy, C.T. (1993) Evidence for large upward trends of ultraviolet-B radiation linked to ozone depletion. *Science*, 262: 1032-1034.
6. Smith, R.C., Prezelin, B.B., Baker, K.S., Bidigare, R.R., Boucher, N.P., Coley, T., Karentz, D., MacIntyre, S., Matlick, H.A., Menzies, D., Ondrusek, M., Wan, Z. and Waters, K.J. (1992) Ozone depletion: ultraviolet radiation and phytoplankton biology in Antarctic waters. *Science*, 255: 952-959.
7. Coohill, T.P., Häder, D.-P. and Mitchell, D.L. (1996) Environmental ultraviolet photobiology: Introduction. *Photochem. Photobiol.*, 64: 401-402.
8. Häder, D.-P., Kumar, H.D., Smith, R.C. and Worrest, R.C. (1998) Effects on aquatic ecosystems. In *UNEP Environmental Effects Panel Report*, UNEP, Nairobi, pp. 86-112.
9. Häder, D.-P., Worrest, R.C., Kumar, H.D. and Smith, R.C. (1995) Effects of increased solar ultraviolet radiation on aquatic ecosystems. *AMBIO*, 24: 174-180.
10. Häder, D.-P. and Worrest, R.C. (1997) Consequences of the effects of increased solar ultraviolet radiation on aquatic ecosystems. In *The Effects of Ozone Depletion on Aquatic Ecosystems* (D.-P. Häder, ed.), Acad. Press, R.G. Landes Company, Austin, pp. 11-30.
11. Takahashi, T., Feely, R.A., Weiss, R.F., Wanninkhof, R.H., Chipman, D.W., Sutherland, S.C. and Takahashi, T. (1997) Global air-sea flux of CO_2: An estimate based on measurements of sea-air pCO_2 difference. *Proc. Natl. Acad. Sci.*, 94: 8282-8299.
12. Thomson, D.J. (1997) Dependence of global temperatures on atmospheric CO_2 and solar irradiance. *Proc. Natl. Acad. Sci. USA*, 94: 8370-8377.
13. Sinha, R.P. and Häder, D.-P. (1997) Impacts of UV-B irradiation on rice-field cyanobacteria. In *The Effects of Ozone Depletion on Aquatic Ecosystems* (D.-P. Häder, ed.), Acad. Press, R.G. Landes Company, Austin, pp. 189-198.
14. Banerjee, M. and Häder, D.-P. (1996) Effects of UV radiation on the rice field cyanobacterium, *Aulosira fertilissima*. *Environ. Experim. Bot.*, 36: 281-291.
15. Donkor, V.A. and Häder, D.-P. (1997) Ultraviolet radiation effects on pigmentation in the cyanobacterium *Phormidium uncinatum*. *Acta Protozool.*, 36: 49-55.
16. Sinha, R.P., Singh, N., Kumar, A., Kumar, H.D., Häder, M. and Häder, D.-P. (1996) Effects of UV irradiation on certain physiological and biochemical processes in cyanobacteria. *J. Photochem. Photobiol. B: Biol.*, 32: 107-113.
17. Araoz, R. and Häder, D.-P. (1997) Ultraviolet radiation induces both degradation and synthesis of phycobilisomes in *Nostoc* sp.: a spectroscopic and biochemical approach. *FEMS Microbiol. Ecol.*, 23: 301-313.
18. Kumar, A., Sinha, R.P. and Häder, D.-P. (1996) Effect of UV-B on enzymes of nitrogen metabolism in the cyanobacterium *Nostoc calcicola*. *J. Plant Physiol.*, 148: 86-91.
19. Vincent, W.F. and Roy, S. (1993) Solar ultraviolet-B radiation and aquatic primary production: damage, protection, and recovery. *Environ. Rev.*, **1**: 1-12.
20. Ignatiades, L. (1990) Photosynthetic capacity of the surface microlayer during the mixing period. *J. Plankton Res.*, 12: 851-860.

21. Herndl, G.J. (1997) Role of ultraviolet radiation on bacterioplankton activity. In *The Effects of Ozone Depletion on Aquatic Ecosystems* (D.-P. Häder, ed.), Acad. Press, R.G. Landes Company, Austin, pp. 143-154.
22. Häder, D.-P., Kumar, H.D., Smith, R.C. and Worrest, R.C. (1998) Effects on aquatic ecosystems. *J. Photochem. Photobiol. B: Biol.*, 46: 53-68.
23. Neale, P.J., Cullen, J.J. and Davis, R.F. (1998) Inhibition of marine photosynthesis by ultraviolet radiation: Variable sensitivity of phytoplankton in the Weddell-Scotia Sea during austral spring. *Limnol. Oceanogr.*, 43: 433-488.
24. Cullen, J.J. and Neale, P.J. (1994) Ultraviolet radiation, ozone depletion, and marine photosynthesis. *Photosynth. Res.*, 39: 303-320.
25. Figueroa, F.L., Jimenez, C., Lubian, L.M., Montero, O., Lebert, M. and Häder, D.-P. (1997) Effects of high irradiance and temperature on photosynthesis and photoinhibition in *Nannochloropsis gaditana* Lubian (Eustigmatophyceae). *J. Plant Physiol.*, 151: 6-15.
26. Gieskes, W.W.C. and Buma, A.G.J. (1997) UV damage to plant life in a photobiologically dynamic environment: the case of marine phytoplankton. *Plant Ecol.*, 128: 16-25.
27. Herrmann, H., Häder, D.-P. and Ghetti, F. (1997) Inhibition of photosynthesis by solar radiation in *Dunaliella salina*: relative efficiencies of UV-B, UV-A and PAR. *Plant, Cell Environ.*, 20: 359-365.
28. Häder, D.-P. (1997) Effects of UV radiation on phytoplankton. In *Adv. Microbial Ecol.* (J.G. Jones, ed.), Plenum Press, New York, pp. 1-26.
29. Döhler, G. and Hagmeier, E. (1997) UV effects on pigments and assimilation of ^{15}N-ammonium and ^{15}N-nitrate by natural marine phytoplankton of the North Sea. *Bot. Acta*, 110: 481-488.
30. Buma, A.G.J., Engelen, A.H. and Gieskes, W.W.C. (1997) Wavelength-dependent induction of thymine dimers and growth rate reduction in the marine diatom *Cyclotella* sp. exposed to ultraviolet radiation. *Mar. Ecol. Prog. Ser.*, 153: 91-97.
31. Klisch, M. and Häder, D.-P. (2000) Mycosporine-like amino acids in the marine dinoflagellate *Gyrodinium dorsum*: induction by ultraviolet irradiation. *J. Photochem. Photobiol. B: Biol.*, 55: 178-182.
32. Klisch, M. and Häder, D.-P. (2001) Wavelength dependence of mycosporine-like amino acid synthesis in *Gyrodinium dorsum*. *J. Photochem. Photobiol. B: Biol.*, 66: 60-66.
33. Gröniger, A., Sinha, R.P., Klisch, M. and Häder, D.-P. (2000) Photoprotective compounds in cyanobacteria, phytoplankton and macroalgae - a database. *J. Photochem. Photobiol. B: Biol.*, 58: 115-122.
34. Dunlap, W.C. and Shick, J.M. (1998) Ultraviolet radiation-absorbing mycosporine-like amino acids in coral reef organisms: a biochemical and environmental perspective. *J. Phycol.*, 34: 418-430.
35. Lüning, K. (1985) *Seaweeds. Their environment, biogeography and ecophysiology*, Wiley, New York.
36. Trebst, A. (1991) A contact site between the two reaction center polypeptides of photosystem II is involved in photoinhibition. *Z. Naturforsch.*, 46: 557-562.
37. Krause, G.H. and Weis, E. (1991) Chlorophyll fluorescence and photosynthesis: the basics. *Ann. Rev. Plant Physiol.*, 42: 313-349.
38. Franklin, L.A., Seaton, G.G.R., Lovelock, C.E. and Larkum, A.W.D. (1996) Photoinhibition of photosynthesis on a coral reef. *Plant Cell Environ.*, 19: 825-836.
39. Häder, D.-P., Herrmann, H. and Santas, R. (1996) Effects of solar radiation and solar radiation deprived of UV-B and total UV on photosynthetic oxygen production and pulse amplitude modulated fluorescence in the brown alga *Padina pavonia*. *FEMS Microbiol. Ecol.*, 19: 53-61.
40. Häder, D.-P., Herrmann, H., Schäfer, J. and Santas, R. (1996) Photosynthetic fluorescence induction and oxygen production in corallinacean algae measured on site. *Bot. Acta*, 109: 285-291.
41. Häder, D.-P., Lebert, M., Mercado, J., Aguilera, J., Salles, S., Flores-Moya, A., Jimenez, C. and Figueroa, F.L. (1996) Photosynthetic oxygen production and PAM fluorescence in the brown alga *Padina pavonica* (Linnaeus) Lamouroux measured in the field under solar radiation. *Mar. Biol.*, 127: 61-66.
42. Häder, D.-P., Porst, M., Herrmann, H., Schäfer, J. and Santas, R. (1996) Photoinhibition in the Mediterranean green alga *Halimeda tuna* Ellis et Sol measured in situ. *Photochem. Photobiol.*, 64: 428-434.
43. Dring, M.J., Wagner, A., Boeskop, J. and Lüning, K. (1996) Sensitivity of intertidal and subtidal red algae to UV-A and UV-B radiation, as monitored by chlorophyll fluorescence measurements: influence of collection, depth and season and length of irradiation. *Europ. J. Phycol.*, 31: 293-302.

44. Vicent, W.F., Castenholz, R.W., Dowes, M.T. and Howard-Williams, C. (1993) Antarctic cyanobacteria: light, nutrients, and photosynthesis in the microbial mat environment. *J. Phycol.*, 29: 745-755.
45. Schreiber, U., Schliwa, U. and Bilger, W. (1986) Continuous recording of photochemical and non-photochemical chlorophyll fluorescence quenching with a new type of modulation fluorometer. *Photosynth. Res.*, 10: 51-62.
46. Büchel, C. and Wilhelm, C. (1993) In vivo analysis of slow chlorophyll fluorescence induction kinetics in algae: progress, problems and perspectives. *Photochem. Photobiol.*, 58: 137-148.
47. Schreiber, U. and Bilger, W. (1993) Progress in chlorophyll fluorescence research: major developments during the past years in retrospect. In *Progr. Bot.* (U. Lüttge and H. Ziegler, eds.), Springer Verlag, Berlin, pp. 151-153.
48. Hanelt, D. (1992) Photoinhibition of photosynthesis in marine macrophytes of the South China Sea. *Mar. Ecol. Prog. Ser.*, 82: 199-206.
49. Falkowski, P.G., Greene, R. and Kolber, Z. (1994) Light utilization and photoinhibition of photosynthesis in marine phytoplankton. In *Photoinhibition of Photosynthesis from Molecular Mechanisms to the Field* (N.R. Baker and J.R. Bowyer, eds.), Bios. Scientific Publishers, Oxford, pp. 407-432.
50. Osmond, C.B. (1994) What is photoinhibition? Some insights from comparisons of shade and sun plants. In *Photoinhibition of Photosynthesis from Molecular Mechanisms to the Field* (N.R. Baker and J.R. Bowyer, eds.), Bios. Scientific Publishers, Oxford, pp. 1-24.
51. Demmig-Adams, B. and Adams, W.W.I. (1992) Photoprotection and other responses of plants to high light stress. *Ann. Rev. Plant Physiol.*, 43: 599-626.
52. Larkum, A.W.D. and Wood, W.F. (1993) The effect of UV-B radiation on photosynthesis and respiration of phytoplankton, benthic macroalgae and seagrasses. *Photosynth. Res.*, 36: 17-23.
53. Herrmann, H., Ghetti, F., Scheuerlein, R. and Häder, D.-P. (1995) Photosynthetic oxygen and fluorescence measurements in *Ulva laetevirens* affected by solar irradiation. *J. Plant Physiol.*, 145: 221-227.
54. Häder, D.-P. and Schäfer, J. (1994) In-situ measurement of photosynthetic oxygen production in the water column. *Environmental Monitoring Assessment*, 32: 259-268.
55. Häder, D.-P. and Schäfer, J. (1994) Photosynthetic oxygen production in macroalgae and phytoplankton under solar irradiation. *J. Plant Physiol.*, 144: 293-299.
56. Lüning, K. (1985) *Meeresbotanik: Verbreitung, Ökophysiologie und Nutzung der marinen Makroalgen.* Thieme Verlag, Stuttgart, New York
57. Häder, D.-P., Porst, M. and Santas, R. (1998) Photoinhibition by solar radiation in the Mediterranean alga *Peyssonnelia squamata* measured on site. *Plant Ecology*, 139: 167-175.
58. Häder, D.-P., Porst, M. and Lebert, M. (2000) On site photosynthetic performance of Atlantic green algae. *J. Photochem. Photobiol. B: Biol.*, 57: 159-168.
59. Häder, D.-P., Porst, M. and Lebert, M. (2001) Photoinhibition in common Atlantic macroalgae measured on site in Gran Canaria. *Helgol. Mar. Res.*, 55: 67-76.
60. Häder, D.-P., Porst, M. and Lebert, M. (2001) Photosynthetic performance of the Atlantic brown macroalgae, *Cystoseira abies-marina*, *Dictyota dichotoma* and *Sargassum vulgare*, measured in Gran Canaria on site. *Env. Exp. Bot.*, 45: 21-32.
61. Häder, D.-P., Lebert, M. and Helbling, E.W. (2000) Photosynthetic performance of the chlorophyta *Ulva rigida* measured in Patagonia on site. *Recent Res. Devel. Photochem. Photobiol.*, 4: 259-269.
62. Häder, D.-P., Lebert, M. and Helbling, E.W. (2001) Photosynthetic performance of marine macroalgae measured in Patagonia on site. *Trends Photochem. Photobiol.*, 8: 145-151.
63. Niyogi, K.K. (1999) Photoprotection revisited: genetics and molecular approaches. *Ann. Rev. Plant Physiol.*, 50: 333-359.
64. Schubert, H., Kroon, B.M.A. and Matthijs, H.C.P. (1994) In vivo manipulation of the xanthophyll cycle and the role of zeaxynthin in the protection against photodamage in the green alga *Chlorella pyrenoidosa*. *J. Biol. Chem.*, 269: 7267-7272.
65. Franklin, L.A. (1994) The effects of temperature acclimation on the photoinhibitory responses of *Ulva rotundata* Blid.. *Planta*, 192: 324-331.
66. Grevby, C. (1996) *Organisation of the light harvesting complex in fucoxanthin containing algae*, PhD Thesis Göteborg University Sweden.
67. Uhrmacher, S., Hanelt, D. and Nultsch, W. (1995) Zeaxanthin content and the degree of photoinhibition are linearly correlated in the brown alga *Dictyota dichotoma*. *Mar. Biol.*, 123: 159-165.
68. Karentz, D. (1994) Ultraviolet tolerance mechanisms in Antarctic marine organisms. *Antarctic Research*, 62: 93-110.
69. Wood, W.F. (1987) Effect of solar ultraviolet radiation on the kelp *Eklonia radiata*. *Marine Biology*, 96: 143-150.

70. Karentz, D., McEuen, F.S., Land, M.C. and Dunlap, W.C. (1991) A survey of mycosporine-like amino acid compounds in Antarctic marine organisms: potential protection from ultraviolet exposure. *Marine Biol.*, 108: 157-166.
71. Beach, K.S. and Smith, C.M. (1996) Ecophysiology of tropical rhodophytes. I. Microscale acclimation in pigmentation. *J. Phycol.*, 32: 701-710.
72. Smith, R.C., Baker, K.S., Byers, M.L. and Stammerjohn, S.E. (1998) Primary productivity of the Palmer long-term ecological research area and the southern ocean. *J. Mar. Syst.*, 529:
73. Neale, P.J., Davis, R.F. and Cullen, J.J. (1998) Interactive effects of ozone depletion and vertical mixing on photosynthesis of Antarctic phytoplankton. *Nature*, 392: 585-589.
74. Villafañe, V.E., Helbling, E.W., Holm-Hansen, O. and Chalker, B.E. (1995) Acclimatization of Antarctic natural phytoplankton assemblages when exposed to solar ultraviolet radiation. *J. Plankt. Res.*, 17: 2295-2306.
75. Helbling, E.W., Chalker, B.E., Dunlap, W.C., Holm-Hansen, O. and Villafañe, V.E. (1996) Photoacclimation of Antarctic marine diatoms to solar ultraviolet radiation. *J. Exp. Mar. Biol. Ecol.*, 204: 85-101.
76. Helbling, E.W., Villafañe, V.E. and Holm-Hansen, O. (1994) Effects of ultraviolet radiation on Antarctic marine phytoplankton photosynthesis with particular attention to the influence of mixing. *Antarctic Research*, 62: 207-227.
77. Wängberg, S.-Å., Selmer, J.-S., Ekelund, N.G.A. and Gustavson, K. (1996) UV-B effects on Nordic marine ecosystems. *TemaNord, Nordic Council of Ministers*, 515: 1-45.
78. Springer, A.M. and McRoy, C.P. (1993) The paradox of pelagic food webs in the northern Bering Sea. III. Patterns of primary production. *Continental Shelf Research*, 13: 575-599.
79. Döhler, G. (1992) Impact of UV-B radiation (290-320 nm) on uptake of ^{15}N-ammonia and ^{15}N-nitrate by phytoplankton of the Wadden Sea. *Mar. Biol.*, 112: 485-489.

UNDERWATER RADIATION MEASUREMENTS: CONSEQUENCES OF AN INCREASED UV-B RADIATION

BERIT KJELDSTAD
Department of Physics, Norwegian University of Science and Technology, N-7491 Trondheim, Norway

1. Introduction

Almost all outdoor living organisms are exposed to ultraviolet radiation (UV). Ecosystems experience from morning to evening a strong variation of UV intensity due to diurnal changes in solar elevation, which depends on latitude and time of the year. These changes are the most dominant factor causing short term variation in UV radiation on Earth. The amount of UV radiation reaching the surface at any time also depends on atmospheric factors. Components such as ozone and particulate matter in the form of clouds and aerosols absorb solar radiation and cause both short term and long term variability. In the UV range scattering processes are also important and reflection from the ground influences radiation levels measured at the surface. All changes in UV caused by atmospheric factors, such as, for instance, ozone depletion, increased amount of aerosols or increased cloud cover, influence of course the amount of UV in water. To assess the amount of radiation into the water one has to consider radiative transfer in the water and particular components which strongly absorb ultraviolet radiation, such as humic substance, also called gelbstoff or CDOM (Chromophoric Dissolved Organic Matter). Moreover at the air-water boundary the index of refraction changes and must be taken into account when penetration of UV into water is investigated.

In several studies it has been shown that there is a great variability in UV penetration both in fresh-water and marine environments, due to different water quality. The span in diffuse attenuation coefficients retrieved from the clearest ocean water to small lakes can be three order of magnitude[1]. A recent overview of attenuation of ultraviolet irradiance in North European coastal waters indicated a variability of one order of magnitude in open coastal waters[2]. Geographical variability of UV penetration has been shown to be essential, even within the same bay or fjord system[3-4].

This study focuses on seasonal variability of UV penetration at one specific site within Trondheim Fjord, a fjord system in Norway. The total period investigated is four years. The parameters taken into consideration to affect UV penetration are total organic carbon (TOC), chlorophyll (Chla) and salinity.

F. Ghetti et al. (eds.), Environmental UV Radiation: Impact on Ecosystems and Human Health and Predictive Models, 193–201.

2. Methods

UV can be measured underwater in the same way as done in air. The most commonly used instrumentation is broadband meters, which are robust, without any moveable parts and easy to deploy into the water. Spectroradiometers too have been deployed, also used in combination with fibres and an optical head. Both downwards and upwards radiation can be measured and the instruments are deployed at different depth in order to measure attenuation as a function of depth. For measuring profiles a surface unit is strongly recommended to correct for rapid changes in radiation reaching the surface as the profile is carried out.

The instrumentation used in this work was a Biospherical PUV500/510 radiometer with 4 channels in UV (305 nm, 320 nm, 340 nm and 380 nm) and one channel measuring Photosynthetic Active Radiation (PAR) in addition to temperature and depth sensors (Figure 1)

Figure 1. Underwater radiometer measuring downward radiation at 4 UV bands (305 nm, 320 nm, 340 nm and 380 nm) and PAR (Biospherical PUV500) underwater unit (black right) and air unit (Biospherical PUV 510) (white left).

Calibration of underwater radiometers are done in the same way as for instrument measuring in air[5]. Examples of profiles for 4 UV channels are shown in Figure 2 . The difference in UV irradiance at the ground for different wavelength can be observed from the surface values. The absolute irradiance level depended on solar elevation and other atmospheric conditions. Downward irradiance decreased exponentially with depth (note the logarithmic irradiance scale).

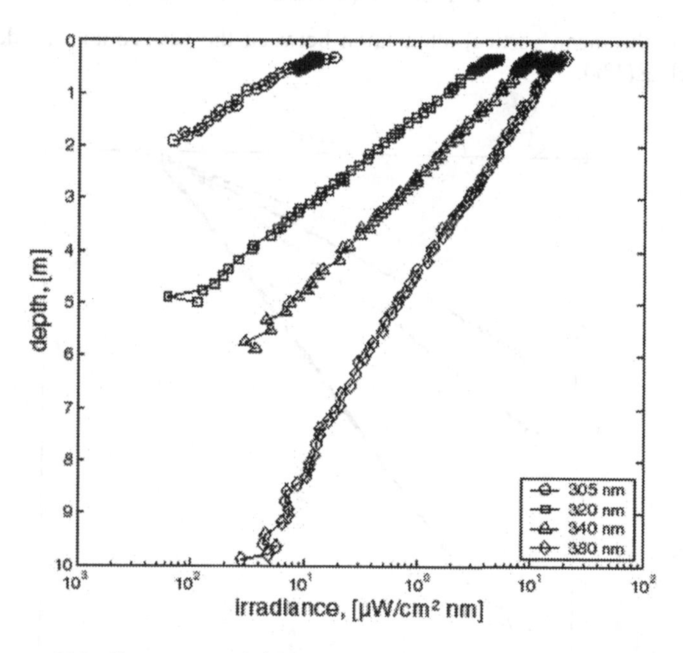

Figure 2. Downward irradiance measured simultaneously at 305 nm, 320 nm, 340 nm and 380 nm as a function of depth. Measured in Trondheim Fjord, Norway (63° 29' N, 10° 18' E)

UV attenuation can be expressed by different quantities. The profiles in Figure 2 show directly how the irradiance decreased with depth for different wavelengths. The attenuation can be described by the slope of the profiles according to equation

$$k_d = -\ln(E_d/E_s)/z \qquad (1)$$

where k_d is the diffuse downwelling attenuation coefficient [m^{-1}], E_d is downwelling irradiance [$\mu W\ cm^{-2}\ nm^{-1}$] measured at depth z, E_s is downwelling irradiance measured at z = 0, approximately 0.2 m below the surface. K_d is calculated according to Equation 1 after a linear regression of the data sampled between two depths, given relatively to the surface measurements. Presenting the underwater profile measurements relative to the surface unit will smoothen the profiles obtained during rapidly changing cloud cover. Diffuse attenuation coefficients can be retrieved from relative profiles with increased accuracy (Figure 3). In a water column with high stratification k_d will change with depth. k_d for ultraviolet radiation is less dependent on solar elevation than k_d in the PAR region, because most of the ultraviolet radiation from the sky is diffuse[2].

Very often attenuation of radiation in a water column is presented as the depth to which 10% (Z(10%)) or 1% (Z(1%)) of the surface irradiance is penetrating, rather than giving the k_d coefficients. According to equation 1 there is a linear dependence between k_d and Z(10%) or Z(1%)

$$Z(10\%) = -\ln(0.10)/k_d = 2.30/k_d \qquad (2)$$

$$Z(1\%) = -\ln(0.01)/k_d = 4.60/k_d \qquad (3)$$

where k_d is the mean vertical attenuation between the surface and the depth Z(10%), respectively Z(1%).

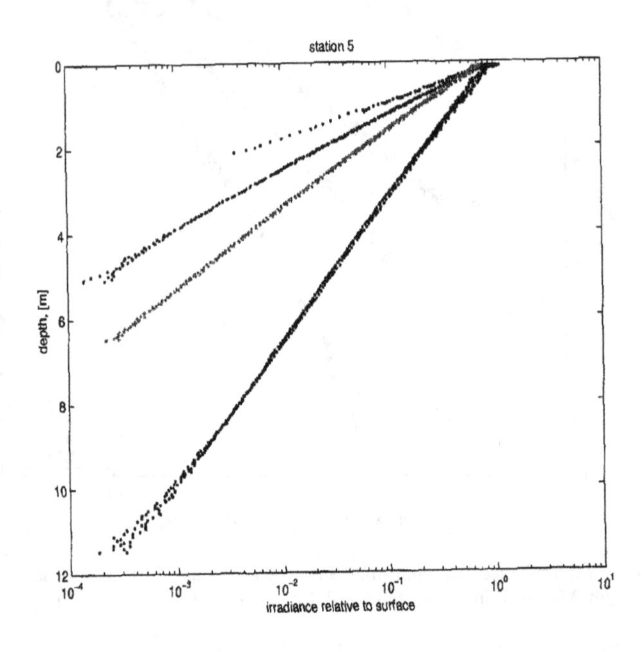

Figure 3. Downward irradiance at 305 nm (upper curve), 320 nm, 340 nm and 380 nm (lower curve) relative to surface irradiance as a function of depth. Measured in Samnanger Fjord, Norway (60° 13' N, 05° 37' E).

For the example given in Figure 3, k_d varied from 2.6 (305 nm) to 0.7 (380 nm) and the corresponding 1% depth values are given in Table 1.

Table 1. Diffuse attenuation coefficient and corresponding 1% depth for data shown in Figure 3

wavelength [nm]	**k_d [m^{-1}]**	**Z(1%) [m]**
305	2.6	1.8
320	1.9	2.4
340	1.4	3.3
380	0.7	6.6

An increase in UV radiation reaching the surface due to for instance ozone depletion will be noticed at all depths in the water. The effect of increased UV will affect the doses measured at certain depths. The profiles given in Figure 1 will be moved parallel to the right (higher irradiances). The Z(1%) will remain the same, because attenuation depends on properties within the water column.

One of the most important factors for UV attenuation in water is the content of CDOM (coloured dissolved organic matter). Bricaud et al.[6] have shown that the absorption coefficient (a) for natural filtered waters in the UV-visible range varies according to the relationship

$$a(\lambda) = a_o^{\exp[-S(\lambda-\lambda o)]} \tag{4}$$

where S is a parameter describing the slope of the exponential curve and a_o is the absorption at a reference wavelength λ_o. Equation 4 explains spectral absorption properties in the water, thus the value of S can be used as an indicator of different absorption properties. In many freshwater systems S lies within the range 0.010 – 0.023 nm^{-1}. The slope values have been used to explain spectral absorption properties within a water system, but care should be taken because S depends on the investigated wavelength interval.

Laurion et al.[7] hypothesised that the diffuse attenuation coefficient (k_d), including both absorption and scattering properties, would follow the same exponential function of wavelength in the UV spectral range as in Equation 4

$$k_d(\lambda)=k_{d440}^{\exp[-S(\lambda-440)]} \tag{5}$$

where S would be a constant and where k_{d440}, the attenuation coefficient at a reference wavelength of 440 nm, would be a function of CDOM variables.

3. Results

In studies performed in Trondheim Fjord (Norway) downward irradiance was measured at the same station (63° 29' N, 10° 18' E) throughout the year. The station was chosen to be representative for the Fjord system with a high circulation rate and directly dominated by input from rivers around (Figure 4, map).

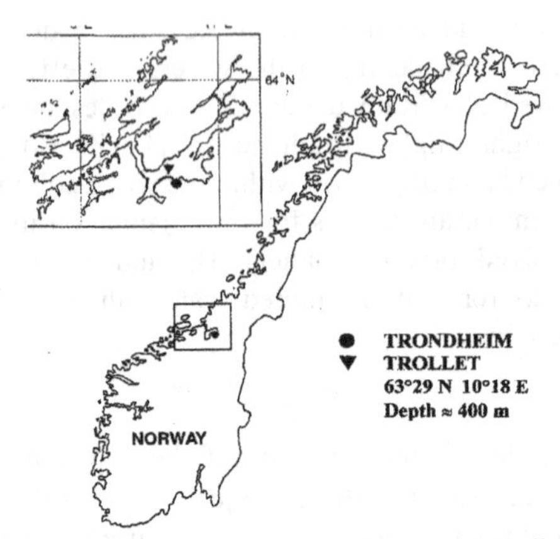

Figure 4. Trondheim Fjord. City of Trondheim (circle) and the station Trollet (triangle).

Diffuse attenuation coefficients were retrieved at four wavelengths in the UV range at different depths. The parameter S was calculated according to equation 5 for λ between 305 and 380 nm. In this fjord S varied between 0.012 to 0.015 in the surface layer (between 0 and 1.0 m) during the year (Figure 5). At one occasion S became much smaller (0.007), which indicated a larger k_d at 380 nm compared with 305 nm. This event was in the spring (Figure 5). K_ds were in general very low after a long winter. When the spring bloom started and more phytoplankton was present absorbing in the blue, the spectral shape of the absorption changes caused a different S. K_ds at 305 nm varied between 1.8 and 9.1 measured in the surface layer. At 380 nm the lowest were 0.9 and the highest 3.1.

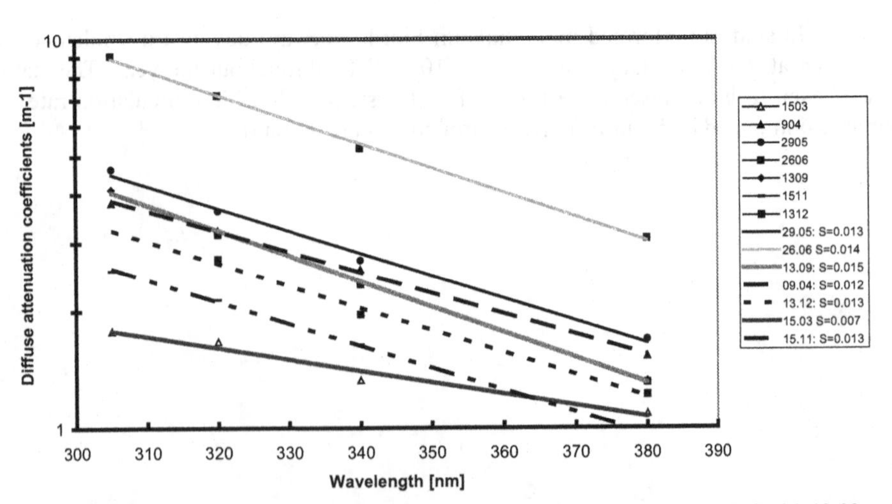

Figure 5. Diffuse attenuation coefficients retrieved during 1999 (15-03, 09-04, 29-05, 26-06, 13-09, 15-11, 13-12) at 305 nm, 320 nm, 340 nm and 380 nm. S was calculated from the interpolated lines shown on the graph.

Penetration of UV into this fjord varied during the year within one order of magnitude. This means that the composition and quantity of absorbing and scattering matter and particles are of great importance for the UV doses relevant for life in the water.

UV transparency in the Fjord depended on the natural environmental cycle during the year. This can be due to snow melting and fresh water flowing into the Fjord containing large amounts of CDOM. From supplementary measurements done at the different depths. Salinity profiles were taken and chlorophyll TOC was measured at 3 m and 10 m depths, as well as chlorophyll measurements. The annual variability of all these factors was compared at different depths. In Figure 6 samples from 3 m depth are shown. Salinity varied between 26 and 32 ppm with a minimum in early July (day number 190, Figure 6) and maximum in winter. TOC varied from 110 μM in winter to maximum 210 μM in late June correlating well with low salinity and low UV penetration (correlation coefficients 0.84 and 0.88, respectively). k_d at 340 nm and 3 m depth increased during spring. The highest values were in early July ($k_d = 4.6$) when salinity was lowest.

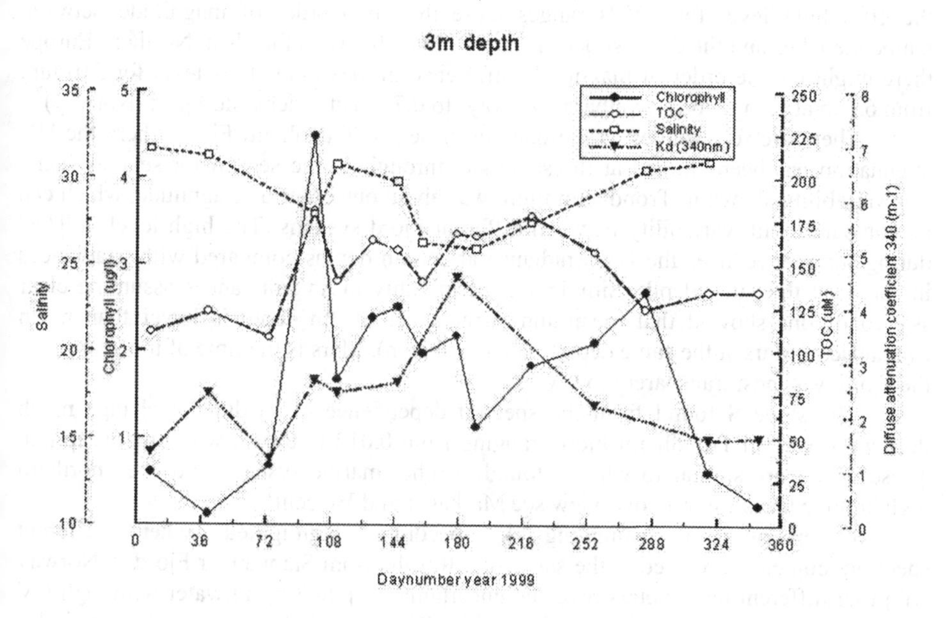

Figure 6. Annual variability at 3 m depth at one location in Trondheim Fjord for salinity (open squares, [ppm]), chlorophyll (closed circles,[μg/ℓ]), total organic carbon (open diamonds, [μM]) and diffuse attenuation coefficients at 340 nm (closed triangles, [m^{-1}]).

There was a high correlation between TOC and diffuse attenuation coefficients in UV. In this Fjord system a high correlation between TOC and salinity were found as well, indicating that organic carbon brought with river water was an important source for the TOC.

4. Discussion

It has been shown in several papers that by far the most successful predictor of UV attenuation in aquatic systems is dissolved organic carbon (DOC) concentration[8]. Models have been developed to establish this dependence[7]. However, it is not clear how general they are, mostly because the chemical nature of the DOC varies tremendously[9]. An empirical model developed in one system might not be valid for another water system. For temperate lakes a linear relationship between UV attenuation and DOC fluorescence was found, whereas UV attenuation was a power function of DOC concentration, but the reason for that was not clear[10]. Going to the extremes, highly coloured ponds, it was shown that UVB attenuates in the top 10-20 cm of the pond. A lack of reliable relationship between DOC and attenuation resulted from differences throughout the season in the fraction of DOC capable of absorbing radiation and the fraction capable of fluorescing[11]. Thus, seasonal variability or organic substances strongly influence the UV doses at different depths.

Attenuation coefficients in the UV-B vary from 20 m^{-1} in strong coloured ponds or lakes to less than 0.2 m^{-1} for the clearest ocean water. Likewise, for these examples, the 10% light level for UV-B ranges more than two order of magnitude between temperate lakes and the clearest ocean water[12]. Even between Fjords in Northern Europe there might be one order of magnitude difference in maximum 10% level for 310 nm, from 6.1 m at Kongsfjord (Svalbard, Norway) to 0.72 m at Inner Oslo Fjord (Norway)[2].

There are few studies as the one performed in Trondheim Fjord where the UV attenuation has been studied at the same site throughout the season for several years. The variability shown in Trondheim Fjord was about one order of magnitude, which can be compared with variability between different Fjord systems. The high level of TOC during summer reduces the UV irradiance at certain depths compared with irradiances in spring at the same depth. Simulations of monthly mean irradiances assuming clear sky conditions showed that mean June irradiances at 3m depth is lower than mean irradiance in Mars at the same depth (data not shown). Mars is the time of the year when the Fjord was most transparent to UV.

The slope S found from the spectral dependence of k_d did not change much during the year in Trondheim Fjord, ranging from 0.012 to 0.015, with one exception. These values are similar to what is found in other marine systems and even northern high latitude lakes[7]. For an overview see Markager and Vincent[13].

Uncertainty in estimating k_d becomes highlighted when different spectroradiometers are used at the same site. Results from Samnanger Fjord in Norway with four different radiometers revealed uncertainties up to 30% in water with high UV attenuation. An intercomparison performed by Kirk et al.[14] showed uncertainties in k_d, specially for short UV wavelengths in optical thick lakes, of the same order of magnitude as the Norwegian study. The same type of instrumentation participated in both intercomparisons.

All these data indicates how important it is to take into account the attenuation of UV in the water column and the annual variability in different aquatic systems when consequences of climate change is discussed.

References

1. Booth, C.R. and Morrow, J.H. (1997) The penetration of natural UV into natural waters, *Photochem. Photobiol.,* 65: 254-257.
2. Aas, E. and Højerslev, N.K. (2001) Attenuation of ultraviolet irradiance in North European coastal waters, *Oceanologia*, 43: 139-168.
3. Kuhn, P., Browman, H.I., McArthur, B. and St-Pierre, J.-F. (1999) Penetration of ultraviolet radiation in the waters of the estuary and Gulf of St. Lawence. *Limn. Oceanogr.*, 44: 710-716
4. Kjeldstad, B., Frette, Ø., Erga, S.R., Browman, H., Kuhn, P., Davis, R., Miller, W. and Stamnes, J.J. (2000) Penetration of solar ultraviolet radiation in a Norwegian fjord, intercomparison. *Manuscript*
5. Zerefos, C.S. and Bais, A.F. (1997) *Solar Ultraviolet Radiation Modelling, Measurements and Effects*, NATO ASI Series I, Global Environmental Change vol. 52, Springer, Berlin.
6. Bricaud, A., Morel, A. and Prieur, L. (1981) Absorption by dissolved organic matter of the sea (yellow substance) in the UV and visible domains, *Limnol. Oceanogr.*, 26: 43-53.
7. Laurion, I., Vincent, W.F. and Lean, D.R.S. (1997) Underwater ultraviolet radiation: development of spectral models for northern high latitude lakes, *Photochem. Photobiol.*, 65: 107-114.
8. Arts, M.T., Robarts, R.D., Kasai, F., Waiser, M.J., Tumber, V.P., Plante, A.J., Rai, H. and del Lange, H.J. (2000) The attenuation of ultraviolet radiation in high dissolved organic carbon waters of wetlands and lakes on the northern Great Plains, *Limnol. Oceanogr.*, 45: 292-299.
9. Morris, D.P., Zagarese, H., Williamson, C.E., Balseiro, E.G., Hargreaves, B.R., Modenutti, B., Moeller, R. and Queimalinos, C. (1995) The attenuation of solar UV radiation in lakes and the role of dissolved organic carbon. *Limnol. Oceanogr.*, 40: 1381-1391.
10. Scully, N.M. and Lean, D.R.S. (1994) The attenuation of ultraviolet radiation in temperate lakes. *Arch. Hydorbiol. Belh. Ergebn. Limnol.*, 43: 135-144.
11. Crump, D., Lean, D., Berrill, M., Coulson, D. and Toy, L. (1999) Spectral irradiance in pond water: influence of water chemistry, *Photochem. Photobiol.*, 70: 893-901.
12. Diaz, S.B., Morrow, J.H. and Booth, C.R. (2000) UV physics and optics. In *The effects of UV radiation in the marine environment* (S. de Mora, S. Demers and M. Vernet, eds), Cambridge Environmental Chemistry Series 10, Cambridge University Press, pp. 35-71.
13. Markager, S. and Vincent, W.F. (2000) Spectral light attenuation and the absorption of UV and blue light in natural waters, *Limnol. Oceanogr.*, 45: 642-650
14. Kirk, J.T.O., Hargreaves, B.R., Morris, D.P., Coffin, R.B., David, B., Frederickson, D., Karentz, D., Lean, D.R.S., Lesser, M.P., Madronich, S., Marrow, J.H., Nelson, N.B. and Scully, N.M. (1994) Measurements of UV-B radiation in two freshwater lakes: an instrument intercomparison, *Arch. Hydorbiol. Belh. Ergebn. Limnol.*, 43: 71-99.

INFLUENCE OF ULTRAVIOLET RADIATION ON THE CHROMOPHORIC DISSOLVED ORGANIC MATTER IN NATURAL WATERS

R. DEL VECCHIO AND N. V. BLOUGH
Department of Chemistry and Biochemistry, University of Maryland, College Park, 20742, MD, USA.

1. Abstract

Colored or chromophoric dissolved organic matter (CDOM) is by definition that portion of the dissolved organic matter (DOM) capable of absorbing light (i.e. contains chromophores). It represents a dominant absorbing species in natural waters and therefore it plays a critical role in controlling the light distribution in aquatic environments. CDOM shows a featureless absorption spectrum that increases exponentially with decreasing wavelength. Under light exposure CDOM loses its optical properties (photobleaching), altering the aquatic light field. Field and laboratory studies indicate that the CDOM photobleaching can represent a quite significant sink of this material over a short time scale.

2. Introduction

Dissolved organic matter (DOM) represents one of the largest reservoirs of organic carbon on the earth[1]. The amount of dissolved organic carbon (DOC) in aquatic environments is comparable to that of carbon (as CO_2) in the atmosphere. DOM has therefore the potential to influence the global carbon cycle and climate change[2]. CDOM, also referred to as Gelbstoff, yellow substances or humic substances, is a complex mixture of species originating from the decay of photosynthetically produced organic matter and found ubiquitously in the environment. Because CDOM is one of the primary absorbing species in aquatic environment, its light absorption is important in determining the aquatic light field. Due to its strong absorption in the ultraviolet (UV) spectral range, CDOM limits the penetration of biologically damaging UV-B radiation in the water column, so shielding the living organisms. At higher levels, CDOM absorption extends well into the visible regime, affecting the quality and quantity of UV-B and photosynthetically active radiation (PAR) available to the phytoplankton, with potentially very important impacts on the ecosystems[3-10].

3. Optical properties

Because CDOM absorbs and emits light in the ultraviolet and visible, studies on this material often employ absorbance and fluorescence spectroscopy. The featureless

F. Ghetti et al. (eds.), Environmental UV Radiation: Impact on Ecosystems and Human Health and Predictive Models, 203–216.

CDOM absorption spectrum decays exponentially from the UV to the visible regime (Figure 1).

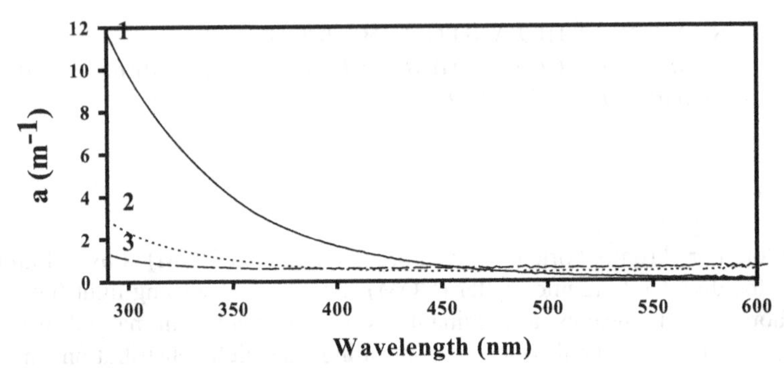

Figure 1. CDOM absorption spectra along the middle Atlantic Bight (MAB) during September 1998: 1. Delaware River at 39°41.33'N and 75°31.11'W (——); 2. Delaware Bay mouth at 38°48.86'N and 75°4.91'W (···); 3. Gulf Stream Western Edge at 37°6.95'N and 72°55.42'W (– –).

An exponential decay function has therefore been used to fit the decay of CDOM absorbance with wavelength:

$$a(\lambda) = a(\lambda_0)e^{-S(\lambda-\lambda_0)} \tag{1}$$

where $a(\lambda)$ and $a(\lambda_0)$ are the absorption coefficients at the wavelength λ and at the reference λ_0[11-15]. The absorption coefficient is obtained as follows:

$$a(\lambda) = 2.303 \times A(\lambda) / L \tag{2}$$

where $A(\lambda)$ is the absorbance at the wavelength λ and L is the cell path length in meters. The spectral slope, S, indicates the rate at which the CDOM absorption coefficient decreases with increasing wavelength and it has been used to discriminate among CDOM of different nature. S increases with decreasing CDOM absorption coefficient, aromatic content and molecular weight[16]. Moreover, S also increases from coastal and fresh waters (~0.014 nm^{-1}) to offshore locales (>0.02 nm^{-1})[13-15,17-19]. This trend may be attributed to the replacement of terrestrial CDOM with marine CDOM or to transformation of the terrestrial CDOM as it moves along the shelf[20].

Since 1949, it is known that seawater fluoresces blue when hit by UV light[21]. Because fluorescence spectroscopy is a more sensitive and easier tool to employ than absorption spectroscopy, it is often preferred when examining the distribution of CDOM in natural waters. Unfortunately, fluorescence is only representative of a portion of the total chromophore pool and not of the whole CDOM. As true for absorption, CDOM fluorescence spectra are broad and unstructured; the emission intensity decreases and its maximum is red shifted with increasing excitation wavelength, indicative of the presence of numerous emitting species or of interactions of a few chromophores that produce long wavelength emission. For this reason, excitation/emission matrix spectra (EEMS) have become a useful tool for studying this complex material, allowing for different classes of fluorophores to be detected[22-25]. EEMS are obtained by collecting a series of emission spectra at successive excitation wavelengths, and then merging them into three-dimensional plots in which excitation and emission wavelengths are plotted against fluorescence intensity (Figure 2)[16].

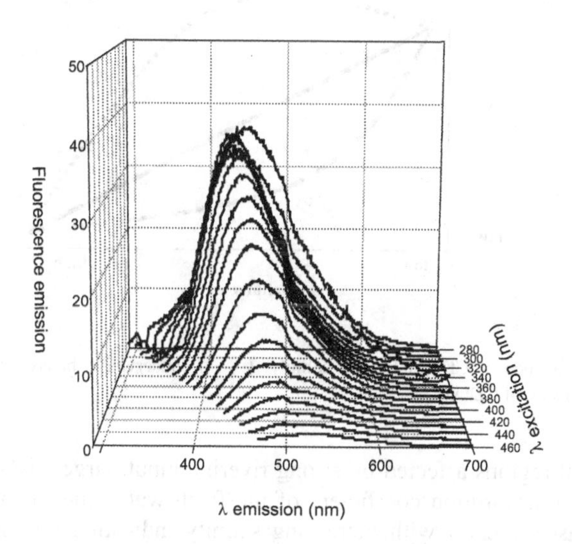

Figure 2. EEMS of Suwanne River fulvic acid (SRFA) (10 mg/L in MilliQ water). [λ_{ex}: 280-460 nm, every 10 nm; λ_{em}: λ_{ex}+10 nm to700 nm]. [The emission spectra have not been corrected for the instrument response].

Unfortunately, EEMS requires long time for collection, time that is not often available in field studies. In this case, acquiring single excitation/emission spectra becomes more advantageous, permitting large field mapping of CDOM over shorter timescales[26-28].

4. CDOM origin and fate

The origin and fate of CDOM is still a matter of discussion. The results from field studies conducted at very different geographical locales are now available (although comparisons among these different studies are often impossible because of the different protocols employed to measure CDOM properties), but there remain many unanswered questions about the factors controlling the seasonal and spatial distributions of CDOM.

Salinity (a conservative tracer) can be employed to investigate sources and sinks of CDOM. CDOM of terrestrially origin that is simply diluted in waters of increasing salinity will follow an inverse linear relationship against salinity; in the presence of sources or sinks this linear relationship will deviate upward or downward, respectively (Figure 3).

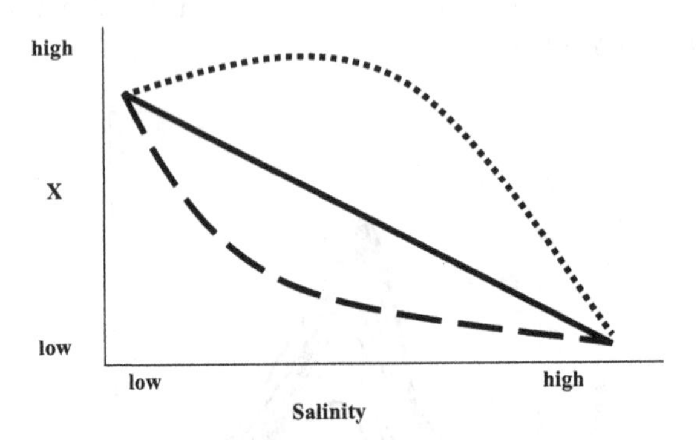

Figure 3. Dependence of CDOM (X) on salinity: (——) conservative behavior; (······) source of CDOM; (– – –) sink of CDOM.

In coastal regions affected by strong riverine input, large CDOM optical signals are observed; the absorption coefficient of the fresh water end members (5-15 m^{-1} at 355 nm) decreases linearly with increasing salinity indicating no sources or sinks as well as a strong terrestrial origin for this material;[14,20,28-32]; the inverse linear dependence of CDOM absorption on salinity is often seasonally and spatially dependent because of CDOM river discharge and/or of CDOM photodegradation[18,30]; at some locations, however, a slight upward curvature of the salinity to absorption dependence is observed at lower salinity, suggesting the possibility of some in situ production of CDOM (Figure 4).

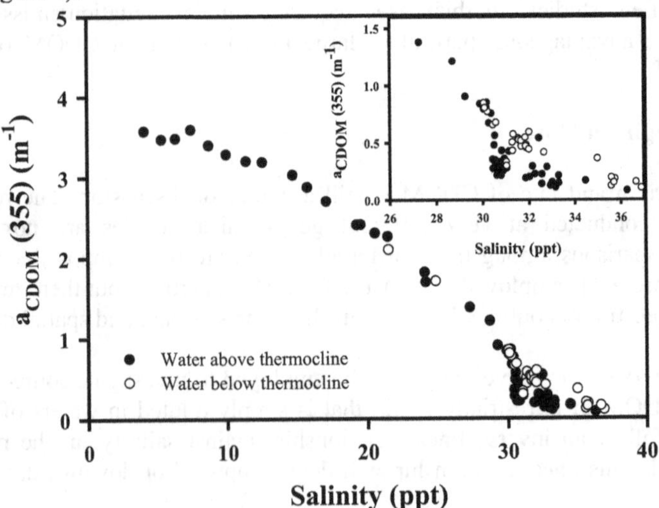

Figure 4. Dependence of $a_{CDOM}(355)$ on salinity for a transect from the Delaware Bay to the MAB during September 1998. Inset: same data on expanded axes.

In coastal regions not influenced by large riverine inputs, lower values of the CDOM absorption coefficient are observed (~0.1 m^{-1} at 355 nm) that do not correlate with salinity[30,33], suggesting that sources other than the terrestrial ones are involved in the CDOM production.

CDOM may also be produced by the release from phytoplankton or by diffusion from sediment porewater[34-35] as well as through atmospheric deposition of CDOM, a possibility not yet investigated. The lack of correlation between CDOM and phytoplankton chlorophyll in both laboratory experiments[36] and in field measurements in coastal regions[19,37] suggests that other sources beside direct phytoplankton input may be important for the CDOM budget.

On the contrary, in the open ocean, low values of CDOM absorption coefficient are observed (~0.05 m^{-1} at 355 nm). At these locations, a correlation between CDOM and chlorophyll-a is sometimes, but not always, observed suggesting that CDOM could be produced *in situ*[32,37], although the clear mechanism of this production is not yet known.

Although rivers are large sources of terrestrial DOM, this material does not seem to contribute significantly to the DOM in seawater and in marine sediments[38]. In fact, a recent field study on the lignin distribution in open oceans has shown that lignin represents only a few percent of the total DOM in the ocean and has a shorter ocean residence time than the marine DOM[39]. Similar conclusions have been reached by Blough and Del Vecchio[20] for CDOM, based on the differences of CDOM absorption between freshwater and central gyre end-members. Both of these results indicate the presence of a quite effective sink of terrestrial material during its transit from inshore to offshore waters. Losses of terrestrial CDOM due to flocculation appear to be inconsequential, based on the conservative or quasi-conservative mixing that has been observed for many estuaries. Further, bacterial consumption alone does not appear to be a substantial sink of CDOM[40]. Other possible sinks are photochemical degradation[19,28,41-43] and/or photodegradation coupled to bacterial uptake[40,44-45].

5. Field evidence of CDOM photobleaching

As described above, for regions influenced by high riverine input, CDOM absorption coefficients often decline linearly with salinity from inshore to off-shore locations, indicating conservative behavior[14,20,28-32]. However, departures from this behavior have been also reported[14,46], due to input or depletion of CDOM or simply to mixing of water masses with different CDOM end-members.

At some locations (as it occurs in the MAB during the summertime) the vertical mixing depth decreases, concomitant with the development of a seasonal stratification, while the CDOM exposure to sunlight increases. Under these conditions, a net decrease of CDOM absorption is observed in surface waters over a narrow salinity range; in waters below the thermocline the CDOM absorption is higher and maintains an inverse linear relationship with salinity (Figure 4). The seasonal surface sink of CDOM has been attributed to photodegradation[19,28,41]. Studies of CDOM absorbance at high temporal and spatial resolution have also provided evidence for a surface sink of CDOM due to photobleaching[43].

Changes of CDOM content in surface water modify the light distribution within the water column and therefore potentially affect the entire ecosystem[4,47]. For example, a two- to three-fold reduction of surface CDOM (Figure 5A) can increase the amount of light penetrating down the water column by several order of magnitude depending on the wavelength of observation and original level of CDOM (Figure 5B). Thus, small changes of CDOM can dramatically increase the amount of radiation potentially damaging living organisms.

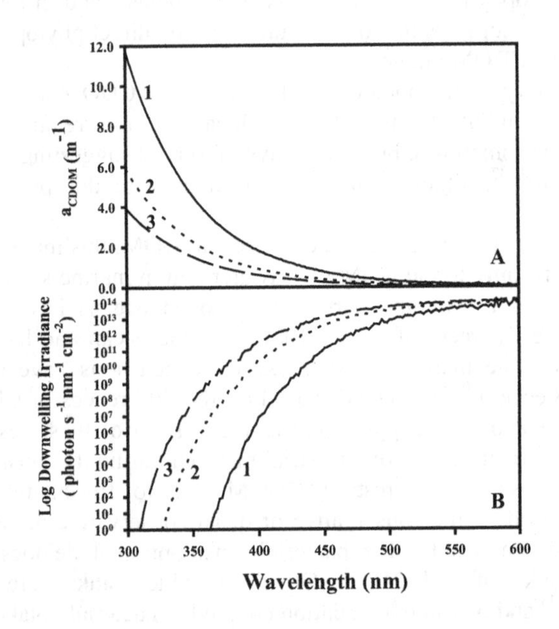

Figure 5. CDOM absorption spectra (A) and calculated downwelling irradiance at 10 m depth (B) in the Delaware River (DR) at 39°41.33'N and 75°31.11'W. A: 1. DR (——); 2. DR diluted two fold (……); 3. DR diluted three fold (– – –). B: Calculated downwelling irradiance on a logarithmic scale in presence of CDOM 1-2-3.

Field and laboratory studies have also shown an increase in the CDOM spectral slope (*S*) induced by light exposure[14,20,28,40,48-49]. Other studies, however, have shown the opposite trend[41,50-51]; the reason for this discrepancy is still unknown, but it may be due to the different protocols employed in calculating the *S* values.

An increase in *S* is also observed at higher salinity (~32 ppt)[20], concomitant with a blue-shift in the CDOM emission maximum[52-55]. The increase in *S* and the blue-shift of the emission maximum are observed at high salinity values where presumably the terrestrial CDOM is diluted enough to allow photobleaching to become evident.

6. Laboratory evidence of CDOM photobleaching

Upon light exposure, CDOM undergoes a complex series of reaction that also can potentially affect the entire aquatic system (Figure 6)[47].

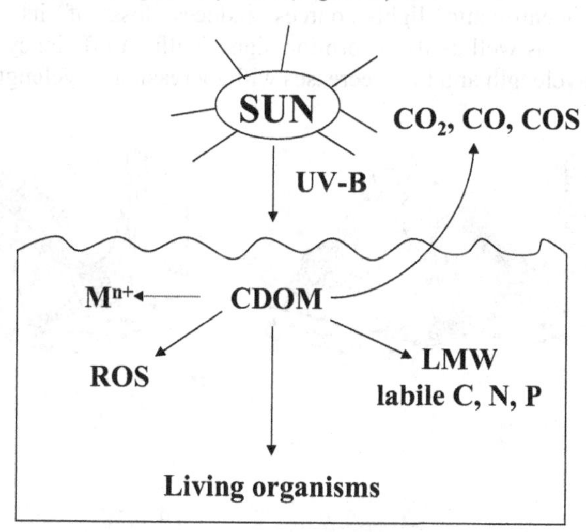

Figure 6. Scheme of CDOM photodegradation (Modified from Zepp et al.[47]).

Because CDOM is one of the primary light absorbing species in aquatic systems, its breakdown affects the optical properties of the whole water body[56]. Its breakdown further produces potentially important substrates for the plankton growth[57]. Photochemical degradation of CDOM releases carbon monoxide[58], carbon dioxide[51,59] and a variety of low molecular weight organic compounds, as well as inorganic nitrogen[60-61]. Most of these compounds are however released with very low quantum yields[62]. Biologically available substrates such as pyruvate, other low molecular weight (LMW) carbonyl compounds are released from the photolytic breakdown of the biologically refractive CDOM[63-65]. These photochemical products seem to stimulate the bacterial growth, although to different extents depending upon the DOM source (e.g., surface versus deep water DOM)[66-67]. Photochemical and microbial degradation combined also result in a more efficient consumption of CDOM[40,45,68].

Overall, absorption of photons by the CDOM can initiate photochemical reactions in which primary products are produced as well as intermediates. Photochemical reactions ultimately lead not only to a destruction of the chromophores and hence to a loss of the CDOM optical properties (called photobleaching)[20,34,49,69-71] but also to transformation of the terrestrial DOM into a new form[42].

Considerable new research (field and laboratory) has been conducted during the last few years to further understand the CDOM photobleaching. However, the nature of the CDOM chromophores and the mechanisms of their photochemical reactions need to be understood to quantify the importance of photochemistry in natural waters.

The availability of a complete and consistent set of experiments devoted to the loss of the optical properties of CDOM in the UV-visible range becomes an important issue in order to estimate the CDOM photobleaching rates in the field. Several laboratory studies are now available on CDOM photobleaching. Exposure of CDOM to different monochromatic light sources induces loss of its fluorescence signal (Figure 7)[49,69-72] as well as its absorption signal with an efficiency that is largest at the irradiation wavelength and that decreases with increasing wavelength[49].

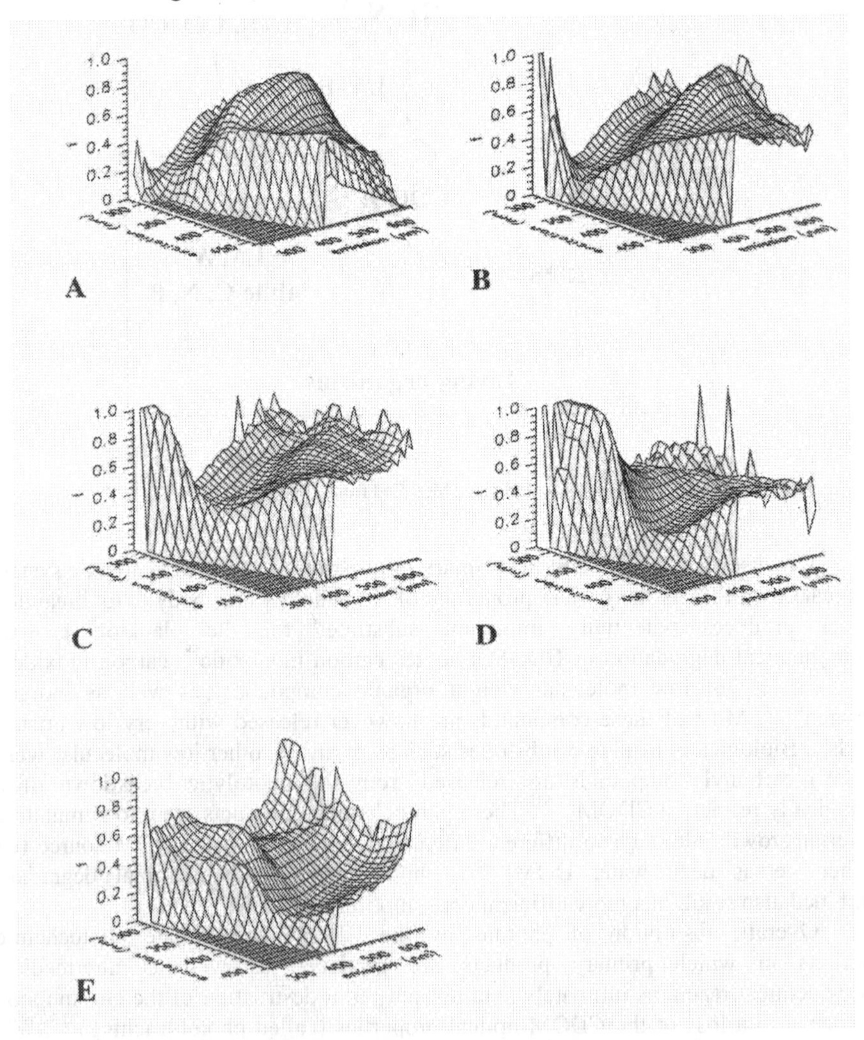

Figure 7. Fraction of the original fluorescence intensity (*f*) remaining after monochromatic irradiation of SRFA at 296 nm (Fig. A; t=83 hr; E=$5.53 \cdot 10^{+15}$ photons cm^{-2} s^{-1}), at 313 nm (Fig. B; t=83 hr; E=$12.67 \cdot 10^{+15}$ photons cm^{-2} s^{-1}), at 334 nm (Fig. C; t=185 hr; E=$4.63 \cdot 10^{+15}$ photons cm^{-2} s^{-1}), at 366 nm (Fig. D; t=103 hr; E=$11.92 \cdot 10^{+15}$ photons cm^{-2} s^{-1}) and at 407 nm (Fig. E; t=233 hr; E=$5.43 \cdot 10^{+15}$ photons cm^{-2} s^{-1}). The *f* is defined as the ratio of the EEMS at irradiation time t to the EEMS at time zero.

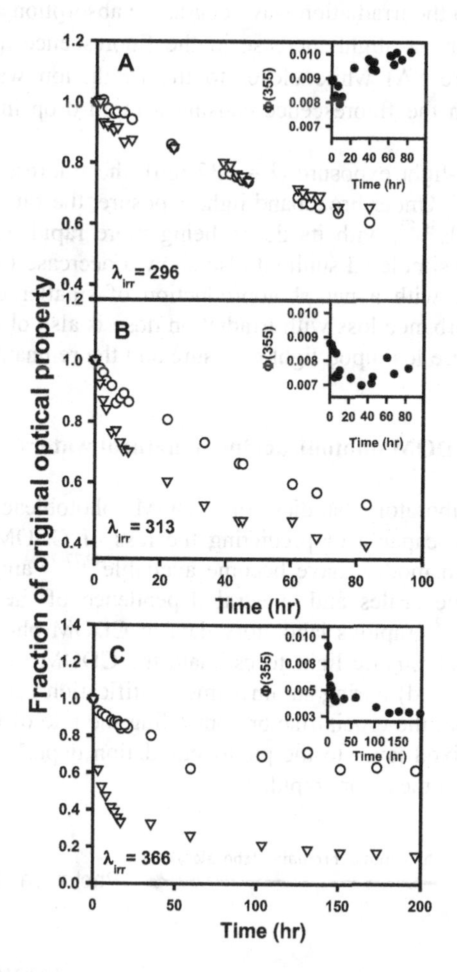

Figure 8. Fractional loss of a(355) (O) and F/R (∇) with time (hours) during monochromatic irradiation of SRFA at 296 nm (Figure A), 313 nm (Figure B) and 366 nm (Figure C). Insets: Dependence of Φ(355) (λ) on irradiation time. [F/R is the fluorescence intensity (F) (λ_{ex} 355 nm) normalized to the Raman intensity (R) (λ_{ex} 355) as reported in Hoge et al.[73]. Φ(355) is the fluorescence quantum yield as defined in Green and Blough[15].

Further, under monochromatic light exposure, the CDOM absorption (at 355 nm) and fluorescence (λ_{ex} 355 nm) decay as a single exponential away from the irradiation wavelength (Figure 8A) indicating that all the chromophores are decaying with the same rate constant. On the other hand, closer to the irradiation wavelength the CDOM absorption and fluorescence decay as a sum of two exponentials (Figures 8B, 8C), suggesting the presence of at least two pools of chromophores decaying with different rate constants. These and other observations suggested that the CDOM absorption spectrum cannot be solely due to a superposition of independent chromophores[49].

Further, away from the irradiation wavelength, the absorption and fluorescence decay at similar rates causing a slight increase in the fluorescence quantum yield (ϕ(355) at λ_{ex} 355 nm) (Figure 8A) while closer to the irradiation wavelength the absorption decays slower than the fluorescence causing a rapid drop in the fluorescence ϕ(355) (Figures 8B, 8C).

Under laser-light exposure (λ = 337 nm), the fluorescence loss depends on the irradiation dose[74-75]. Under broadband light exposure, the fluorescence decays as a sum of two exponentials[34,49], with its decay being more rapid than the absorption loss[49]. Experiments using simulated sunlight also show a decrease in the molecular weight of the organic matter with a net photoproduction of smaller compounds[31,76]. A strong correlation of absorbance loss with irradiation dose is also observed[48]. Overall, CDOM optical properties are lost upon light exposure and the remnant CDOM shows modified optical properties.

7. Modeling the CDOM photobleaching in natural waters

Detailed laboratory studies of CDOM photobleaching are necessary for developing models capable of predicting the loss of CDOM due to photobleaching. Very recently, such models have become available[49,77-78] and have been employed to investigate the time scales and spectral dependence of the CDOM photobleaching. Briefly, our model[49] employs laboratory data of CDOM photodegradation induced by exposure to monochromatic light to estimate the CDOM loss that occurs under solar irradiation (broad-band) during summertime stratification across the shelf in the MAB. Results indicate that the crucial factor controlling the rate of CDOM photobleaching is the ratio of the mixed layer to the photodegradation depth[49]. By decreasing this ratio, the CDOM loss becomes more rapid.

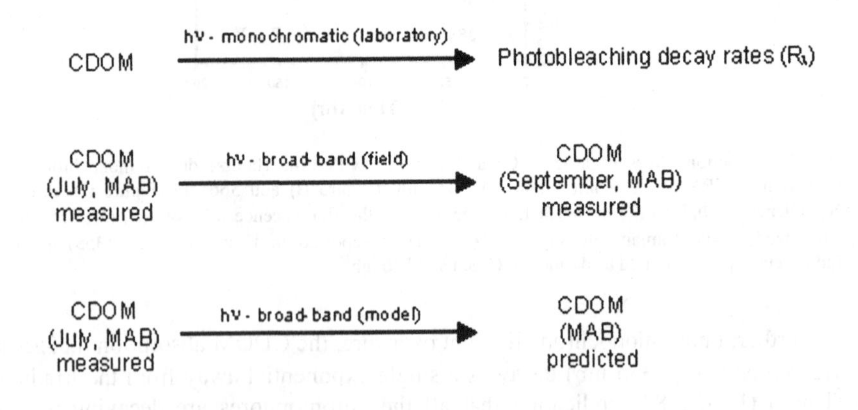

Figure 9. Scheme of CDOM photobleaching model (modified from Del Vecchio and Blough[49]).

Furthermore, using this model, the indirect (or dark) photobleaching (i.e. that occurring outside the irradiation wavelength) has been shown to be important to the overall rate of CDOM absorption loss[49]. Ignoring this indirect bleaching leads to a substantial underestimate of the loss of CDOM absorption upon light exposure.

A theoretical model, based on the a superposition of independent chromophores undergoing photobleaching independently[49], has also been shown to be inconsistent with the experimental photobleaching results, suggesting that the CDOM absorption spectrum is not simply the result of a superposition of the absorption of numerous, distinct chromophores. Further experiments are in progress to address this question.

References

1. Hedges, J.I. (1992) Global biogeochemical cycles: progress and problems, *Mar. Chem.*, 39: 67-93.
2. Farrington, J. (1992) Overview and key recommendations. Marine organic geochemistry workshop, January 1990, *Mar. Chem.*, 39: 5-9.
3. Arrigo, K.R. and Brown, C.W. (1996) Impact of chromophoric dissolved organic matter on UV inhibition of primary productivity in the sea, *Mar. Ecol. Prog. Ser.*, 140: 207-216.
4. Williamson, C.E., Stemberger, R.S., Morris, D.P., Frost, T.M. and Paulsen, S.G. (1996) Ultraviolet radiation in North American lakes: Attenuation estimates from DOC measurements and implications for plankton communities, *Limnol. Oceanogr*,. 41: 1024-1034.
5. Häder, D.P., Kumar, H.D., Smith R.C. and Worrest, R.C. (1998) Effects of UV-B radiation on aquatic ecosystems, *J. Photochem. Photobiol. B:Biol.*, 46: 53-68.
6. Kuhn, P., Browman, H., McArthur, B. and St-Pierre, J.-F. (1999) Penetration of ultraviolet radiation in the waters of the estuary and Gulf of St. Lawrence, *Limnol. Oceanogr.*, 44: 710-716.
7. West, L.J.A., Greenberg, B.M. and Smith, R.E.H. (1999) Ultraviolet radiation effects on a microscopic green alga and the protective effects of natural dissolved organic matter, *Photochem. Photobiol.*, 69: 536-544.
8. De Mora, S., Demers, S. and Vernet, M. (2000) *The Effects of UV Radiation in the Marine Environment*, Cambridge University Press, Cambridge.
9. Gibson, J.A.E., Vincent, W.F., Nieke, B. and Pienitz, R. (2000) Control of biological exposure to UV radiation in the Arctic Ocean: Comparison of the roles of ozone and riverine dissolved organic matter, *Arctic*, 53: 372-382.
10. Kuwahara, V.S., Ogawa, H., Toda, T., Kikuchi, T. and Taguchi, S. (2000) Variability of bio-optical factors influencing the seasonal attenuation of ultraviolet radiation in temperate coastal waters of Japan, *Photochem. Photobiol.*, 72: 193-199.
11. Bricaud, A., Morel, A. and Prieur, L. (1981) Absorption by dissolved organic matter in the sea (yellow substance) in the UV and visible domains, *Limnol. Oceanogr.*, 26: 43-53.
12. Zepp, R.G. and Schlotzhauer, P.F. (1981) Comparison of the photochemical behavior of various humic substances in water. III. Spectroscopic properties of humic substances, *Chemosphere*, 10: 479-486.
13. Carder, K.L., Steward, R.G., Harvey, G.R. and Ortner. P.B. (1989) Marine humic and fulvic acids: their effects on remote sensing of chlorophyll, *Limnol. Oceanogr.*, 34: 68-81.
14. Blough, N.V., Zafiriou, O.C. and Bonilla, J. (1993) Optical absorption spectra of water from the Orinoco River outflow: terrestrial input of colored organic matter to the Caribbean, *J. Geophys. Res.*, 98: 2271-2278.
15. Green, S.A. and Blough, N.V. (1994) Optical absorption and fluorescence properties of chromophoric dissolved organic matter in natural waters, *Limnol. Oceanogr.*, ***39***: 1903-1916.
16. Blough, N.V. and Green, S.A. (1995) Spectroscopic characterization and remote sensing of non-living organic matter. In *The role of non-living organic matter in the earth's carbon cycle* (R.G. Zepp and C. Sonntag, eds.), John Wiley & Sons, pp. 23-45.
17. Brown, M. (1977) Transmission spectroscopy examinations of natural waters. C. Ultraviolet spectral characteristic of the transition from terrestrial humus to marine yellow substance, *Estuar. Coast. Mar. Sc.*, 5: 309-317.

18. Nelson, J.R. and Guarda, S. (1995) Particulate and dissolved spectral absorption on the continental shelf of the southeastern United States, *J. Geophys. Res.*, 100: 8715-8732.
19. Nelson, N.B., Siegel, D.A. and Michaels, A.F. (1998) Seasonal dynamics of colored dissolved organic matter in the Sargasso Sea, *Deep-Sea Res. I*, 45: 931-957.
20. Blough, N.V. and Del Vecchio, R. (2002) Chromophoric DOM in the Coastal Environment. In *Biogeochemistry of marine dissolved organic matter* (D. Hansell and C. Carlson, eds.), Academic Press, Cambridge (MA), pp. 509-546.
21. Kalle, K. (1949) Fluoreszenz und gelbstoff im Bottnischen und Finnishchen Meerbusen, *Deutsche Hydrographische Zietschrift*, 2: 9-124.
22. Coble, P.G., Green, S.A., Blough, N.V. and Gagosian, R.B. (1990) Characterization of dissolved organic matter in the Black Sea by fluorescence spectroscopy, *Nature*, 348: 432-435.
23. Coble, P.G., Del Castillo, C.E. and Avril, B. (1998) Distribution and optical properties of CDOM in the Arabian Sea during the 1995 Southwest Monsoon, *Deep-Sea Res. II*, 45: 2195-2223.
24. Mopper, K. and Schultz, C.A. (1993) Fluorescence as a possible tool for studying the nature and water column distribution of DOC components, *Mar. Chem.*, 41: 229-238.
25. Coble, P.G. (1996) Characterization of marine and terrestrial DOM in seawater using excitation-emission matrix spectroscopy, *Mar. Chem.*, 51: 325-346.
26. Hoge, F.E., Vodacek, A., Swift, R.N., Yungel, J.K. and Blough, N.V. (1995) Inherent optical properties of the ocean: retrieval of the absorption coefficient of chromophoric dissolved organic matter from airborne spectral fluorescence measurements, *Appl. Opt.*, 34: 7033-7038.
27. Vodacek, A., Hoge, F., Swift, R.N., Yungel, J.K., Peltzer, E.T. and Blough N.V. (1995) The use of in situ and airborne fluorescence measurements to determine UV absorption coefficients and DOC concentrations in surface waters, *Limnol. Oceanogr.*, 40: 411-415.
28. Vodacek, A., Blough, N.V., DeGrandpre, M.D., Peltzer, E.T. and Nelson, R.K. (1997) Seasonal variation of CDOM and DOC in the Middle Atlantic Bight: terrestrial inputs and photooxidation, *Limnol. Oceanogr.*, 42: 674-686.
29. Nieke, B., Reuter, R., Heuermann, R., Wang, H., Babin, M. and Therriault, J.C. (1997) Light absorption and fluorescence properties of chromophoric dissolved organic matter (CDOM) in the St. Lawrence Estuary (Case 2 waters), *Cont. Shelf Res.*, 17: 235-252.
30. Ferrari, G.M. and Dowell, M.D. (1998) CDOM absorption characteristics with relation to fluorescence and salinity in coastal areas of the southern Baltic Sea, *Est. Coast. Shelf Sci.*, 47: 91-105.
31. Seritti, A., Russo, D., Nannicini, L. and Del Vecchio, R. (1998) DOC, absorption and fluorescence properties of estuarine and coastal waters of the northern Tyrrhenian Sea, *Chem. Spec. Bioavail.*, 10: 95-105.
32. Kowalczuk, P. (1999) Seasonal variability of yellow substance absorption in the surface layer of the Baltic Sea, *J. Geophys. Res.*, 104: 30047-30058.
33. Obernosterer, I. and Herndl, G.J. (2000) Differences in the optical and biological reactivity of the humic and non-humic dissolved organic carbon component in two contrasting coastal marine environments; *Limnol. Oceanogr.*, 45: 1120-1129.
34. Skoog, A., Wedborg, M. and Fogelqvist, E. (1996) Photobleaching of fluorescence and organic carbon concentration in a coastal environment, *Mar. Chem.*, 55: 333-345.
35. Boss, E., Pegau, W.S., Zaneveld, J.R.V. and Barnard, A.H. (2001) Spatial and temporal variability of absorption by dissolved material at a continental shelf, *J. Geophys. Res.-Ocean*, 106: 9499-9508.
36. Rochelle-Newall, E.J., Fisher, T.R., Fan, C. and Glibert, P.M. (1999) Dynamics of chromophoric dissolved organic matter and dissolved organic carbon in experimental mesocosms, *Int. J. Remote Sensing*, 20: 627-641.
37. DeGrandpre, M.D., Vodacek, A., Nelson, R.K., Bruce, E.J. and Blough, N.V. (1996) Seasonal seawater optical properties of the U.S. Middle Atlantic Bight, *J. Geophys. Res.*, 101: 22722-22736.
38. Hedges, J.I., Keil, R. and Benner, R. (1997) What happens to terrestrially-derived organic matter in the sea?, *Org. Geochem.*, 27: 195-212.
39. Opsahl, S. and Benner, R. (1997) Distribution and cycling of terrigenous dissolved organic matter in the ocean, *Nature*, 386: 480-482.
40. Moran, M.A., Sheldon, W.M. and Zepp, R.G. (2000) Carbon loss and optical property changes during long-term photochemical and biological degradation of estuarine dissolved organic matter, *Limnol. Oceanogr.*, 45: 1254-1264.

41. Morris, D.P. and Hargreaves, B.R. (1997) The role of photochemical degradation of dissolved organic carbon in regulationg the UV transparency of three lakes on Pocono Plateau, *Limnol. Oceanogr.*, 42: 239-249.
42. Opsahl, S. and Benner, R. (1998) Photochemical reactivity of dissolved lignin in river and ocean waters, *Limnol. Oceanogr.*, 43: 1297-1304.
43. Twardowsky, M.S. and Donaghay, P.L. (2001) Separating in situ and terrigenous sources of absorption by dissolved materials in coastal waters, *J. Geophys. Res.*, 106: 2545-2560.
44. Amon, R.M.W. and Benner, R. (1996) Photochemical and microbial consumption of dissolved organic carbon and dissolved oxygen in the Amazon River system, *Geochim. Cosmochim. Acta*, 60: 1783-1792.
45. Miller, W.L. and Moran, M.A. (1997) Interaction of photochemical and microbial processes in the degradation of refractory dissolved organic matter from a coastal environment, *Limnol. Oceanogr.*, 42: 1317-1324.
46. Højerslev, N.K., Holt, N. and Aarup, T. (1996) Optical measurements in the North Sea-Baltic Sea transition zone. I. On the origin of the deep water in the Kattegat, *Contin. Shelf. Res.*, 16: 1329-1342.
47. Zepp, R.G., Callaghan, T.V. and Erickson, D.J. (1998) Effects of enhanced solar ultraviolet radiation on biogeochemical cycles; *J. Photochem. Photobiol. B: Biol.*, 46: 69-82.
48. Whitehead, R.F., de Mora, S., Demers, S., Gosselin, M., Monfort, P. and Mostajir, B. (2000) Interactions of ultraviolet-B radiation, mixing, and biological activity on photobleaching of natural chromophoric dissolved organic matter: a mecocosm study, *Limnol. Oceanogr.*, 45: 278-291.
49. Del Vecchio, R. and Blough, N.V. (2002) Photobleaching of Chromophoric Dissolved Organic Matter in Natural Waters: Kinetics and Modeling, *Marine Chem.*, 78: 231-253
50. Miller, W.L. (1994) Recent advances in the photochemistry of natural dissolved organic matter. In *Aquatic and Surface Photochemistry* (G.R. Helz, R.G. Zepp and D.G. Crosby, eds.), CRC Press, pp. 111-128.
51. Gao, H.Z. and Zepp, R.G. (1998) Factors influencing photoreactions of dissolved organic matter in a coastal river of the southern United States, *Environ. Sci. Technol.*, 32: 2940-2946.
52. De Souza Sierra, M.M., Donard, O.F.X. and Lamotte, M. (1997) Spectral identification and behaviour of dissolved organic fluorescent material during estuarine mixing processes, *Mar. Chem.*, 58: 51-58.
53. De Souza Sierra, M.M., Donard, O.F.X., Lamotte, M., Belin, C. and Ewald, M. (1994) Fluorescence spectroscopy of coastal and marine waters, *Mar. Chem.*, 47: 127-144.
54. Del Castillo, C.E., Gilbes, F., Coble, P.G. and Müller-Karger, F.E. (2000) On the dispersal of riverine colored dissolved organic matter over the West Florida Shelf, *Limnol. Oceanogr.*, 45: 1425-1432.
55. Del Castillo, C.E., Coble, P.E., Morell, J.M., Lopez, J.M. and Corredor, J.E. (1999) Analysis of the optical properties of the Orinoco River plume by absorption and fluorescence spectroscopy, *Mar. Chem.*, 66: 35-51.
56. Blough, N.V. and Zepp, R.G. (1995) Reactive oxygen species in natural waters. In *Active Oxygen in Chemistry* (C.S. Foote, J.S. Valentine, A. Greenberg and J.F. Liebman, eds.), Chapman and Hall, New York., pp. 280-333.
57. Kieber, D.J. (2000) Photochemical production of biological substrates. In *The effect of UV radiation in the marine environment* (S.J. de Mora, S. Demers and M. Vernet, eds.), Cambridge University Press, Cambridge, pp. 130-148.
58. Valentine, R.L. and Zepp, R.G. (1993) Formation of carbon monoxide from the phtoodegradation of terrestrial dissolved organic carbon in natural waters, *Environm. Sci. Technol*, 27: 409-412.
59. Miller, W.L. and Zepp, R.G. (1995) Photochemical production of dissolved inorganic carbon from terrestrial organic matter: significance to the oceanic carbon cycle, *Geophys. Res. Lett.*, 22: 417-420.
60. Carlsson, P. and Graneli, E. (1993) Availability of humic bound nitrogen for coastal phytoplankton, *Estuar. Coast. Shelf Sci.*, 36: 433-447.
61. Bushaw, K.L., Zepp, R.G., Tarr, M.A., Schulz-Jander, D., Bourbonniere, R.A., Hodson, R.E., Miller, W.L., Bronk, D.A. and Moran, M.A. (1996) Photochemical release of biologically available nitrogen from aquatic dissolved organic matter, *Nature*, 381: 404-407.
62. Blough, N.V. (2001) Air-Sea Interaction: Photochemical Processes. In *Encyclopedia of Ocean Sciences* (J. Steele, S. Thorpe and K. Turekian, eds.), Academic Press, pp. 2162-2172.
63. Kieber, D.J., Daniel, J.M. and Mopper, K. (1989) Photochemical source of biological substrates in seawater: implications for carbon cycling, *Nature*, 341: 637-639.

64. Kieber, R.J., Zhou, X. and Mopper, K. (1990) Formation of carbonyl compounds from UV-induced photodegradation of humic substances in natural waters: Fate of riverine carbon in the sea, *Limnol. Oceanogr.*, 35: 1503-1515.
65. Wetzel, R.G., Hatcher, P.G. and Bianchi, T.S. (1995) Natural photolysis by ultraviolet irradiance of recalcitrant dissolved organic matter to simple substrates for rapid bacterial metabolism, *Limnol. Oceanogr.*, 40: 1369-1380.
66. Mopper, K., Zhou, X., Kieber, R.J., Kieber, D.J., Sikorski, R.J. and Jones, R.D. (1991) Photochemical degradation of dissolved organic carbon and its impact on the oceanic carbon cycle, *Nature*, 353: 60-62.
67. Benner, R. and Biddanda, B. (1998) Photochemical transformations of surface and deep marine dissolved organic matter: effects on bacterial growth, *Limnol. Oceanogr.*, 43: 1373-1378.
68. Moran, M.A. and Zepp, R.G. (1997) Role of photoreactions in the formation of biologically labile compounds from dissolved organic matter, *Limnol. Oceanogr.*, 42: 1307-1316.
69. Kouassi, A.M., and Zika, R.G. (1990) Light-induced alteration of the photophysical properties of dissolved organic matter in seawater, *Netherlands J. Sea Res.*, 27: 25-32.
70. Kouassi, A.M., R.G. Zika, and J.M.C. Plane. (1990) Light-induced alteration of the photophysical properties of dissolved organic matter in seawater. Part II. Estimates of the environmental rates of the natural water fluorescence, *Neth. J. Sea Res.*, 27: 33-41.
71. Patsayeva, S.V., Fadeev, V.V., Filippova, E.M., Chubarov, V.V. and Yuzhakov V.I. (1991) The effect of temperature and ultraviolet radiation influence on the luminescence spectrum characteristics of dissolved organic substance, *Moscow Univ. Physics Bull.*, 46: 66-69.
72. Kouassi, A.M. and Zika, R.G. (1992) Light-induced destruction of the absorbance property of dissolved organic matter in seawater, *Toxicol. Environ. Chem.*, 35: 195-211.
73. Hoge, F.E., Vodacek, A. and Blough, N.V. (1993) Inherent optical properties of the ocean: retrieval of the absorption coefficient of chromophoric dissolved organic matter from fluorescence measurements, *Limnol. Oceanogr.*, 38: 1394-1402.
74. Patsayeva, S.V., Filippova, E.M., Chubarov, V.V. and Yuzhakov V.I. (1991) Variation in the fluorescence band of a dissolved natural organic substances induced by UV laser radiation, *Moscow Univ. Physics Bull.*, 46: 73-76.
75. Patsayeva, S.V. (1996) New methodological aspects of the old problem. Laser diagnostic of dissolved organic matter, *EARSeL, Advanced in Remote Sensing.*, 3: 66-70.
76. Frimmel, F.H. (1998) Impact of light on the properties of aquatic natural organic matter, *Environm. Internat.*, 24: 559-571.
77. Del Vecchio, R., Vodacek, A. and Blough N.V. (1998) The photobleaching of the Colored Dissolved Organic Matter in Natural Waters, SETAC 19th annual meeting, The natural connection: environmental integrity and human health, Charlotte, North Carolina, USA, November 15-19, 1998.
78. Miller, W.L. and Kuhn, P. (2000) A quantitative model for photochemical fading of aquatic chromophoric dissolved organic matter, Pacifichem 2000 International chemical congress of pacific basin society. Honolulu, Hawaii, December 14-19, 2000.

IMPACT OF UV RADIATION ON RICE-FIELD CYANOBACTERIA: ROLE OF PHOTOPROTECTIVE COMPOUNDS

RAJESHWAR P. SINHA AND DONAT-P. HÄDER
Friedrich Alexander Universität, Institut für Botanik und Pharmazeutische Biologie, Staudtstr. 5, 91058 Erlangen, Germany

1. Abstract

Members of the cyanobacteria are cosmopolitan in distribution, forming a prominent component of microbial populations in aquatic as well as terrestrial ecosystems. They play a central role in successional processes, global photosynthetic biomass production and nutrient cycling. In addition, N_2-fixing cyanobacteria are often the dominant microflora in wetland soils, especially in rice paddy fields, where they significantly contribute to fertility as a natural biofertilizer. Recent studies have shown a continuous depletion of the stratospheric ozone layer, as a result of anthropogenically released atmospheric pollutants such as chlorofluorocarbons (CFCs) and the consequent increase in solar UV-B radiation reaching the Earth's surface. Considering the vital role of cyanobacteria in crop production, the fluence rates of UV-B radiation impinging on the natural habitats are of major concern since, being photoautotrophic organisms, cyanobacteria depend on solar radiation as their primary source of energy. UV-B radiation causes reduction in growth, survival, protein content, heterocyst frequency and fixation of carbon and nitrogen, bleaching of pigments, disassembly of phycobilisomal complexes, DNA damage and alteration in membrane permeability in cyanobacteria. However, a number of cyanobacteria have developed photoprotective mechanisms to counteract the damaging effects of UV-B which includes synthesis of water soluble colourless mycosporine-like amino acids (MAA) and the lipid soluble yellow-brown coloured sheath pigment, scytonemin. Cyanobacteria, such as *Anabaena* sp., *Nostoc commune*, *Scytonema* sp. and *Lyngbya* sp. were isolated from rice paddy fields and other habitats in India and screened for the presence of photoprotective compounds. Spectroscopic and biochemical analyses revealed the presence of only shinorine, a bisubstituted MAA containing both a glycine and a serine group with an absorption maximum at 334 nm in all cyanobacteria except *Lyngbya* sp. There was a circadian induction in the synthesis of this compound by UV-B. A polychromatic action spectrum for the induction of MAAs in *Anabaena* sp. and *Nostoc commune* also shows the induction to be UV-B-dependent and peaking at around 290 nm. Another photoprotective compound, scytonemin, with an absorption maximum at 386 nm (which also absorbs significantly at 300, 278, 252 and 212 nm) was detected in all cyanobacteria except *Anabaena* sp. In addition, two unidentified, water-soluble, yellowish (induced by high white light) and brownish (induced by UV-B) compounds with an absorption maximum at 315 nm were recorded only in *Scytonema* sp. In

F. Ghetti et al. (eds.), Environmental UV Radiation: Impact on Ecosystems and Human Health and Predictive Models, 217–230.

conclusion, cyanobacteria having photoprotective mechanisms may be potent candidates as biofertilizers for crop plants.

2. Introduction

The stratospheric ozone layer shields the Earth from the biologically most hazardous short-wavelength solar radiation. There is mounting evidence that the solar flux of UV-B radiation has increased at the Earth's surface due to the depletion of the ozone layer by anthropogenically released atmospheric pollutants, such as chlorofluorocarbons (CFCs), chlorocarbons (CCs) and organo-bromides (OBs)[1-2]. Ozone depletion has been reported in the Antarctic, where ozone levels commonly have declined by more than 70 % in late winter and early spring during the last few decades, as well as in the Arctic and subarctic regions[3]. This decline in ozone level is commonly attributed to a unique combination of extreme cold temperatures and stratospheric circulation (the polar vortex) which results in conditions that are favourable for the CFC-ozone reactions. Polar stratospheric clouds play important roles in the formation of the springtime Antarctic "ozone hole" by activating chlorine and denitrifying the stratosphere. Recent TOMS (Total Ozone Mapping Spectrometer) data indicate an Antarctic 'ozone hole' that is three times larger than the entire land mass of the United States. The 'hole' had expanded to a record size of approximately 28.3 million square kilometres in the year 2000[4]. Moreover, ozone depletion has been predicted to continue throughout this century[5-6]. Recent reports indicate that widespread severe denitrification could enhance future Arctic ozone loss by up to 30 %[6].

Light is one of the most important factors determining the growth of photosynthetic organisms in their natural habitats. As biologically effective doses of UV-B radiation can penetrate deep into ecologically significant depths in natural waters, the observed increases in surface UV-B radiation may adversely affect the productivity of aquatic organisms[7]. UV-B has the potential to cause wide ranging effects, including alteration in the structure of proteins, DNA and other biologically relevant molecules, chronic depression of key physiological processes and acute physiological stress leading to either reduction in growth and cell division rates or death of the organism. Studies have shown a wide variation in tolerance to UV-B among species and taxonomic groups[8-10].

Although almost all aquatic organisms including cyanobacteria are more or less susceptible to UV-B, they are not defenceless. As a first line of defence, organisms can limit the damage of DNA and other chromophoric molecules by using photoprotective mechanisms[11-12]. Photoprotective compounds such as scytonemin and mycosporine-like amino acids (MAAs) play an important role in screening UV-B radiation in cyanobacteria[11-13]. An additional defence against UV-B radiation includes the induction of carotenoids[14] and the production of enzymes such as superoxide dismutases, catalases and peroxidases and scavengers such as vitamins C, B and E or cysteine and glutathione which can quench or scavenge UV-induced excited states and reactive oxygen species[15-16]. Certain organisms show diurnal rhythms and thereby protect themselves during the period of high UV-B irradiances. In the case that photoprotective mechanisms are insufficient to prevent damage, organisms have developed certain repair mechanisms for reversing UV-B-induced damage. DNA repair by the enzyme photolyase in the presence of light (photoreactivation) and nucleotide excision repair (NER) are the two major pathways to remove UV-induced DNA lesions from the genome, thereby preventing mutagenesis and cell death[17-18]. This chapter deals with the effects of UV-B radiation on rice-field cyanobacteria and the defence mechanisms operative against the negative effects of UV-B in these organisms.

3. Impacts of UV-B radiation on rice-field cyanobacteria

Cyanobacteria are phylogenetically a primitive group of gram-negative prokaryotes possessing higher plant-type oxygenic photosynthesis. Fossil evidence dates their appearance to the Precambrian era (between 2.8 and 3.5 x 10^9 years ago). At that time they might have been the main primary producers of organic matter and the first organisms to release oxygen into the then oxygen-free atmosphere. The advent of oxygenic photosynthesis on Earth probably have increased global biological productivity by a factor of 100 – 1000 and thereby profoundly affected both geochemical and biological evolution[19]. Thus, cyanobacteria were responsible for a major global evolutionary transformation leading to the development of aerobic metabolism and the subsequent rise of higher plant and animal forms[20-21]. Cyanobacteria are thought to have survived from a wide spectrum of environmental stresses such as extreme temperatures, drought, salinity, nitrogen starvation, photooxidation, anaerobiosis, osmotic and ultraviolet radiation stress[10,22].

Cyanobacteria have a cosmopolitan distribution ranging from hot springs to Arctic and Antarctic regions showing high variability in adapting to diverse environmental factors. They not only colonize oceans, lakes, rivers and various soils (especially rice paddy fields where cyanobacteria contribute significantly to fertility as natural biofertilizer), but, also make symbiotic associations with several organisms[10,23]. Members of cyanobacteria also possess a central position in the nutrient cycling largely due to their inherent capacity to fix atmospheric N_2 with the help of the enzyme nitrogenase, directly into ammonium (NH_4^+), a form through which nitrogen enters into the food chain[10]. Free-living as well as symbiotic cyanobacteria, particularly *Azolla-Anabaena* symbiosis, have been recognized as one of the most promising biofertilizer systems for wetland soils[10,24-25]. In nature the process of nitrogen fixation is of utmost importance because it counterbalances losses of nitrogen from the environment by denitrification[26].

Considering the vital role of cyanobacteria as a biofertilizer in rice and other crop production, the fluence rate of UV-B radiation impinging on their natural habitats, is of major concern since UV radiation has been reported not only to impair motility and photoorientation[27] but also to affect growth, survival, pigmentation, heterocyst differentiation, enzymes of nitrogen metabolism and total protein profiles[10,28-31]. In the following we discuss the effects of UV-B on some of the important physiological and biochemical processes in cyanobacteria.

Effects on growth and survival

The growth and survival of several cyanobacteria have been reported to be severely affected following UV irradiation for different durations. Growth and survival was arrested within 120 - 180 min of UV-B irradiation, depending upon the species type. Strains such as *Scytonema* sp. and *Nostoc commune*, filaments of which are embedded in mucilaginous sheaths, have been reported to be more tolerant in comparison to filaments which do not contain such coverings, such as *Anabaena* sp. and *Nostoc* sp.[28-31]. Several research groups have suggested that the cellular constituents absorbing radiation in the range 280 - 315 nm are destroyed by UV-B radiation, which may further affect the cellular membrane permeability and damage proteins, eventually resulting in the death of the cell[9-10,28-32].

Effects on pigmentation and phycobiliproteins

Photosynthetic organisms use various types of photoreceptor pigments to efficiently harvest the full spectrum of the light energy. In cyanobacteria, light harvesting is primarily carried out by a group of pigmented (accessory) proteins, known as phycobiliproteins. which are the constituents of macromolecular complexes called phycobilisomes (PBSs). There are three major classes of phycobiliproteins present in the PBSs: phycoerythrin (PE, red colored, λ_{max} 560 nm), phycocyanin (PC, blue colored, λ_{max} 620 nm) and allophycocyanin (APC, bluish green, λ_{max} 650 nm), which are soluble in aqueous media and constitute up to 50 % of the soluble proteins and about 24 % of the dry weight of the cell[33-34]. The radiant energy captured by the phycobiliproteins is transferred both within the PBSs as well as from the PBSs to the chlorophylls of the thylakoid membrane with about 100 % efficiency[35]. Each phycobiliprotein is made up of a heteromonomer of two different subunits, α (12 – 20 kDa) and β (15 – 22 kDa), each of which contains one to four chromophores. The aggregation state of phycobiliproteins suggest that the PBSs are composed of trimeric [($\alpha\beta_3$), single disc] and hexameric [($\alpha\beta_6$), double disc] phycobiliprotein assemblages. In addition to the pigmented proteins, the PBSs also contain about 15 % of proteins, mostly lacking in chromophores, referred to as linker polypeptides, arbitrarily divided into two groups, A (70 – 120 kDa) and B (25 – 70 kDa). These polypeptides play an important role in stabilization of PBSs assembly and unidirectional energy flow within the PBSs and from the PBSs to the photosynthetic reaction centers[33-35].

The study of the effects of UV-B on pigmentation of various cyanobacteria has revealed that all types of photosynthetic pigments such as chlorophyll *a* (λ_{max} at 437 and 672 nm) and the accessory light harvesting pigments (PC and PE) are affected by UV-B. In some cases the accessory light harvesting pigment phycocyanin was bleached more rapidly and drastically than any other pigment, such as the carotenoids (λ_{max} 480 nm) or chlorophyll *a*[28,30]. This shows that UV radiation can photooxidize and thereby bleach all types of photosynthetic pigments and may also cause a depression of Chl *a* and carotenoids via reduced rate of biosynthesis. A decrease in the phycobiliprotein content and the disassembly of phycobilisomal complex following UV-B irradiation have been reported in a number of cyanobacteria, indicating impaired energy transfer from the accessory pigments to the photosynthetic reaction centers[32,36-40]. Spectroscopic analyses have shown a drastic decline in both absorption and fluorescence, as well as a shift of the fluorescence peak towards shorter wavelengths, which are indicative of the disassembly of phycobilisomes and hence the impaired energy transfer from the phycobilisomes to the photosynthetic reaction centers[36-37]. This notion is further supported by SDS-PAGE analysis of the phycobiliproteins which shows a loss in the low molecular mass proteins between 16 and 22 kDa (phycobiliprotein α and β subunits) and high molecular mass rod and rod-core linker polypeptides (molecular masses between 24 and 45 kDa) as well as core membrane linker polypeptides (molecular masses around 66 kDa). It seems probable that the supramolecular organization of the phycobilisomes is disassembled largely due to the loss of linker polypeptides after UV irradiation, as has been shown by electrophoretic analyses[32,36-37,39].

Effects on heterocyst differentiation and enzymes of nitrogen metabolism

Differentiation of vegetative cells into heterocysts has also been reported to be severely affected by UV-B irradiation in a number of cyanobacteria[41]. Processes such as heterocyst differentiation and nitrogen fixation are metabolically very expensive and during extreme stress conditions may require cellular resources for their maintenance. Since heterocysts are important primarily in supplying fixed nitrogen to the vegetative cells, apparently vegetative cells growing in areas without external nitrogen sources may be seriously affected. Most probably the C:N ratio is severely affected following UV-B irradiation, which in turn affects the spacing pattern of heterocysts in a filament[41]. In addition, major heterocyst polypeptides of around 26, 54 and 55 kDa have also been shown to decrease following UV-B irradiation, suggesting that the multilayered thick wall of heterocysts may be disrupted resulting in the inactivation of the nitrogen fixing enzyme nitrogenase[41]. UV-B-induced membrane disruption, leading to changes in membrane permeability and release of ^{14}C-labelled compounds, has been observed in a number of cyanobacteria[32]. UV-B-induced inactivation of the nitrogen fixing enzyme nitrogenase has been reported in many cyanobacteria[41-42]. The process of nitrogen fixation has also been reported to be depressed during midday, when visible and UV irradiance are higher[43]. Nitrogenase is a molybdoenzyme and requires ATP and reductants for its activity. There is a possibility that the inhibition of the nitrogenase activity might be due to reduced supply of reductants and ATP following UV treatment. However, the loss of reductants and ATP does not occur immediately in cyanobacteria since many N_2 fixing species are capable to drive nitrogenase activity at the cost of an endogenous pool. It has been suggested that UV irradiation causes inactivation/denaturation of the nitrogenase enzyme and this appears to be a novel phenomenon[41-42]. Similarly, the primary ammonia assimilating enzyme, glutamine synthetase (GS), is also affected by UV-B irradiation, whereas nitrate reductase (NR) activity is stimulated[28,42].

Effects on total protein profiles

Total protein profiles of several cyanobacteria as evidenced by SDS-PAGE show a decrease in protein content with increasing UV-B exposure time, indicating that cellular proteins are among the main targets of UV-B[28,39,41], which is known to damage proteins and enzymes, especially those rich in aromatic amino acids such as tryptophan, tyrosine, phenylalanine and histidine, all of which show strong absorption in the UV range from 270 to 290 nm[9,28,39,41-42].

Effects on photosynthetic enzymes

The activity of the ribulose 1,5-bisphosphate carboxylase (RuBISCO), the primary CO_2 fixing enzyme, has been reported to be inhibited by UV-B irradiation in a number of cyanobacteria which may be due to protein destruction or enzyme inactivation[32]. The control of RuBISCO biosynthesis is strongly influenced by the prevailing light environment. During UV-B irradiation, proteins may undergo a number of modifications including photodegradation, increased aqueous solubility of membrane proteins, and fragmentation of the peptide chain, leading to inactivation of proteins (enzymes) and disruption of their structural entities[32,39]. UV-induced inhibition of $^{14}CO_2$ uptake in various rice-field cyanobacteria has been reported which could be due to the effect on the photosynthetic apparatus leading to the reduction in the supply of ATP and $NADPH_2$[41-42]. A disruption of the cell membrane and/or alteration in thylakoid integrity as a result of UV-B irradiation may partly or wholly destroy the components required for photosynthesis and may thus affect the rate of CO_2 fixation[32,41]. UV-induced loss of photosystem II activity[44] and opening of the membrane-bound calcium channels have been demonstrated in cyanobacteria[45].

Effects on DNA

The peak absorption of DNA lies in the UV-C range that is absorbed in the upper layers of the atmosphere and does not reach the Earth's surface. However, the absorption of UV-B radiation by DNA is sufficient to induce severe damage to the DNA in cyanobacteria. Absorbed quanta of UV can induce changes in the molecular structure of the DNA[46]. The two major classes of mutagenic DNA lesions induced by UV radiation are *cis-syn* cyclobutane-pyrimidine dimers (CPDs) and pyrimidine (6-4) pyrimidone photoproducts (6-4PPs). 6-4PPs are formed at 20 – 30 % of the yields of CPDs[18,47]. Both classes of lesions distort the DNA helix. CPDs and 6-4PPs induce a bend or kink of 7 – 9° and 44°, respectively. The ability of UV radiation to damage a given base is determined by the flexibility of the DNA. Sequences that facilitate bending and unwinding are favourable sites for damage formation, e.g. CPDs are formed at higher yields in single-stranded DNA, at the flexible ends of poly(dA)(dT) tracts, but not in their rigid centre. Bending of DNA towards the minor groove reduces CPD formation. One of the transcription factors having a direct effect on DNA damage formation and repair is the TATA-box binding protein (TBP). TBP promotes the selective formation of 6-4PPs in the TATA-box, where the DNA is bent, but CPDs are formed at the edge of the TATA-box and outside, where DNA is not bent. These DNA lesions interfere with DNA transcription and replication and can lead to misreadings of the genetic code and cause mutations and death[17-18,47]. UV-induced thymine dimer and other photoproducts formation have been reported in a number of cyanobacteria[48-50].

4. Impacts of UV-B radiation on cyanobacteria in polar ecosystems

In addition to the tropics, cyanobacteria are also common organisms in the freshwater and terrestrial ecosystems of Antarctica and also in the north polar regions[51]. Cyanobacteria have colonized a wide variety of polar environments including soils, the surface and interior of rocks, lakes, ponds, streams, moss beds, melt pools on glaciers and ice shelves, and littoral marine sediments. In some of these habitats they constitute the ecosystem dominant species in terms of total biomass and productivity. Cyanobacteria are also a conspicuous element of mature microbial communities across the surface of Antarctic rocks, as well as under and within translucent rocks[16]. Cyanobacterial communities form crusts, films and spectacular mats up to several centimeters in thickness, often intensely coloured with pigments. Mat-forming cyanobacteria are especially widespread in high latitude ponds and streams. In these environments the cyanobacteria often occupy shallow-water habitats that are exposed to full sunlight. Even moderate levels of UV-B radiation can have a major physiological impact on Antarctic cyanobacteria, but there are substantial differences between closely related species in their ability to escape the damaging effects of this high-energy waveband[51]. It has been concluded that the phototrophic organisms living in cold environments may be especially prone to the damaging effects of UV-B radiation. These findings are relevant to the perennially cold waters found in the north and south polar zones, where stratospheric ozone depletion and the associated increase in ambient UV-B radiation are proceeding most rapidly[6,52]. Thus, in natural habitats avoidance of UV-B radiation seems to be of utmost importance for cyanobacterial growth and nitrogen fixation.

5. Photoprotective compounds in cyanobacteria

Cyanobacteria may protect themselves from photodamage by adopting one or more of the following strategies: (a) production of ultraviolet-absorbing/screening substances such as mycosporine-like amino acids (MAAs) and scytonemin[11-13]; (b) escape from ultraviolet radiation by migration into habitats having reduced bright light exposure. Such strategies include phototactic, photokinetic and photophobic responses[53-54], vertical migration into deeper strata of mat communities[55] and sinking and floating behaviour by a combination of gas vacuoles and ballast[56]; (c) production of quenching agents such as carotenoids[15] or enzyme systems such as those containing superoxide dismutase that react with and thereby neutralize the highly toxic reactive oxygen species produced by UV radiation[16]; (d) availability of a number of repair mechanisms such as photoreactivation and light-independent nucleotide excision repair of DNA[18] and UV-A/blue light-guided repair of the photosynthetic apparatus[57]; and (e) chromatic adaptation (variation in phycocyanin/phycoerythrin ratio), which allows regulation of the balance of wavelengths of absorbed light [58]. Below we discuss only the MAAs and scytonemin and their evolutionary history.

Mycosporine-like Amino Acids (MAAs)

MAAs are water-soluble substances characterized by a cyclohexenone or cyclohexenimine chromophore conjugated with the nitrogen substituent of an amino acid or its imino alcohol, having absorption maxima ranging from 310 to 360 nm and an average molecular weight of around 300[11-13,59-61]. Accumulation in large quantities of colorless UV-absorbing substances (now known as MAAs) in cyanobacterial cells were reported in 1969 by Shibata[62]. Synthesis of MAAs probably takes place from the first part of the shikimate pathway[63-64]. A number of cyanobacteria isolated from freshwater, marine or terrestrial habitats contain MAAs[12,61]. Presence of MAAs has also been reported in Antarctic[51] as well as halophilic[65] cyanobacteria. The occurrence of high concentrations of MAAs in the organisms exposed to high levels of solar radiation has been described to provide protection as a UV-absorbing sunscreen[12,60], but there is no conclusive evidence for the exclusive role of MAAs as sunscreen, and it is possible that they play more than one role in the cellular metabolism in all or some organisms[9,62,66]. MAAs may act as antioxidants to prevent cellular damage resulting from UV-induced production of active oxygen species[67].

Studies with cyanobacteria have shown that MAAs prevent 3 out of 10 photons from hitting cytoplasmic targets. Cells with high concentrations of MAAs are approximately 25 % more resistant to ultraviolet radiation centred at 320 nm than those with no or low concentrations[60]. This protection is unlikely to be effective for thin, solitary trichomes, but may be especially important in some mat communities or large phytoplankton. The MAAs in *Nostoc commune* have been reported to be extracellular and linked to oligosaccharides in the sheath[68]. These glycosylated MAAs represent perhaps the only known example of MAAs that are actively excreted and accumulated extracellularly and therefore act as a true screen[69]. Experiments with rice paddy field cyanobacteria, *Anabaena* sp., *Nostoc commune* and *Scytonema* sp. have revealed the existence of a circadian induction by UV-B radiation of shinorine, having an absorption maximum at 334 nm[70-71]. The single-cell sunscreen effect of intracellular MAAs in cyanobacteria is modest and only 10 - 30 % of incident photons were intercepted in a fairly large-celled cyanobacterium *Gloeocapsa* sp.[60], but the screening efficiency may be substantially increased in colony- and mat-forming cyanobacteria[66]. There may be physiological limitations to the accumulation of osmotically active compounds such as MAAs within the cell, and it seems probable that the maximal specific content of MAAs in the cell is regulated by osmotic mechanisms which is reflected by the fact that field populations of halotolerant cyanobacteria contain unusually high concentration of MAAs[65]. UV and osmotic stress have been reported to induce and regulate the synthesis of MAAs in the cyanobacterium *Chlorogloeopsis*[72-73]. The polychromatic action spectra for the induction of shinorine in *Anabaena* sp. and *Nostoc commune* show a pronounced peak at 290 nm[74-75]. In an action spectrum of MAA synthesis in another cyanobacterium, *Chlorogloeopsis*, a distinct peak at 310 nm has been reported, and reduced pterin has been proposed as a putative candidate for the induction of shinorine[73]. It seems that the induction of MAAs in cyanobacteria is under photocontrol (particularly UV-B).

Scytonemin

Scytonemin is a yellow-brown, lipid soluble dimeric pigment located in the extracellular polysaccharide sheath of some cyanobacteria. The pigment was first observed in some terrestrial cyanobacteria by Nägeli[76] and later termed “scytonemin”. The occurrence of scytonemin is restricted to cyanobacteria, but it is widespread among this diverse group, and more than 300 species with sheaths coloured with yellow to brown pigments have been described. Scytonemin has an *in vivo* absorption maximum at 370 nm[77] with a molecular mass of 544 Da and a structure based on indolic and phenolic subunits[78]. Purified scytonemin has an absorption maximum at 386 nm, but it also absorbs significantly at 252, 278 and 300 nm[12,78-79]. Strong evidence for the role of scytonemin as ultraviolet shielding compound has been presented in several cyanobacterial isolated and collected materials from various harsh habitats, mostly exposed to high light intensities[61,77,80]. It was also found that the well-known red and green forms of scytonemin were related and could be interconverted by mild oxidation and reduction[80]. Its role as a sunscreen was clearly demonstrated in a terrestrial cyanobacterium *Chlorogloeopsis* sp.[77]. Scytonemin was also found to be responsible for the screening of ultraviolet radiation in a number of cyanobacterial lichens, such as *Collema*, *Gonohymenia* and *Petulla*, all of which were collected from high light intensity habitats[81]. However, intense illumination may not always be necessary for the production of scytonemin, as the pigment has been reported to occur in a cyanobacterium *Calothrix* deficient in Fe or Mg and exposed to low levels of irradiation[82].

In cyanobacterial cultures, as much as 5 % of the cellular dry weight may be accumulated as scytonemin. Naturally occurring cyanobacteria may have an even higher specific content[66]. The desiccation tolerant cyanobacterium *Lyngbya* sp., collected from the bark of mango trees, has also been reported to contain scytonemin[79]. The correlation between UV flux and scytonemin content in populations of shaded freshwater *Scytonema* and in colonies of *Rivularia* sp. were examined by Pentecost[83]. He found a variable and even negative correlation between UV and scytonemin in *Scytonema*, but a positive correlation in *Rivularia*. However, it was still probable that scytonemin functioned as a radiation shield in *Scytonema*, even in shaded sites, when water relations and cell division localization were taken into account. The correlation between UV protection and presence of scytonemin has been established under solar irradiance in a naturally occurring monospecific population of a cyanobacterium, *Calothrix* sp., and it was shown that high scytonemin content is required for uninhibited photosynthesis under high UV flux[84]. Studies indicate that the incident UV-A radiation entering the cells may be reduced by around 90 % due to the presence of scytonemin in the cyanobacterial sheaths[77,80,84]. Once synthesized, it remains highly stable and carries out its screening activity without further metabolic investment from the cell. Rapid photodegradation of scytonemin does not occur which is evidenced by its long persistence in terrestrial cyanobacterial crusts or dried mats[77,79,84]. This strategy may be invaluable to several scytonemin containing cyanobacteria inhabiting harsh habitats, such as intertidal marine mats or terrestrial crusts, where they experience intermittent physiological inactivity (e.g., desiccation). Under these metabolically inactive periods, other ultraviolet protective mechanisms such as active repair or biosynthesis of damaged cellular components would be ineffective[69,79,84].

In addition to the above reports, another UV protective agent with absorption maxima at 312 and 330 nm has been reported for the terrestrial cyanobacterium *Nostoc commune*, a species that also produces scytonemin[85]. A brown *Nostoc* sp. that produces three UV-absorbing compounds with absorption maxima at 256, 314 and 400 nm, has been reported to be resistant to high light intensity and UV radiation[86]. The shielding role of certain cyanobacterial pigments (a brown-colored pigment from *Scytonema hofmanii* and a pink extract from *Nostoc spongiaeforme*) against UV-B induced damage has been demonstrated[87].

6. Evolutionary history of photoprotective compounds

The possibility that organic molecules might have acted as a UV screen in aquatic environments of the early Earth was first considered by Sagan[88]. The nature and evolutionary origin of the first specific photoprotective compounds on the Achaean Earth is unknown, but according to one assumption early aromatic containing reaction centres were some of the earliest UV screens that over the period altered from a nonproductive dissipative UV screen to a light harvesting role in photosynthesis[89]. MAAs play a vital role as osmotic regulators in some cyanobacteria[65] and such alternative role may have given rise to the first UV-screening MAAs. MAAs evolution as specific UV-protectants may represent an early innovation in dealing with Achaean UV-B fluxes. Certain MAAs such as mycosporine-glycine specifically absorb in the UV-B range of the spectrum. It has been postulated that later, as oxygen levels increased, UV-A screening MAAs became more important since many of the effects of UV-A are mediated through oxygen free radicals and thus the role of UV-A as a damaging agent in the biosphere increased. In these compounds, the nitrogen atom replacing the ketone function has a greater mesomeric effect on the benzene ring and the absorbance is shifted towards the UV-A. Most probably, a mutation in the earliest UV-B screening compounds resulted in a UV-A screen which became physiologically advantageous. Since MAAs have been reported in several eukaryotic algae, it is likely that they were passed to the eukaryotic algae by cyanobacteria in the plastidic line[13]. It has been suggested that the protection against UV radiation provided by scytonemin may have been an important ecophysiological factor in cyanobacterial evolutionary history[90]. Scytonemin would have facilitated the ecological expansion of cyanobacterial mat communities into exposed shallow-water and terrestrial habitats during the early to middle Precambrian, despite the high levels of UV radiation impinging on the Earth's surface at that time.

7. Conclusions

Increases in the level of UV-B radiation are likely to induce changes in cyanobacterial community composition since there are great differences in susceptibility of species to UV-induced damage. Species having the ability to accumulate UV-screening substances or with more effective repair mechanisms will likely be favoured. The ecological significance of photoprotective compounds in diverse organisms as screening agents against UV-induced damage has yet to be fully elucidated. The spatial distribution of MAAs within the cells is not well known. However, photoprotective compounds may serve at least three different functions: (a) protection of the cell from UV photodamage by playing the role of sunscreen[11-13], (b) transfer radiant energy to the photosynthetic reaction centres, which is supported by the fact that the emitted fluorescence spectrum of MAAs peaks at a wavelength near the Soret band of chlorophyll absorption[70], and (c) aid in osmotic regulation[65,72].

Acknowledgement. The work outlined here was financially supported by the European Union (DG XII, Environmental Program, ENV4-CT97-0580).

References

1. Crutzen, P.J. (1992) Ultraviolet on the increase. *Nature*, 356: 104-105.
2. Lubin, D. and Jensen, E.H. (1995) Effects of clouds and stratospheric ozone depletion on ultraviolet radiation trends. *Nature,* 377: 710-713.
3. Kerr, J.B. (1994) Decreasing ozone cause health concern. *Environ. Sci. Technol.*, 28: 514-518.
4. News in brief (2000) NASA encounters biggest-ever Antarctic ozone hole. *Nature*, 407: 122.
5. Toon, O.B. and Turco, R.P. (1991) Polar stratospheric clouds and ozone depletion. *Sci. Am.*, 264: 68-74.
6. Tabazadeh, A., Santee, M.L., Danilin, M.Y., Pumphrey, H.C., Newman, P.A., Hamill, P.J. and Mergenthaler, J.L. (2000) Quantifying denitrification and its effect on ozone recovery. *Science*, 288: 1407-1411.
7. Häder, D.-P., Kumar, H.D., Smith, R.C. and Worrest, R.C. (1998) Effects on aquatic ecosystems. *J. Photochem. Photobiol. B: Biol.*, 46: 53-68.
8. Karentz, D., Cleaver, J.E. and Mitchell, D.L. (1991) Cell survival characteristics and molecular responses of Antarctic phytoplankton to ultraviolet-B radiation. *J. Phycol.*, 27: 326-341.
9. Vincent, W.F. and Roy, S. (1993) Solar ultraviolet-B radiation and aquatic primary production: damage, protection, and recovery. *Environ. Rev.*, 1: 1-12.
10. Sinha, R.P. and Häder, D.-P. (1996) Photobiology and ecophysiology of rice field cyanobacteria. *Photochem. Photobiol.*, 64: 887-896.
11. Garcia-Pichel, F. and Castenholz, R.W. (1993) Occurrence of UV-absorbing, mycosporine-like compounds among cyanobacterial isolates and an estimation of their screening capacity. *Appl. Environ. Microbiol.*, 59: 163-169.
12. Sinha, R.P., Klisch, M., Gröniger, A. and Häder, D.-P. (1998) Ultraviolet-absorbing/screening substances in cyanobacteria, phytoplankton and macroalgae. *J. Photochem. Photobiol. B: Biol.*, 47: 83-94.
13. Cockell, C.S. and Knowland, J. (1999) Ultraviolet radiation screening compounds. *Biol. Rev.*, 74: 311-345.
14. Götz, T., Windhövel, U., Böger, P. and Sandmann, G. (1999) Protection of photosynthesis against ultraviolet-B radiation by carotenoids in transformants of the cyanobacterium *Synechococcus* PCC7942. *Plant Physiol.*, 120: 599-604.
15. Burton, G.W. and Ingold, K.U. (1984) β-Carotene: an unusual type of lipid antioxidant. *Science*, 224: 569-573.
16. Vincent, W.F. and Quesada, A. (1994) Ultraviolet radiation effects on cyanobacteria: implication for Antarctic microbial ecosystems. *Antarctic Research Series*, 62: 111-124.
17. Britt, A.B. (1996) DNA damage and repair in plants. *Annu. Rev. Plant Physiol. Plant Mol. Biol.*, 47: 75-100.
18. Thoma, F. (1999) Light and dark in chromatin repair: repair of UV-induced DNA lesions by photolyase and nucleotide excision repair. *EMBO J.*, 18: 6585-6598.
19. Hoehler, T.M., Bebout, B.M. and Des Marais, D.J. (2001) The role of microbial mats in the production of reduced gases on the early earth. *Nature*, 412: 324-327.
20. Schopf, J.W. (1993) Microfossils of the early archean chert: new evidence of the antiquity of life. *Science*, 260: 640-646.
21. Pace, N.R. (1997) A molecular view of microbial diversity and the biosphere, *Science*. 276: 734-740.
22. Tandeau de Marsac, N. and Houmard, J. (1993) Adaptation of cyanobacteria to environmental stimuli: new steps towards molecular mechanisms. *FEMS Microbiol. Rev.*, 104: 119-190.
23. Rai, A.N., Söderbäck, E. and Bergman, B. (2000) Cyanobacterium-plant symbioses. *New Phytol.*, 147: 449-481.
24. Vaishampayan, A., Sinha, R.P. and Häder, D.-P. (1998) Use of genetically improved nitrogen-fixing cyanobacteria in rice paddy fields: prospects as a source material for engineering herbicide sensitivity and resistance in plants. *Bot. Acta*, 111: 176-190.
25. Sinha, R.P., Vaishampayan, A. and Häder, D.-P. (1998) Plant-cyanobacterial symbiotic somaclones as a potential bionitrogen-fertilizer for paddy agriculture: biotechnological approaches. *Microbiol. Res.*, 153: 297-307.
26. Kuhlbusch, T.A., Lobert, J.M., Crutzen, P.J. and Warneck, P. (1991) Molecular nitrogen emission. Trace nitrification during biomass burning. *Nature*, 351: 135-137.
27. Donkor, V. and Häder, D.-P. (1991) Effects of solar and ultraviolet radiation on motility, photomovement and pigmentation in filamentous, gliding cyanobacteria. *FEMS Microbiol. Ecol.*, 86: 159-168.
28. Sinha, R.P., Kumar, H.D., Kumar, A. and Häder, D.-P. (1995) Effects of UV-B irradiation on growth, survival, pigmentation and nitrogen metabolism enzymes in cyanobacteria. *Acta Protozool.*, 34: 187-192.

29. Sinha, R.P., Singh, S.C. and Häder, D.-P. (1999) Photoecophysiology of cyanobacteria. *Recent Res. Devel. Photochem. & Photobiol.*, 3: 91-101.
30. Sinha, R.P. and Häder, D.-P. (2000) Effects of UV-B radiation on cyanobacteria. *Recent Res. Devel. Photochem. & Photobiol.*, 4: 239-246.
31. Sinha, R.P., Sinha, J.P. and Häder, D.-P. (2000) Effects of solar UV-B radiation on aquatic ecosystems. *Res. Adv. in Photochem. & Photobiol.*, 1: 95-103.
32. Sinha, R.P., Singh, N., Kumar, A., Kumar, H.D. and Häder, D.-P. (1997) Impacts of ultraviolet-B irradiation on nitrogen-fixing cyanobacteria of rice paddy fields. *J. Plant Physiol.*, 150: 188-193.
33. Glazer, A.N. (1985) Light harvesting by phycobilisomes. *Annu. Rev. Biophy. Chem.*, 14: 47-77.
34. Grossman, A.R., Schaefer, M.R., Chiang, G.G. and Collier, J.L. (1993) The phycobilisome, a light harvesting complex responsive to environmental conditions. *Microbiol. Rev.*, 57: 725-749.
35. Glazer, A.N. (1989) Directional energy transfer in photosynthetic antenna. *J. Biol. Chem.*, 264: 1-4.
36. Sinha, R.P., Lebert, M., Kumar, A., Kumar, H.D. and Häder, D.-P. (1995) Disintegration of phycobilisomes in a rice field cyanobacterium *Nostoc* sp. following UV irradiation. *Biochem. Mol. Biol. Int.*, 37: 697-706.
37. Sinha, R.P., Lebert, M., Kumar, A., Kumar, H.D. and Häder, D.-P. (1995) Spectroscopic and biochemical analyses of UV effects on phycobiliproteins of *Anabaena* sp. and *Nostoc carmium. Bot. Acta*, 180: 87-92.
38. Sinha, R.P. and Häder, D.-P. (1998) Phycobilisomes and environmental stress. In *Advances in Phycology* (B.N. Verma, A.N. Kargupta and S.K. Goyal, eds.), APC Publications Pvt. Ltd., New Delhi, India, pp. 71-80.
39. Sinha, R.P. and Häder, D.-P. (1998) Effects of ultraviolet-B radiation in three rice field cyanobacteria. *J. Plant Physiol.*, 153: 763-769.
40. Lao, K. and Glazer, A.N. (1996) Ultraviolet-B photodestruction of a light-harvesting complex. *Proc. Natl. Acad. Sci. USA*, 93: 5258-5263.
41. Sinha, R.P., Singh, N., Kumar, A., Kumar, H.D., Häder, M. and Häder, D.-P. (1996) Effects of UV irradiation on certain physiological and biochemical processes in cyanobacteria. *J. Photochem. Photobiol. B: Biol.*, 32: 107-113.
42. Kumar, A., Sinha, R.P. and Häder, D.-P. (1996) Effect of UV-B on enzymes of nitrogen metabolism in the cyanobacterium *Nostoc calcicola. J. Plant Physiol.*, 148: 86-91.
43. Peterson, R.B., Friberg, E.E. and Burris, R.H. (1979) Diurnal variation in N_2 fixation and photosynthesis by aquatic blue-green algae. *Plant Physiol.*, 59: 74-80.
44. Máté, Z., Sass, L., Szekeres, M., Vass, I. and Nagy, F. (1998) UV-B induced differential transcription of *psbA* genes encoding the D1 protein of photosystem II in the cyanobacterium *Synechocystis* 6803. *J. Biol. Chem.*, 273: 17439-17444.
45. Richter, P., Krywult, M., Sinha, R.P. and Häder, D.-P. (1999) Calcium signals from heterocysts of *Anabaena* sp. after UV irradiation. *J. Plant Physiol.*, 154: 137-139.
46. Karentz, D. (1994) Ultraviolet tolerance mechanisms in Antarctic marine organisms. *Antarctic Research Series*, 62: 93-110.
47. Friedberg, E.C., Walker, G.C. and Siede, W. (1995) *DNA Repair and Mutagenesis*, ASM Press, Washington DC.
48. O`Brien, P.A. and Houghton, J.A. (1982) UV-induced DNA degradation in the cyanobacterium *Synechocystis* PCC 6308. *Photochem. Photobiol.*, 36: 417-422.
49. Ng, W.-O., Zentella, R., Wang, Y., Taylor, J.-S.A. and Pakrasi, H.B. (2000) phrA, the major photoreactivating factor in the cyanobacterium *Synechocystis* sp. strain PCC 6803 codes for a cyclobutane-pyrimidine-dimer-specific DNA photolyase. *Arch. Microbiol.*, 173: 412-417.
50. Sinha, R.P., Dautz, M. and Häder, D.-P. (2001) A simple and efficient method for the quantitative analysis of thymine dimers in cyanobacteria, phytoplankton and macroalgae. *Acta Protozool.*, 40: 187-195.
51. Quesada, A. and Vincent, W.F. (1997) Strategies of adaptation by Antarctic cyanobacteria to ultraviolet radiation. *Eur. J. Phycol.*, 32: 335-342.
52. Roos, J.C. and Vincent, W.F. (1998) Temperature dependence of UV radiation effects on Antarctic cyanobacteria. *J. Phycol.*, 34: 118-125.
53. Häder, D.-P. (1987) Photomovement. In *The Cyanobacteria* (P. Fay and C. Van Baalen, eds.), Elsevier, Amsterdam, pp. 325-345.
54. Häder, D.-P. (1987) Photosensory behavior in prokaryotes. *Microbiol. Rev.*, 51: 1-21.
55. Bebout, B.M. and Garcia-Pichel, F. (1995) UV-B induced vertical migrations of cyanobacteria in a microbial mat. *Appl. Environ. Microbiol.*, 61: 4215-4222.
56. Reynolds, C.S., Oliver, R.L. and Walsby, A.E. (1987) Cyanobacterial dominance: the role of buoyancy regulation in dynamic lake environments. *NZ J. Mar. Freshwater Res.*, 21: 379-390.

57. Han, T., Sinha, R.P. and Häder, D.-P. (2001) UV-A/blue light-induced reactivation of photosynthesis in UV-B irradiated cyanobacterium. *Anabaena* sp., *J. Plant Physiol.*, 158: 1403-1413.
58. Tandeau de Marsac, N. (1977) Occurrence and nature of chromatic adaptation in cyanobacteria. *J. Bacteriol.*, 130: 82-91.
59. Dunlap, W.C. and Shick, J.M. (1998) Ultraviolet radiation-absorbing mycosporine-like amino acids in coral reef organisms: a biochemical and environmental perspective. *J. Phycol.*, 34: 418-430.
60. Garcia-Pichel, F., Wingard, C.E. and Castenholz, R.W. (1993) Evidence regarding the UV sunscreen role of a mycosporine-like compound in the cyanobacterium *Gloeocapsa* sp.. *Appl. Environ. Microbiol.*, 59: 170-176.
61. Gröniger, A., Sinha, R.P., Klisch, M. and Häder, D.-P. (2000) Photoprotective compounds in cyanobacteria, phytoplankton and macroalgae - a database. *J. Photochem. Photobiol. B: Biol.*, 58: 115-122.
62. Shibata, K. (1969) Pigments and a UV-absorbing substance in corals and a blue-green alga living on the Great Barrier Reef. *Plant Cell Physiol.*, 10: 325-335.
63. Favre-Bonvin, J., Bernillon, J., Salin, N. and Arpin, N. (1987) Biosynthesis of mycosporine: mycosporine glutaminol in *Trichothecium roseum*. *Phytochemistry*, 26: 2509-2514.
64. Shick, J.M., Romaine-Lioud, S., Ferrier-Pagès, C. and Gattuso, J.-P. (1999) Ultraviolet-B radiation stimulates shikimate pathway-dependent accumulation of mycosporine-like amino acids in the coral *Stylophora pistillata* despite decreases in its population of symbiotic dinoflagellates. *Limnol. Oceanogr.*, 44: 1667-1682.
65. Oren, A. (1997) Mycosporine-like amino acids as osmotic solutes in a community of halophilic cyanobacteria. *Geomicrobiol. J.*, 14: 231-240.
66. Castenholz, R.W. (1997) Multiple strategies for UV tolerance in cyanobacteria. *Spectrum*, 10: 10-16.
67. Dunlap, W.C. and Yamamoto, Y. (1995) Small-molecule antioxidants in marine organisms: antioxidant activity of mycosporine-glycine. *Comp. Biochem. Physiol.*, 112: 105-114.
68. Böhm, G.A., Pfleiderer, W., Böger, P. and Scherer, S. (1995) Structure of a novel oligosaccharide-mycosporine-amino acid ultraviolet A/B sunscreen pigment from the terrestrial cyanobacterium *Nostoc commune*. *J. Biol. Chem.*, 270: 8536-8539.
69. Ehling-Schulz, M., Bilger, W. and Scherer, S. (1997) UV-B-induced synthesis of photoprotective pigments and extracellular polysaccharides in the terrestrial cyanobacterium *Nostoc commune*. *J. Bacteriol.*, 179: 1940-1945.
70. Sinha, R.P., Klisch, M. and Häder, D.-P. (1999) Induction of a mycosporine-like amino acid (MAA) in the rice-field cyanobacterium *Anabaena* sp. by UV irradiation. *J. Photochem. Photobiol. B: Biol.*, 52: 59-64.
71. Sinha, R.P., Klisch, M., Helbling, E.W. and Häder, D.-P. (2001) Induction of mycosporine-like amino acids (MAAs) in cyanobacteria by solar ultraviolet-B radiation. *J. Photochem. Photobiol. B: Biol.*, 60: 129-135.
72. Portwich, A., Garcia-Pichel, F. (1999) Ultraviolet and osmotic stresses induce and regulate the synthesis of mycosporines in the cyanobacterium *Chlorogloeopsis* PCC 6912. *Arch. Microbiol.*, 172: 187-192.
73. Portwich, A., Garcia-Pichel, F. (2000) A novel prokaryotic UVB photoreceptor in the cyanobacterium *Chlorogloeopsis* PCC 6912. *Photochem. Photobiol.*, 71: 493-498.
74. Sinha, R.P., Sinha, J.P., Gröniger, A. and Häder, D.-P. (2002) A polychromatic action spectrum for the induction of a mycosporine-like amino acid (MAA) in a rice-field cyanobacterium, *Anabaena* sp.. *J. Photochem. Photobiol. B: Biol.*, 66: 47-53.
75. Sinha, R.P., Ambasht, N.K., Sinha, J.P. and Häder, D.-P. (2001) Induction of a mycosporine-like amino acid (MAA) in a rice-field cyanobacterium, *Nostoc commune*: a polychromatic action spectrum. *Acta Protozool.*, in press.
76. Nägeli, C. (1849) Gattungen einzelliger Algen, physiologisch und systematisch bearbeitet. *Neue Denkschrift Allg. Schweiz. Natur Ges.*, 10: 1-138.
77. Garcia-Pichel, F., Sherry, N.D. and Castenholz, R.W. (1992) Evidence for a UV sunscreen role of the extracellular pigment scytonemin in the terrestrial cyanobacterium *Chlorogloeopsis* sp.. *Photochem. Photobiol.*, 56: 17-23.
78. Proteau, P.J., Gerwick, W.H., Garcia-Pichel, F. and Castenholz, R.W. (1993) The structure of scytonemin, an ultraviolet sunscreen pigment from the sheaths of cyanobacteria. *Experientia*, 49: 825-829.
79. Sinha, R.P., Klisch, M., Vaishampayan, A. and Häder, D.-P. (1999) Biochemical and spectroscopic characterization of the cyanobacterium *Lyngbya* sp. inhabiting Mango (*Mangifera indica*) trees: presence of an ultraviolet-absorbing pigment, scytonemin. *Acta Protozool.*, 38: 291-298.
80. Garcia-Pichel, F. and Castenholz, R.W. (1991) Characterization and biological implications of scytonemin, a cyanobacterial sheath pigments. *J. Phycol.*, 27: 395-409.

81. Büdel, B., Karsten, U. and Garcia-Pichel, F. (1997) Ultraviolet-absorbing scytonemin and mycosporine-like amino acids derivatives in exposed, rock-inhabiting cyanobacterial lichens. *Oecologia*, 112: 165-172.
82. Sinclair, C. and Whitton, B.A. (1977) Influence of nutrient deficiency on hair formation in the Rivulariaceae. *Br. Phycol. J.*, 12: 297-313.
83. Pentecost, A. (1993) Field relationships between scytonemin density, growth and irradiance in cyanobacteria occurring in low illumination regimes. *Micro. Ecol.*, 26: 101-110.
84. Brenowitz, S. and Castenholz, R.W. (1997) Long-term effects of UV and visible irradiance on natural populations of a scytonemin-containing cyanobacterium (*Calothrix* sp.). *FEMS Microbiol. Ecol.*, 24: 343-352.
85. Scherer, S., Chen, T.W. and Böger, P. (1988) A new UV-A/B protecting pigment in the terrestrial cyanobacterium *Nostoc commune*. *Plant Physiol.*, 88: 1055-1057.
86. de Chazal, N.M. and Smith, G.D. (1994) Characterization of a brown *Nostoc* sp. from Java that is resistant to high light intensity and UV. *Microbiology*, 140: 3183-3189.
87. Kumar, A., Tyagi, M.B., Srinivas, G., Singh, N., Kumar, H.D., Sinha, R.P. and Häder, D.-P. (1996) UVB shielding role of $FeCl_3$ and certain cyanobacterial pigments. *Photochem. Photobiol.*, 64: 321-325.
88. Sagan, C. (1973) Ultraviolet radiation selection pressure on the earliest organisms. *J. Theoret. Biol.*, 39: 195.
89. Mulkidjanian, A. Y. and Junge, W. (1997) On the origin of photosynthesis as inferred from sequence analysis. *Photosynth. Res.*, 51: 27.
90. Dillon, J. G. and Castenholz, R. W. (1999) Scytonemin, a cyanobacterial sheath pigment, protects against UVC radiation: implications for early photosynthetic life. *J. Phycol.*, 35: 673.

EFFECT OF UV-B RADIATION ON CILIATED PROTOZOA

ROBERTO MARANGONI, FABIO MARRONI, FRANCESCO GHETTI, DOMENICO GIOFFRÉ AND GIULIANO COLOMBETTI
CNR Istituto di Biofisica, Pisa, Italy

1. Abstract

This paper reviews some of the results described in the literature on the effect of UV-B radiation on ciliated protozoa, concentrating in particular on the changes induced in motility and photomotility, which are both important in determining the capability of these organisms to survive in their environment. It will be shortly described what ciliates are and why they are an important component of ecological systems. A summary will follow of the early works, where the effects of UV radiation on ciliates were investigated. Finally, it will be described in some more detail the results of studies on a marine ciliate, *Fabrea salina*, and two fresh-water ciliates, *Blepharisma japonicum* and *Ophryoglena flava*.

2. Introduction

The amount of UV-B radiation that reaches the Earth surface is determined in first approximation by the ozone concentration in the stratospheric layer. It is widely acknowledged that the ozone layer has been depleted because of the production of chemicals such as CFCs and methyl bromide which react with ozone molecules turning them into oxygen. A consequence of this depletion of the stratospheric ozone layer might be an increase of the UV-B radiation reaching the Earth.

UVB radiation penetrates to a certain depth in the water (depending on the turbidity of the water itself), so it can affect phyto- and zooplankton microorganisms in both marine and fresh-water ecosystems. Recent studies have shown that UV-B radiation can damage microorganisms in many ways: it can affect growth and metabolic processes such as photosynthesis. In addition to this, UV-B radiation may impair motility and the orientation with respect to various stimuli, such as light, gravity and chemical substances[1]. Microorganisms exposed to UV-B radiation may become unable to orient themselves in the water column and to find suitable condition for growth and survival. The biological effects of a possible increase in UV irradiance on our planet, due to the reduction of the ozone layer, have been investigated in the last twenty years or so. Among the systems that have been examined, a good deal of work has been done on single cell systems, where the complex interactions between cells and organs typical of higher organisms are not present. This makes, in principle, easier to understand the basic biological processes affected by UV radiation.

F. Ghetti et al. (eds.), Environmental UV Radiation: Impact on Ecosystems and Human Health and Predictive Models, 231–248.

Among these systems, ciliates deserve a particular attention, for their importance in many ecosystems. The Ciliata, or Ciliophora, include about 7000 known species of some of the most complex single-cell organisms ever. Ciliates include organisms such as *Tetrahymena*, *Paramecium*, *Stentor*, *Euplotes*, *Fabrea*, *Blepharisma* and *Ophryoglena*. They derive their name from the Latin word *cilium* (eyelash), which gives an immediate idea of the aspect of many ciliates. In fact, some or all of the surface of a ciliate is covered with relatively short, dense and fine hair-like structures, the cilia, that beat rhythmically and in an orderly fashion to propel the ciliate through the water and/or to draw in food particles. Ciliates can be found in almost every environment with liquid water: ocean waters, marine sediments, lakes, ponds and rivers, and even soils. This have a great ecological importance. For instance, ciliates are especially important trophic links in microbial food webs, because they are the major consumers of bacteria, pico- and nano-plankton, diatoms, dinoflagellates, and amoebae. In their turn, they are eaten by higher organisms, such as crustacea in the zooplankton and larval fish.

They also play a role in cleaning water in sewage treatment plants, where some species eat sewage and others feed upon the bacteria that grow on the waste. Because of their role in microbial food network and their sensitivity to pollutants, which may vary greatly from a species to another, they may be used as reliable indicators of the state of the microbial environment and can give quick information on the degree of its pollution.

Even though ciliates are unicellular organisms, they do possess a complex morphology and physiology and are, moreover, extremely sensitive to environmental signals; they can perceive a wide variety of stimuli, from mechanical to thermal, from chemical to optical and gravitational. In many cases, their reaction to these stimuli is shown as an alteration of their motile behavior. These perception functions are all integrated in a single cell and all participate in controlling the motile behavior; in fact, the final step of a stimulus-induced behavior consists in an alteration of the membrane potential and a modulation of the activity of the motile apparatus. The presence of all these control circuits in the same ciliate cell has led some author to speak of ciliates as "walking neurons".

It has been known for a long time that exposure to artificial UV irradiation at high doses can kill these cells; at lower doses, it can destroy (or seriously damage) many of the control circuits and of the ultrastructural elements of the cell itself.

This suggests that also the increase of environmental UV can alter the ciliate physiology, leading to a decrease of their survival[2-4]. Since ciliates are decisive components of the food chain, it is very important, from an ecological point of view, to understand how they are affected by UV irradiation, and UV-B (280-320 nm) in particular, and if they possess repair mechanisms.

Many authors have worked on this subject, exploring a wide range of possible effects of UV exposure (from cell elongation to alterations in the reproduction cycle) often using different investigation methods; the first part of this chapter will give the reader a schematic summary of these works.

3. Early studies and general considerations

The first studies were mainly aimed at determining the relationship between UV exposure and cell survival, trying to establish the minimum dose sufficient to severely damage the cell[5-7]. Most of these experiments were performed by determining the UV dose (often using UV-C wavelengths, 200-280 nm) sufficient to determine an immobilization of the cells or a cytolysis: sometimes the minimum dose sufficient to determine an unusual motile pattern, such as rotation, was assumed as a quantitative assessment of cell damage. Together with these dosimetric assays, the role of environmental conditions, such as pH or salt composition of the medium, has been investigated.

Table 1. Resistance to UV-irradiation as a function of environment and physiological state of different protozoa. The dose for 50% immobilization of *P. multimicronucleatum* was set at 100, and the other values are relative to this (from ref. 5).

Protozoan	**Medium or environment**	**50% rotation**	**50% immobilization**	**50% cytolysis**
Paramecium	Balanced inorganic salt solution	61 ± 6.4	100 ± 2.1	
	Culture medium	84 ± 2.5	114 ± 0	
	Distilled water	72 ± 5.6	89 ± 3.8	
	pH 6.0	75 ± 0	105 ± 1.8	265 ± 12
	pH 7.0	65 ± 0	100 ± 0	230 ± 10
	pH 8.0	50 ± 0	100 ± 0	205 ± 10
	16 °C	39 ± 1.7	82 ± 3.7	
	19 °C	51 ± 4.5	94 ± 1	
	26 °C	79 ± 2.1	100 ± 1	
	4-day-old-culture	62 ± 4.7	100 ± 5.8	2201 ± 12.1
	11-day-old-culture	48 ± 6.7	81 ± 10.0	172 ± 3.1
Blepharisma	pH 6.0		82 ± 2.5	156 ± 1.8
	pH 7.0		58.5 ± 7.5	101 ± 5.3
	pH 8.0		54.5 ± 7.6	121 ± 5.2
Tetrahymena	1-day-old-culture in yeast estract		146 ± 4.4	
	8-day-old-culture in yeast estract		114 ± 7.0	
	3-day-old lettuce culture		48.5 ± 4.2	

These early studies showed that low doses of UV irradiation cause an increase in cell speed, while, after a certain dose, the cell speed decreases, the motion pattern becomes anomalous (cells tend to rotate around their cell body) and, eventually, cells stop and die. Some of these results have been confirmed by our studies (see below).

The data reported in Table 1 show that the effect of UV-irradiation seems to depend on environmental parameters, such as pH or salt concentrations. The nutritional status and the culture age might also play a role.

The resistance against UV-irradiation may greatly vary among ciliates: for example, *F. salina* shows a resistance about tenfold greater than that of *B. undulans*, which belongs to the same order Heterotrichida (Table 2).

Some authors have tried to correlate the UV-resistance with the presence of pigments in the cell cytoplasm; in a white mutant of *Blepharisma incertus*, for example,

Table 2. Relative resistance of different protozoa to UV-irradiation. The dose for 50% immobilization of *P. multimicronucleatum* was set at 100 and the other values are relative to this (from ref. 5).

Protozoan		Size (μm)	50% rotation	50% immobilization
Tetrahymena glaucoformia		25 x 50		146 ± 4.4
Colpidium colpoda		55 x 88		132 ± 6.8
Stylonychia curvata		48 x 98	34 ± 6.4	80 ± 4.7
Paramecium bursaria		43 x 105	87 ± 6.3	103 ± 37.0
Paramecium Aurelia		49 x 120	92 ± 0.3	119 ± 5.7
Paramecium multimicronucleatum				
	N. 1	51 x 175	76 ± 6.9	100 ± 4.2
	N. 2	64 x 123	73 ± 22.0	116 ± 19
Paramecium caudatum		70 x 173	58 ± 4.4	101 ± 3.4
Blepharisma undulans		52 x 175	43 ± 3.2	59 ± 6.1
Spirostomum ambiguum		36 x 369	32 ± 3.3	
Bursaria truncatella		200 x 337	112 ± 6.3	147 ± 14.1
Fabrea salina		169 x 225	555 ± 196	896 ± 126

the pigment deprivation produces an increase in UV-vulnerability[7]. However, UV-resistance is not always correlated with pigmentation, as shown by our results on an apparently colorless strain of *Fabrea salina*, which is more resistant to the same radiation doses than the pigmented protozoa *B. japonicum* (see below). Even if these findings are derived from measures of different parameters and are difficult to compare to those of Giese[5-8], they indicate that there is no direct relationship between pigment content and UV-resistance.

Structural changes in the cell after UV-exposure have been described: cells can elongate, vacuolate or the cell wall can break causing a cytolysis. Particular attention has been given to the so called retardation effect: it has been noticed, in fact, that cells exposed to UV (and UV-B in particular) strongly reduce the frequency of division[9]. The delay in reproduction has been measured as a function of the UV-dose received and of some other environmental factors such as composition of the culture medium, starving period, etc., finding also in this case an interplay of all these variables.

In order to try to identify the molecular targets of UV radiation in ciliates, the action spectrum for the retardation response has been determined in some species; the results are, in certain cases, difficult to interpret: the action spectrum at low doses resembles the absorption spectrum of cytoplasmic proteins, whereas the action spectrum at larger doses is complex, with a contribution surely due to the DNA[8] (Figure 1).

The relationship between UV-damage and nucleic acids has been pointed out by different authors; in general it is suggested that the molecular phenomena which take place after UV exposure are complex, and involve both proteins and nucleic acids[8]. Moreover, there are evidences demonstrating that after UV-exposure there may be some dark reactions, which, in certain cases (for example in *Didinium nasutum*), may amplify the damages produced by UV itself[10]. In fact, a pulsed UV stimulation, at the same dose of a continuous one, is much more efficient in inducing damages. These dark reactions, being independent of light, are of a termochemical nature and, therefore, strongly dependent on temperature[8-10].

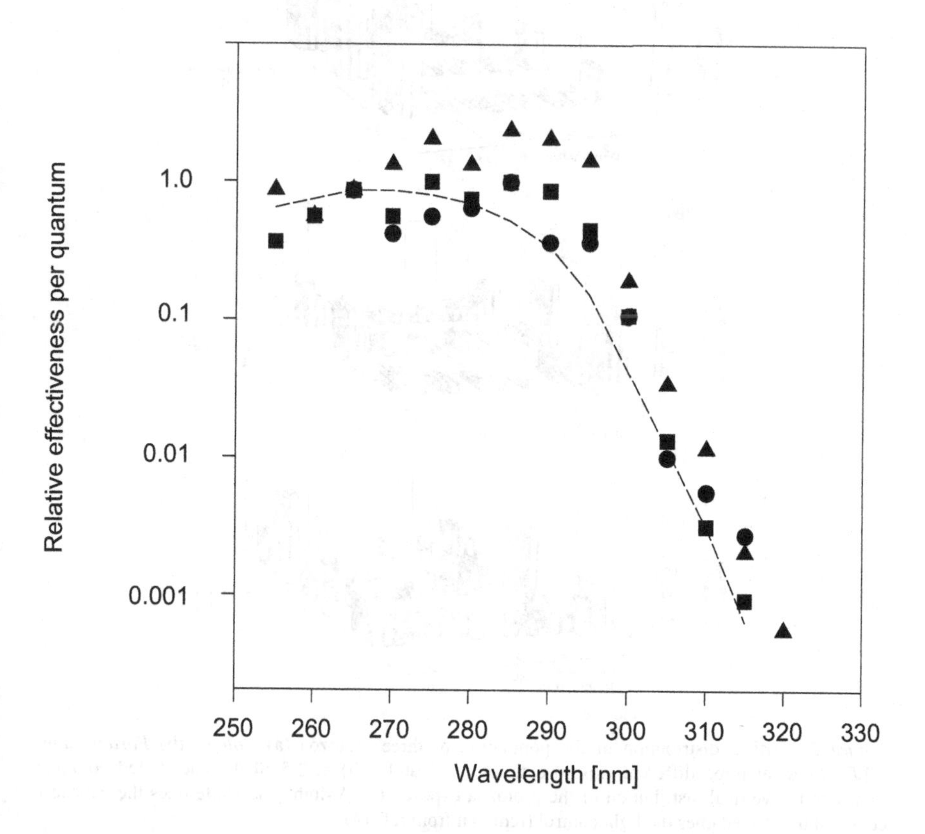

Figure 1. Action spectra of UV-induced photokilling of Tetrahymena as a function of different postirradiation treatments. Squares: no treatment; circles: 20 min photoreactivation; triangles: suspension medium containing 0.02% caffeine. The dotted line represents an "average DNA" action spectrum (from ref. 11).

Probably, but this is not certain yet, these dark reactions are correlated with the repair mechanisms, and, in particular the photo-repair mechanisms, which operate only in the presence of visible light[12]. These findings suggest that UV can affect several different physiological functions; this may happen in a not very specific way, since radiation can affect both proteins and nucleic acids; some researchers, however, have tried to investigate the existence of specific UV receptors in ciliates[11,13-16]. In particular, they tried to characterize the UV sensitivity of some organisms (included little crustaceans), in order to establish if there is a photoreceptor absorbing in the UV region.

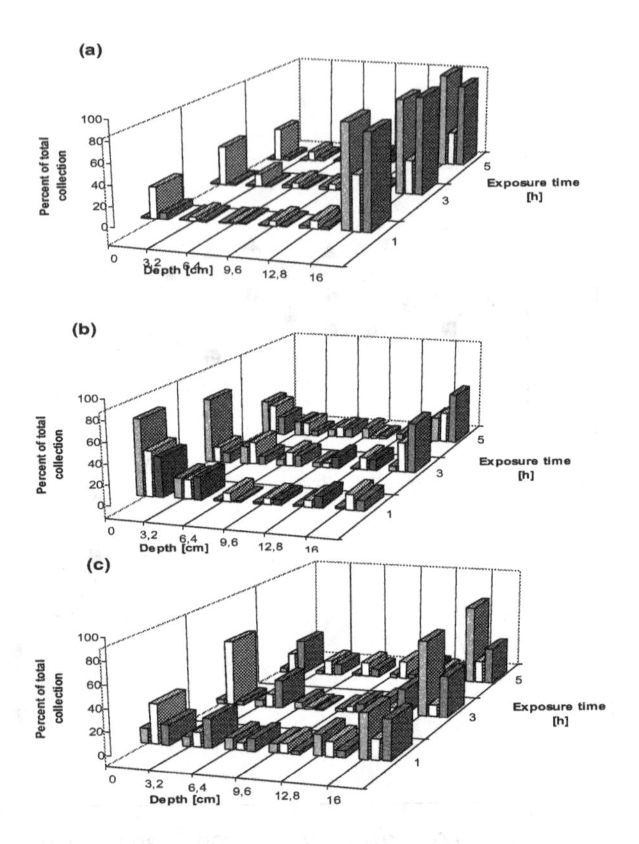

Figure 2. Vertical distribution of the population of three protozoa (a) *Coleps*, (b) *Paramecium*, (c) *Euplotes*, at three different exposure times (1, 3 and 5 h) at 2.5 SU/h. The dotted columns represent the vertical distribution of the protozoa exposed to UV-light; the white ones the dark test cells and the striped ones the light control (redrawn from ref. 14).

These experiments lead to positive results: many microorganisms show a specific UV sensitivity, and can perform an avoiding reaction when exposed to UV radiation; moreover, the strength of this reaction seems to be inversely proportional to the relative resistance against UV radiations, i.e., more resistant the organism, less evident the reaction (see Fig. 2). Recent works seem to indicate that also *B. japonicum* has a specific UV receptor[17].

3. Recent studies on the effects of UV-B on ciliates

The effects of UV-B radiation have been studied in three species of ciliated protozoa, two living in fresh-water, *Blepharisma japonicum* and *Ophryoglena flava*, and a marine one, *Fabrea salina*[18,19]. Motile and photomotile behaviour have been investigated before and after UV exposure, with the aim of finding possible intracellular targets of UV radiation.

B. japonicum possess a great amount of pigment granules, located beneath the cell pellicle, which contain a hypericin-like pigment, called blepharismin, whereas *Ophryoglena flava* is apparently colorless; the *F. salina* strain studied in these experiments also contains a hypericin-like pigment, as shown by fluorescence measurements[20-22] and the pigment granules are very similar to those found in *B. japonicum*, though *Blepharisma* contains a much larger number of them[21]. This explains why *B. japonicum* appears red-pigmented, whilst *F. salina* appears colorless. In *Blepharisma*, the action spectrum of the step-up photophobic response indicates that hypericin is very likely the photoreceptor pigment; in *F. salina* the action spectrum of phototactic reaction[23] and immunocytochemical studies[24] indicate the presence of a rhodopsin-like molecule. It has been suggested that the pigment responsible for phototaxis in *O .flava* might be a rhodopsin-like molecule, as well. Recently, the data of Cadetti et al.[25] have been reevaluated by Foster[26], who concludes that a rhodopsin is indeed the photoreceptor pigment in this ciliate.

Effect of UV-B radiation on motility

In *B. japonicum*, the exposure to UV radiation impairs cell motility, too[18]. Samples were UV irradiated by means of three TL40W/12 Philips UV-B lamps, using a cellulose acetate film to remove the UV-C component of radiation. Irradiations were performed also with a Schott WG1 filter to remove UV-B radiation. In both experimental conditions the total irradiance was kept about the same (~5 W/m^2) by means of neutral density filters. In the presence of UV-B, the spectral distribution was about 52% UV-B, 27% UV-A and 21% visible; UV-B irradiance was about 2.5 W/m^2, comparable to that in the natural environment, whereas UV-A and visible irradiance were about 30 and 400 times lower than the natural ones. In the irradiation condition without UV-B, the spectral distribution was 0% UV-B, 16% UV-A and 84% visible. UV-B irradiation for 30 and 60 minutes caused a gradual rise in the number of cells that swim on circular trajectories and induced a significant decrease in the average speed: from 118 μm/s in control samples to 59 μm/s and 7 μm/s for 30 and 60 min irradiation, respectively (see Fig. 3)[18]. However, recent results showed that short time UV-irradiation (up to 10 min, at 3 W/m^2) causes an increase in cell speed (Lenci, personal communication). On the contrary, cells irradiated without UV-B did not show any movement pattern alteration, but a significant reduction of average speed was observed after 60 min irradiation (see Fig. 3)[18].

In *F. salina* low irradiance UV-B irradiation caused at a meaningful increase in cell speed[19], as also observed in *B. japonicum*. A similar experimental set-up was used: two UV-B fluorescent tubes, TL40W/12 Philips, screened with a cellulose acetate film, and a visible one, TLD36W/54 Philips: irradiance was about 8 W/m^2 in the visible range and about 3 W/m^2 in the UV range (1 W/m^2 UV-A, 2 W/m^2 UV-B). The samples were covered with: a quartz disk transparent to all wavelengths (treated samples), a Schott WG1 filter to remove UV-B (control 1) and with a Schott GG400 filter transparent to visible radiation only (control 2)[19]. UV-B-induced speed increase was already detectable at short irradiation times (3 min 45 s) and reached a plateau after one hour of exposure. Control samples, on the contrary, did not show meaningful alteration of their swimming speed (see Fig. 4). UV-B induced motion alteration was limited to speed increase. In fact, other motion parameters, such as frequency of directional changes and frequency distribution of the angles of directional changes, did not show any dependence on UV-B irradiation[19].

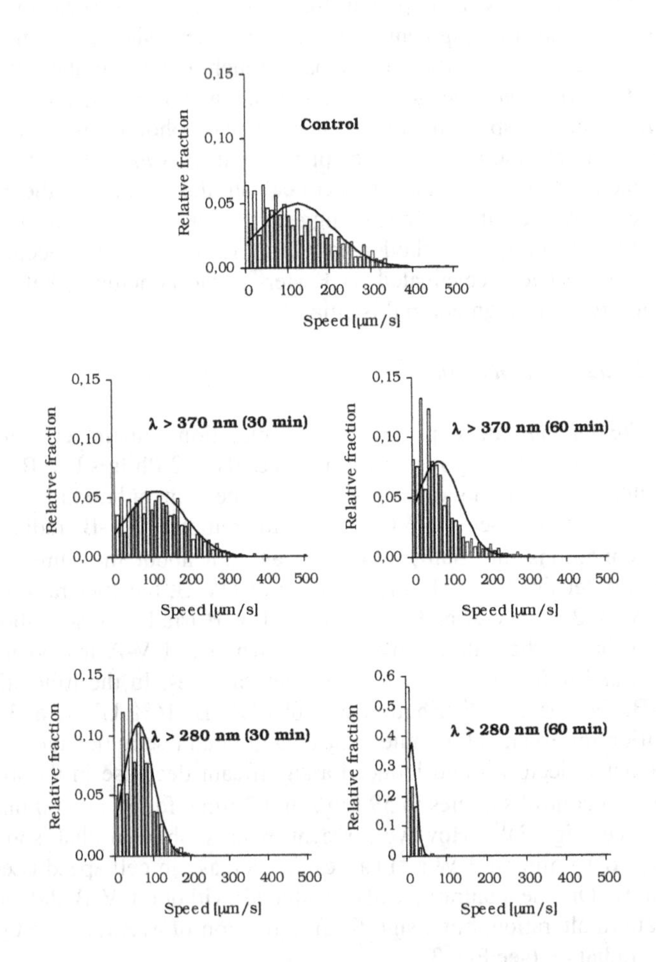

Figure 3. Histograms of the velocity distribution of samples of red B. japonicum cells under different irradiation conditions ("visible"= $\lambda \geq 370$ nm, "UV-B"= $\lambda \geq 280$ nm) and times (30 and 60 min) Redrawn from ref. 18.

The UV-B-induced speed increase was a reversible effect, as shown in Fig. 5, and the time required for speed recovery depended on irradiation time: for short irradiation time (3 min 45 s, 7 min 30 s, 15 min) the recovery occurred within about one hour, for longer irradiation time (30 min, 60 min) it occurred within two hours and for 120 min irradiation speed resumed its initial value after three hours only[19].

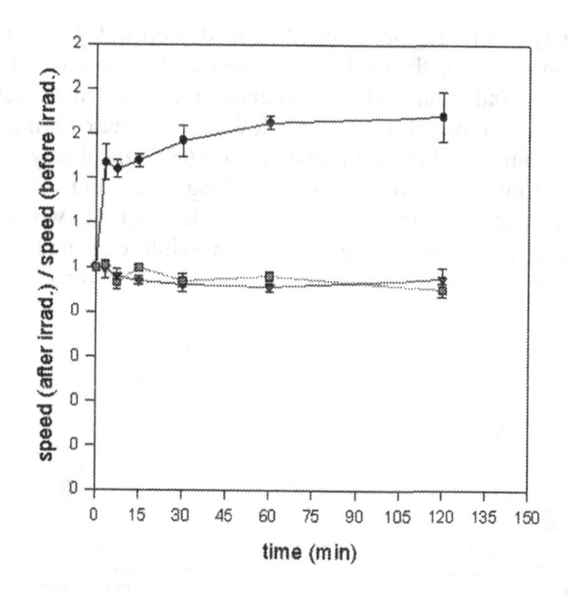

Fig. 4. Dose-effect curves of the speed of *F. salina* cells in samples, irradiated with UV-B, UV-A and visible light (circles), in control 1 samples, irradiated with UV-A and visible light (squares) and in control 2 samples, irradiated with visible light only (triangles) (redrawn from ref. 19).

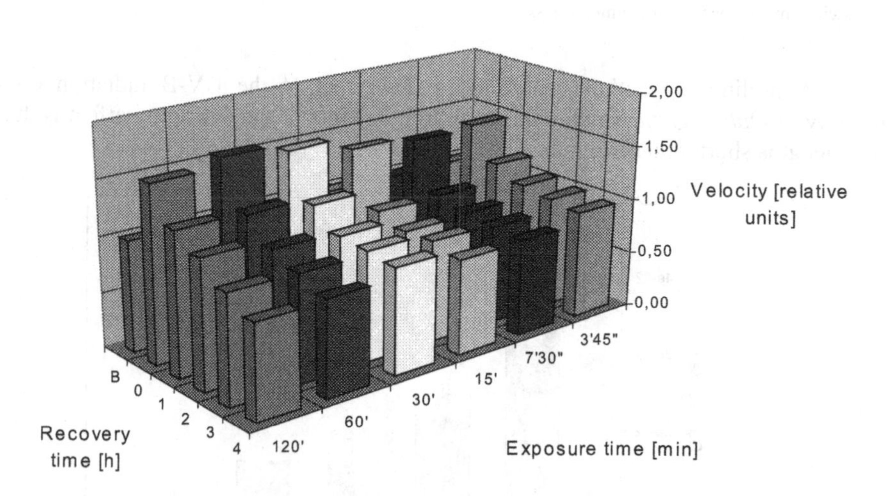

Fig. 5. 3D histogram of the speed of UV-irradiated samples of *F. salina* cells, as a function of both irradiation time and time elapsed after irradiation. The x-axis represents the UV exposure time (in minutes), the y-axis represents the recovery time (in hours), i.e. the time elapsed after the end of irradiation and the z-axis represents the speed of a sample at (t, T), relative to the speed of the same sample before irradiation, shown in the histogram at Recovery time = B (redrawn from ref. 19).

The analysis of the speed distribution demonstrated that the fraction of cells swimming at speed lower than 0.1 mm/s (slow cells) showed a decrease immediately after the end of irradiation and a subsequent increase. Nevertheless, at the longest irradiation time (120 min) the fraction of slow cells increased immediately after the end of irradiation (from 0.05 before irradiation to 0.08 immediately after irradiation) (see Fig. 6). This result could indicate that at long irradiation time a part of the cell population was damaged and swam at very low speed, whereas the majority of population increased its swimming speed. A possible explanation of this result is the existence of a subpopulation of cells more sensitive to UV-B radiation[19].

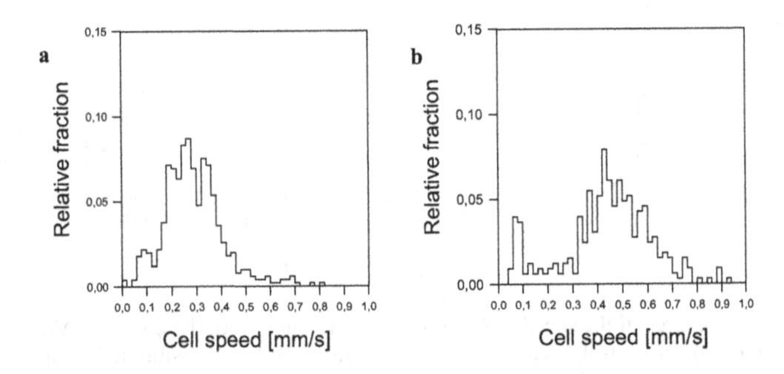

Fig. 6. Histograms of the speed distributions of *F. salina* cells (a) before and (b) after 120 min with UV-B, UV-A and visible light. After irradiation, we observe an increase of the fraction of cells swimming at speed of 0.1 mm/s or less.

A preliminary action spectrum for the effect of the UV-B radiation on the motility of *Ophryoglena flava* is shown in Fig. 7. Most of the damaging effect is due to wavelengths shorter than 300 nm.

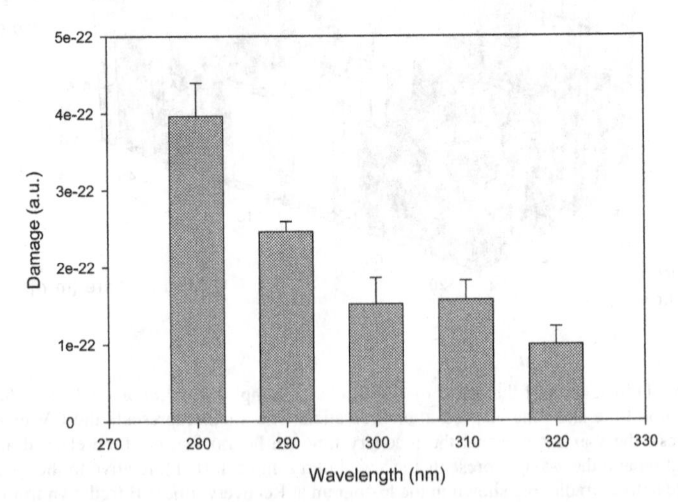

Fig. 7. Preliminary action spectrum for the effect of UV-B wavelength on the motility of *Ophryoglena flava.*

Effect of UV-B radiation on photomotile responses

Photomotile responses of both *B. japonicum* and *F. salina* are impaired by UV-B irradiation[18,19].

B. japonicum reacts to light stimuli exhibiting step-up photophobic responses: upon a sudden increase in light fluence rate the cells stop, turn and start swimming again in a new direction. Action spectra indicated that these responses are mediated by the endogenous pigment blepharismin[27-29]. UV-B irradiation caused a specific inhibition of step-up photophobic responses (see Fig. 8): after 30 min of UV-B irradiation, as described in the previous section, 50% of still motile cells were unable to respond to photic stimuli and after 60 min irradiation the photoresponsiveness was totally suppressed. This inhibition was specifically determined by UV-B radiation; in fact, irradiation with visible light only did not affect in any way the photophobic response[18].

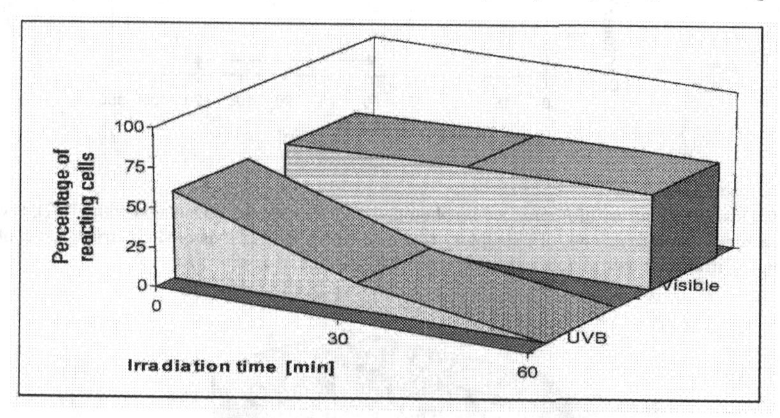

Fig. 8. Fraction of motile cells of *B. japonicum* showing step-up photophobic response as a function of irradiation time in the visible and in the UV-B range (redrawn from ref. 18).

F. salina shows two kinds of photomotile responses: positive phototaxis (cell displacement towards the light source) and step-down photophobic reaction (upon a sudden decrease in light fluence rate the cells stop, turn and start swimming again in a new direction)[30,31]. As previously mentioned it is not clear at present if these responses are mediated by a rhodopsin like pigment and/or by a hypericin-like pigment[32]. Preliminary, qualitative, observations indicate a UV-induced damage to the photophobic step-down reaction (unpublished data).

F. salina phototactic reaction is strongly inhibited by UV-B irradiation[19]. Phototaxis decreased with UV-B irradiation time (at the above-described experimental conditions) and completely disappeared after one hour (Fig. 9). Control samples, as defined in the previous section, show a slight decrease in phototaxis after two hour irradiation, but this decrease is significantly smaller than that of UV-B treated samples. The inhibition of phototactic responsiveness, except for the shortest exposure time, was irreversible (see Fig. 10).

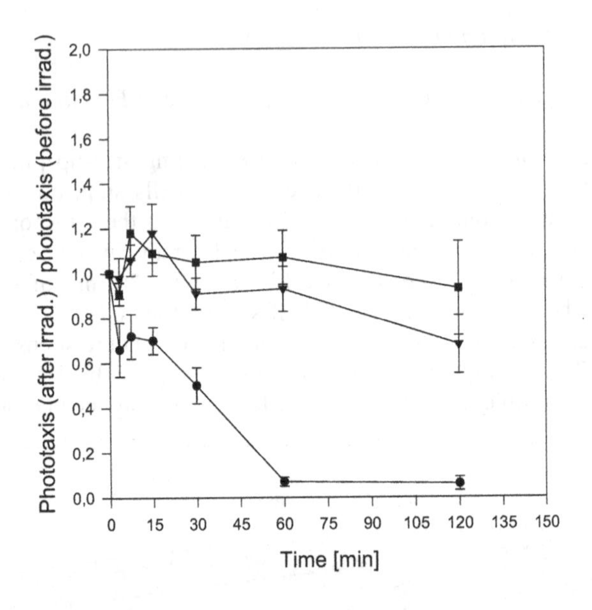

Fig. 9. Dose-effect curves of *Fabrea salina* phototaxis. Circles: samples irradiated with UV-B, UV-A and visible light; squares: control 1 samples, irradiated with UV-A and visible light; triangles: control 2 samples, irradiated with visible light only (redrawn from ref. 19).

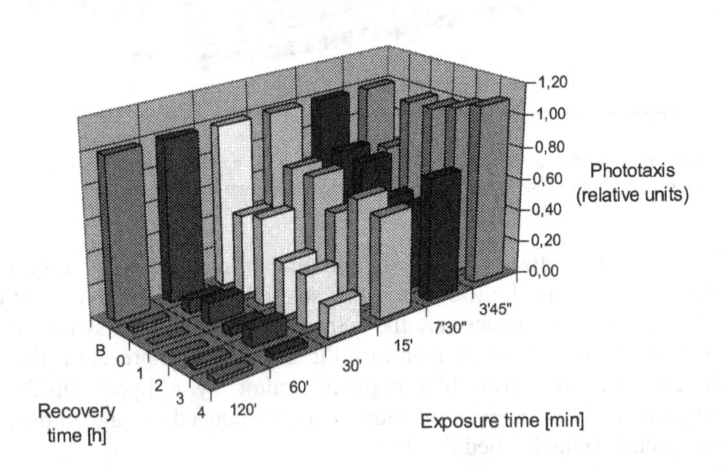

Fig. 10. 3D histogram of the phototaxis of UV-irradiated samples of *F. salina* cells, as a function of both irradiation time and time elapsed after irradiation. Data are reported as described in Fig. 5. (Redrawn from ref. 19).

Figure 11 shows the dose-effect curves of phototactic sensitivity (that is the phototactic response of a sample as a function of increasing phototaxis-stimulating light intensity) for samples irradiated with UV-B, UV-A and visible light for 30 min and for control samples. From these curves, it is clear that the phototactic response of irradiated samples saturated at a much lower level than that of the control and that a higher irradiance is required to obtain half maximum response. This indicates a reduced light sensitivity in irradiated samples.

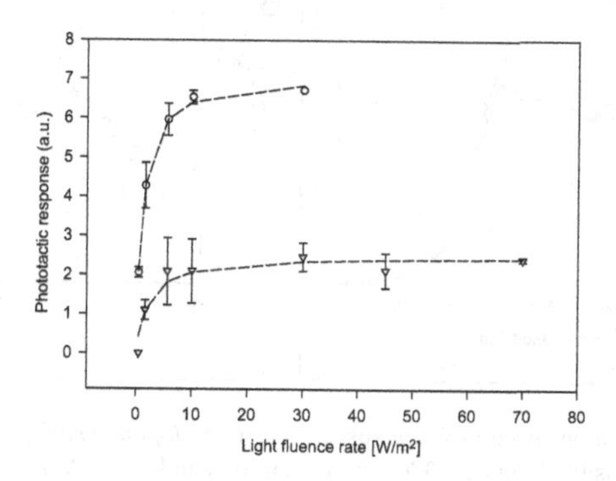

Fig. 11. Dose effect curves of the phototactic sensitivity of *F. salina* cells before (circles) and after (triangles) 30 min irradiation with polychromatic UV-B, UV-A and visible radiation. Phototactic activity in arbitrary units is plotted as a function of the irradiance of the stimulating light at 600nm. Both curves can be fitted by an hyperbolic function (p>95%).

Targets of UV-B radiation

It is well known that UV radiation can also affect living organisms by means of photodynamic reactions[33], in which the excitation energy absorbed by a chromophore molecule is used by either of two different mechanisms, one involving oxygen (type II photodynamic reaction), the other involving other acceptors of the triplet energy (type I photodynamic reaction). Both mechanisms can destroy cellular components[34].

However, photodynamic reactions are unlikely involved in the above-described UV-B effects in *B. japonicum* and *F. salina*, where the target molecules could be intrinsic components of the photoreceptor apparatus and/or the photosensory transduction chain[18,19]. In fact, experiments on *B. japonicum* indicated that the effects of UV-B radiation are about the same in both red and blue *B. japonicum* cells[18], the latter containing an oxidized, less phototoxic form of the pigment and not undergoing photosensitized killing[35].

Absorption spectra of agar suspension of *B. japonicum* cells were not significantly different after irradiation with UV-B or visible light[18] (see Fig. 12). Fluorescence spectra showed an increase of fluorescence quantum yield. In visible-irradiated cells this was probably due to the extrusion of the pigment from the cell, which occurs even in the presence of dim light to avoid the harmful consequences of blepharismin-sensitized photodynamic reactions[36]. In UV-B-irradiated cells the quantum yield increase could be due to a damage of the molecular environment of blepharismin. These spectroscopic measurements seems to indicate that UV-B irradiation does not cause a specific molecular transformation of blepharismin. This was confirmed by recent time-resolved fluorescence measurements on *B. japonicum* showing that the fluorescence lifetimes and the relative amplitudes of the different molecular species emitting at 600 nm are not affected by UV-B irradiation[37].

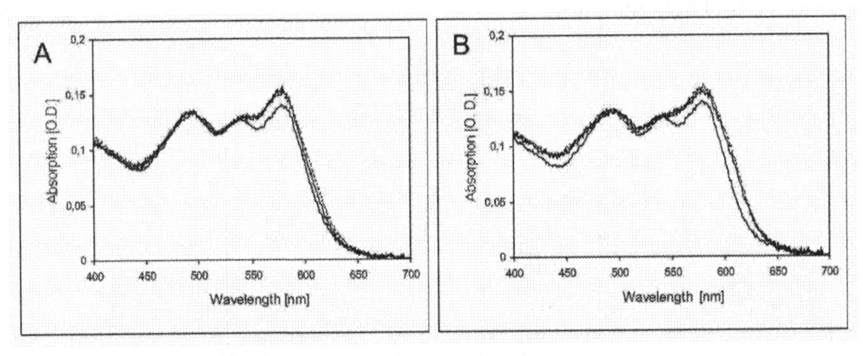

Fig. 12. Optical absorption spectra of agar suspensions of *B. japonicum* kept in the dark (continuous line), irradiated with visible light ($\lambda \geq 370$ nm, broken line) and with UV-B ($\lambda \geq 280$ nm, dotted line) for 30 min (A) and 60 min (B) (redrawn from ref.18).

The UV-B-induced inhibition of photoresponsiveness in *B. japonicum* can, therefore, be explained in terms of a damage to some components of the photosensory transduction chain, such as electron/proton transfer systems and/or membrane channels. The latter hypothesis is supported by the observation that the avoiding reaction to mechanical stimuli are reduced after 30 min and completely suppressed after 60 min of UV-B irradiation. This inhibition, in fact, could be due to a damage at the membrane level[18].

Fabrea salina do not undergo photokilling and the effects of UV-B irradiation cannot be mediated by photodynamic reactions. In agreement with the results obtained in *B. japonicum*, fluorescence spectra of the hypericin-like pigment of *F. salina*, obtained by cell sonication, were unaffected by irradiation with UV-B, UV-A and visible light[19] (Fig. 13). UV-B-caused modifications of the pigment concentration or structure can, therefore, be ruled out.

Interesting data come from the analysis of the dose-effect curves of phototactic sensitivity reported in Fig. 11. The increase, after irradiation, of the fluence rate required in order to obtain half maximum response could indicate a damage to the photoreceptor pigment molecules, with the consequence that a higher number of photons is required in order to obtain the same response. In addition to this reduced light sensitivity, *F. salina* irradiated samples show a lower level of saturation of the phototactic response; this suggests a damage to some parts of the sensory transduction chain, such that even high fluence rate are not sufficient to elicit the same maximum response as in control samples. These results, therefore, point to an impaired photosensory transduction chain due to a possible damage both to the photoreceptor pigment and to some target molecule involved in the dark reactions following light detection[19].

The above-mentioned fluorescence measurements indicate that the hypericin-like pigment of *F. salina* does not appear to be damaged by UV-B radiation; on the basis of these experimental results, the hypothesis of the rhodopsin-like nature of the photoreceptor pigment is strengthened. A damage to the rhodopsin-like pigment, in addition to a damage to some steps of the sensory transduction chain, could, in fact, explain the UV-B induced reduction of the phototactic response.

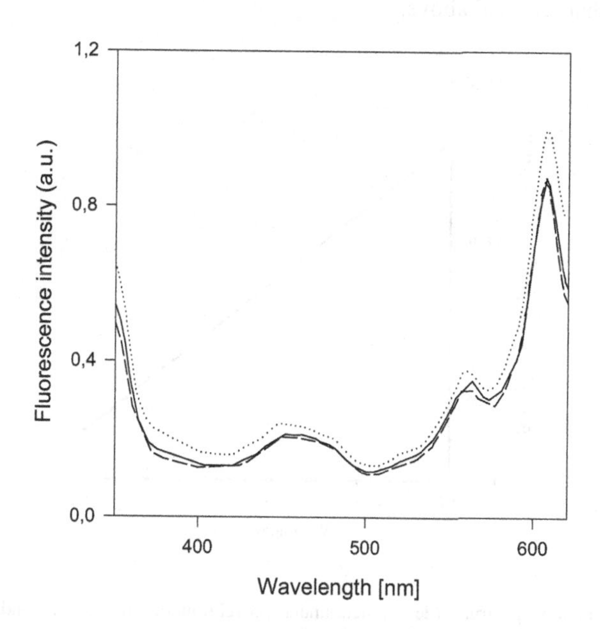

Fig. 13. Fluorescence excitation spectra (emission wavelength = 640 nm) of sonicated suspensions of F. salina cells before (solid line) and after 120 min irradiation with UV-A and visible radiation (dashed line) and with UV-B, UV-A and visible light (dotted line).

In the case of *Ophryoglena flava* an UV action spectrum of positive phototaxis inhibition was determined under polychromatic irradiation conditions. In order not to exclude possible repair mechanisms due to UV-A and/or visible radiation, sample were irradiated with radiation from 280 to about 700 nm (TL40W/12 Philips and TDL36W/54 Philips fluorescent lamps, emitting in the UV and visible, respectively), removing increasingly larger portions of UV-B and short wavelength UV-A, by means of coloured glass filters (see Ghetti et al., this volume). Moreover UV irradiance was modulated by using one or three UV fluorescent lamps. UV-B irradiance was comparable to that recorded outside by an ELDONET dosimeter (www.eldonet.org and Häder and Lebert, this volume) in summer days.

Even these irradiance values damaged the cells and, if the exposure lasted long enough, they were killed. However, in natural conditions the organisms might be shaded by the organic matter usually present in lakes and ponds. Moreover, in the above-described experimental set-up cells are continuously irradiated, whereas in the natural environment, after being damaged by UV radiation, they can start swimming in random direction and disperse in the water column, far from the surface.

It should be noted that the ratios between UV-B irradiance and UV-A and visible one in the above-described experiments were much higher than those of the solar radiation, so that we cannot exclude the hypothesis that in natural conditions the repair mechanisms could work more efficiently. However, it was observed that light-dependent photolyase DNA repair can take place under an irradiance of 15 W/m^2 [38], comparable to that reported above.

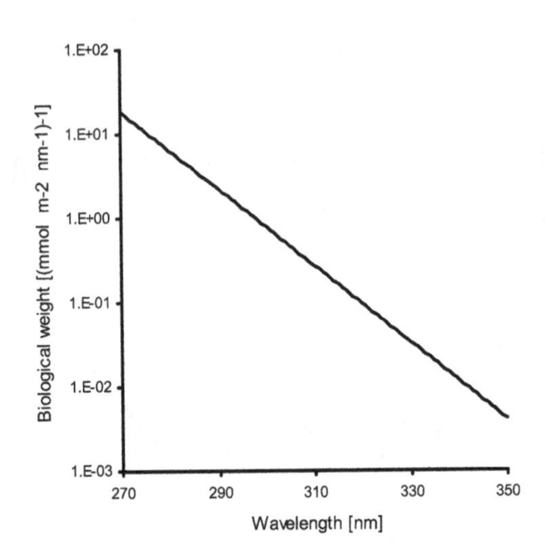

Fig. 14. UV action spectrum (determined under polychromatic irradiation conditions) for the inhibition of positive phototaxis in *Ophryoglena flava.*

The UV action spectrum for the phototaxis inhibition in *Ophryoglena flava* (see Fig. 14) clearly shows that the damage increases exponentially with decreasing wavelength. Thus, a further depletion of the ozone hole, causing shorter wavelengths to reach the Earth surface, could significantly increase the risk of damage to aquatic microorganisms in natural ecosystems.

References

1. Marangoni R, Martini B, Colombetti G (1997) Effects of UV on ciliates. In *The Effects of Ozone Depletion in Aquatic Ecosystems* (D-P Häder, ed.), R.G. Landes Company, Georgetown, and Academic Press, San Diego, pp. 229-246.
2. Dahlback A, Henriksen T, Larsen SH, Stamnes K (1989) Biological UV-doses and the effect of an ozone layer depletion. *Photochem. Photobiol.,* 49: 621-625.
3. Frederick JE (1990) Trends in atmospheric ozone and ultraviolet radiation: mechanisms and observations for the northern hemisphere. *Photochem Photobio.* 51: 757-763.
4. Frederick JE (1993) Ultraviolet sunlight reaching the earth's surface: a review of recent research. *Photochem Photobiol* 57: 175-178.
5. Giese AC (1938) Differential scusceptibility of a number of protozoans to ultraviolet radiations. *J Cell Comp Physiol* 12: 129-138.
6. Giese AC (1945) The ultraviolet action spectrum for retardation of division of *Paramecium. J Cell Comp Physiol* 26: 47-55.
7. Hirshfield H, Giese AC (1953) Ultraviolet radiation effects on growth processes of *Blepharisma undulans. Exp Cell Res* 4: 283-294.

8. Giese,A.C. (1971) Nucleic acid and protein synthesis, as measured by incorporation of tracers, during regeneration in ultraviolet-treated *Blepharisma*. *Exp Cell Res* 64: 218-226.
9. Giese AC, McCaw B, Cornell R (1963) Retardation of division of three ciliates by intermittent and continuous ultraviolet radiations at different temperatures. *J Gen Physiol* 46: 1095-1108.
10. Giese AC, Shepard DC, Bennett J, Farmanfarmaian A, Brandt CL (1956) Evidence for thermal reactions following exposure of *Didinium* to intermittent ultraviolet radiations. *J Gen Physiol* 40: 311-325.
11. Calkins J, Colley E, Wheeler J (1987) Spectral dependence of some UV-B and UV-C responses of *Tetrahymena pyriformis* irradiated with dye laser generated UV. Photochem Photobiol. 45: 389-398.
12. Lakhanisky T, Hendrickx B, Mouton RF, Cornelis JJ (1979) A serological study of removal of UV-induced photoproducts in the DNA of *Tetrahymena pyriformis* GL: influence of caffeine, quinacrine and chloroquine. *Photochem Photobiol* 29: 851-853.
13. Barcelo JA, Calkins J (1980) The kinetics of avoidance of simulated solar UV radiation by two arthropods. *Biophys J* 32: 921-929.
14. Barcelo JA, Calkins J (1979) Positioning of aquatic microorganisms in response to visible light and simulated solar UV-B irradiation. *Photochem Photobiol* 29: 75-83.
15. Calkins J, Selby C, Enoch HG (1987) Comparison of UV action spectra for lethality and mutation in *Salmonella typhimurium* using a broad band source and monochromatic radiations. *Photochem Photobiol* 45: 631-636.
16. Gutierrez JC Perez-Silva J (1983) Effects of UV-Radiation on kinetics of encystment and on morphology of resting cysts of the ciliate *Laurentiella acuminata*. *J Protozool* 30: 715-718,
17. Lenci F, Checcucci G, Ghetti F, Gioffré D, Sgarbossa A (1997) Sensory perception and transduction of UV-B radiation by the ciliate *Blepharisma japonicum*. Biochim Biophys Acta 1336: 23-27.
18. Sgarbossa A, Lucia S, Lenci F, Gioffré D, Ghetti F, Checcucci G (1995) Effects of UV-B irradiation on motility and photoresponsiveness of the coloured ciliate *Blepharisma japonicum*. *J Photochem Photobiol B: Biol* 27: 243-249.
19. Martini B, Marangoni R, Gioffré D, Colombetti G (1997) Effects of UV-B irradiation on the motility and photomotility of the marine ciliate *Fabrea salina*. *J Photochem Photobiol B: Biol* 39: 197-203.
20. Marangoni R, Cubeddu R, Valentini G, Taroni P, Sorbi R, Gioffré D, Batistini A, Colombetti G (1994) Fluorescence microscopy of an endogenous pigment in the marine ciliate *Fabrea salina*. *Med Biol Environ* 22: 85-89.
21. Marangoni R, Cubeddu R, Taroni P, Valentini G, Sorbi R, Lorenzini E, Colombetti G (1996) Microspectrofluorometry, fluorescence imaging and confocal microscopy of an endogenous pigment of the marine ciliate *Fabrea salina*. *J Photochem Photobiol B: Biol* 34: 183-189.
22. Marangoni R, Gobbi L, Verni F, Albertini G, Colombetti G (1996) Pigment granules and hypericin-like fluorescence in the marine ciliate *Fabrea salina*. *Acta Protozool* 35: 177-182.
23. Marangoni R, Puntoni S, Favati L, Colombetti G (1994) Phototaxis in *Fabrea salina*. I. Action spectrum determination. *J Photochem Photobiol B: Biol* 23: 149-154.
24. Podestà A, Marangoni R, Villani C, Colombetti G (1994) A rhodopsin-like molecule on the plasma membrane of *Fabrea salina*. *J Euk Microbiol* 41: 565-569.
25. Cadetti L, Marroni F, Marangoni R, Gioffré D, Kuhlmann HW, Colombetti G (2000) Positive phototaxis in the ciliate protozoan *Ophryoglena flava*: dose-effect curves and action spectrum determination. *J Photochem Photobiol B: Biol* 57: 41-46.
26. Foster KW (2001) Action Spectroscopy of Photomovements. In *Photomovements* (DP Häder, ed.), Elsevier, Amsterdam, pp. 51-115.
27. Scevoli P, Bisi F, Colombetti G, Ghetti F, Lenci F, Passarelli V (1987) Photomotile responses of *Blepharisma japonicum* I: Action spectra determination and time-resolved fluorescence of photoreceptor pigments. *J Photochem Photobiol B: Biol* 1: 75-84.
28. Matsuoka T, Matsuoka S, Yamaoka Y, Kuriu T, Watanabe Y, Takayanagi M, Kato Y, Taneda K (1992) Action spectra for step-up photophobic response in *Blepharisma*. *J Protozool* 39: 498-502.
29 Checcucci G, Damato G, Ghetti F, Lenci F (1993) Action spectra of the photophobic response of blue and red forms of *Blepharisma japonicum*. *Photochem Photobiol* 57: 686-689.
30. Colombetti G, Marangoni R, Machemer H (1992) Phototaxis in *Fabrea salina*. *Med Biol Environ* 20: 93-100.
31. Colombetti G, Bräucker R, Machemer H (1992) Photobehavior of *Fabrea salina*: responses to directional and diffused gradient-type illumination. *J Photochem Photobiol B: Biol* 15: 253-257.
32. Marangoni R, Puntoni S, Colombetti G (1998) A model system for photosensory perception in Protozoa: the marine ciliate *Fabrea salina*. In *Biophysics of photoreception: molecular and phototransductive events* (C. Taddei-Ferretti, ed.), World Scientific, Singapore, pp. 83-92.

33. Ito T (1983) Photodynamic agents as tools for cell biology. In *Photochemical and Photobiological Reviews* (KC Smith, ed.), Plenum, New York, pp. 141-188.
34. Häder DP, Worrest RC (1991) Effects of enhanced solar ultraviolet radiation on aquatic ecosystems. *Photochem Photobiol* 53: 717-725.
35. Ghetti F, Heelis PF, Checcucci G, Lenci F (1992) A laser flash photolysis study of the triplet states of the red and the blue forms of *Blepharisma japonicum* pigment. *J Photochem Photobiol B: Biol* 13: 315-321.
36. Giese AC (1981) The photobiology of *Blepharisma*. In *Photochemical and Photobiological Reviews* (KC Smith, ed.), Plenum, New York, pp. 139-180.
37. Angelini N, Cubeddu R, Ghetti F, Lenci F, Taroni P, Valentini G (1995) In vivo spectroscopic study of photoreceptor pigments of *Blepharisma japonicum* red and blue cells. *Biochim Biophys Acta* 1231: 247-254.
38. Malloy KD, Holman MA, Mitchell D, Detrich III HW (1997) Solar UVB-induced DNA damage and photoenzymatic DNA repair in Antarctic zooplankton. *Proc Nat Acad Sci USA* 94: 1258-1263.

UV RADIATION, DNA DAMAGE, MUTATIONS AND SKIN CANCER

FRANK R. DE GRUIJL
Department of Dermatology, Leiden University Medical Centre, NL-2333 AL Leiden, The Netherlands.

1. Introduction

About UV radiation, ozone and life

More than blue light, UV radiation is scattered in the earth's atmosphere: going from a wavelength (λ) of 400 down to 340 nm, the scattering goes up almost 2 fold (Rayleigh scattering $\propto \lambda^{-4}$). Although we do not see it, the sky is more violet-ultraviolet than it is blue. Blocking the UV radiation from the direction of the solar disc - e.g. by a small parasol - would still leave a substantial part of the UV radiation, i.e. the part reaches us through scattering from the blue sky (and possibly by reflection from sand or snow). UV radiation can initiate a great variety of photochemical reactions. We are all familiar with examples such as "photochemical smog", photodegradation of plastics and bleaching of pigments in paints. Although high-energy UV radiation form the sun may have contributed to the early synthesis of organic molecules in the primitive anoxic atmosphere (wavelengths below 200 nm are absorbed by CH_4, H_2O and NH_3), it needed to be blocked out in early evolution because of its very detrimental effects on highly evolved organic molecules such as proteins and DNA, essential building blocks of living organisms. Most of the sun's harmful UV radiation could pass through the earth's primordial anoxic atmosphere, and life could not sustain itself on the earth's bare surface. Early life on earth must have evolved shielded from the UV radiation by barriers, e.g., in oceans below a layer of dissolved organic compounds[1] or very deep near hot vents and/or in the earth's crust, and/or a few millimeters deep in soil[2]. By a very fortunate evolutionary event plant-like organisms started to produce oxygen from photosynthesis in substantial amounts over 2 billion years ago, and this oxygen provided protection against UV radiation from the sun.

Up in the stratosphere, oxygen (O_2) is bombarded by high-energy UV radiation, absorbs this radiation, splits up (in O) and recombines with oxygen to form ozone (O_3). This stratospheric layer of ozone is spread over a wide range in altitude (from 10 to 50 km), but it is extremely rarified and would amount to a sheath of only about 3 mm thick if it were compressed at ground level conditions. This modest amount of ozone shields off most of the harmful UV radiation in the UVB band that is not absorbed by oxygen. Life on earth has thus generated its own "stratospheric sunscreen". However, the UVB radiation is only partially absorbed, and the fraction that reaches ground level is still capable of damaging and killing unprotected cells. To cope with this residual UV

F. Ghetti et al. (eds.), Environmental UV Radiation: Impact on Ecosystems and Human Health and Predictive Models, 249–258.

radiation organisms at the earth's surface have acquired certain adaptive features: e.g., UV-absorbing surface layers, repair of damaged cells or replacing damaged or killed cells entirely. Because of the high absorption in organic material, UV radiation does not penetrate any deeper than the skin, and this organ clearly shows all of the mentioned adaptive features. Despite these adaptations, solar UV radiation still affects human health, but the full extent of these effects remain largely unknown. Any increase in ambient UVB radiation due to loss of stratospheric ozone can therefore be expected to have important public health impacts.

About DNA, genes and cancer

Basic cancer research has shown that cancer is a disease stemming from disturbances in signaling pathways that control the cell cycle and differentiation. The most persistent disturbance is introduced by synthesis of dysfunctional signaling proteins or by a complete lack of synthesis of such proteins from miscoding or lost genes. Two types of genes are directly relevant to carcinogenesis: the oncogene (dominant) whose protein actively contributes to the cancerous progression, and the tumor suppressor gene (recessive) whose protein should counter such an uncontroled progression. A combination of such genes need to be affected in order for a cell to become cancerous. Such altered genes are passed on to daughter cells, thus propagating the problem of controlling cell growth. From the above it is immediately obvious that the ubiquitous solar UV radiation can damage the DNA of genes in exposed skin cells. Thus solar UV radiation poses a continuous threat to the genomic integrity of skin cells. The fact that healthy humans do not readily develop skin cancer attests to an impressively adequate adaptation of the human skin, it has several lines of defence ranging from increasing UV absorption to protect germinative basal cells, to DNA repair or removal of damages or transformed cells.

2. UV radiation, DNA damage, repair and mutation

UV absorption and DNA damage

Typical UV absorbing features in organic molecules are conjugated bonds: alternating single and double bonds, absorbing generally radiation of wavelengths between 200 and 250 nm, and in a ring between 250 and 300 nm (see reference 3). (Feynman et al.[4] give an elegantly simplified, quantummechanical introductory analysis of energy states of double bonds). Absorption maxima at longer wavelengths (300 - 450 nm) are found in some molecules with three (e.g., riboflavin) or four rings (e.g., porphyrines), and in long-chain repeats (e.g., carotenoids) of conjugated bonds. The collective protein fraction from cells shows a maximum absorption around 280 nm, whereas the DNA fraction shows a maximum around 260 nm. Major absorbers in the protein fraction are the tryptophan and tyrosine amino acids. In the DNA all nucleic acids are aromatic and contribute to the absorption maximum. DNA appears to contribute appreciably to the total absorption of UVC (200 - 280nm) radiation by a cell. Although absorption by DNA in the UVB around 300 nm is far less than in the UVC (10 to 100 fold lower), sun exposure causes significant levels of DNA damage.

In DNA the pyrimidines are more sensitive than purines (quantum yield about 10 times greater). The dominant DNA damage caused by UVC and UVB exposures is targeted at neighbouring pyrimidines that become dimerized; either through a four-carbon cyclobutane ring at the 5 and 6 positions in the successive bases or a binding of the 6th position to the 4th in the next base, where the former adduct (the cyclobutane pyrimidine dimer, CPD) is formed more frequently than the latter (the 6-4 pyrimidine-pyrimidone photoproduct, 6-4PP). CPD distort the DNA helix less than 6-4PP.

Other DNA lesions are known to be caused by UV radiation, e.g., protein-DNA crosslinks, oxidative base damage (e.g. 8-oxo-7,8-dihydroxyguanine, 8-oxo-G for short) and single strand breaks. Especially, in the longwave UVA spectral region (340 - 400 nm) these lesions become relatively more important[5]. There the DNA absorption becomes vanishingly small, but the DNA can still be damaged by radicals generated through absorption of the radiation by other (unidentified) molecules ("endogenous sensitizers", possibly riboflavins or porphyrins) in the cell. Reactive oxygen species are known to play an important role in cellular damage by UVA exposure (evidenced by oxygen dependency), which includes lipid peroxidation (membrane damage) and oxidation of single bases in the DNA (for an overview see De Laat and De Gruijl[6]).

DNA repair and UVB vs UVA mutations

Pyrimidine dimers are the predominant DNA lesions caused by solar UV radiation (the effect of UVB), and they are mainly repaired enzymatically by Nucleotide Excision Repair (NER). Xeroderma pigmentosum (XP) patients lack this form of repair, and run a dramatically increased risk of skin cancer[7]. This indicates that NER is a main line of defence against UV carcinogenesis, which has been experimentally verified in transgenic animals with deficiencies in NER[8-9]. The increased risk in patients with XP-variant shows that besides NER, an other form of DNA repair is important. The XP-variant patients are proficient in NER but lack functional DNA polymerase-η[10]. This is a "sloppy" polymerase which can bypass a (UV-)damaged base, frequently inserting an adenine. When mutated, the dysfunctional polymerase causes replication problems and an increase in sister chromatide exchange by recombinational repair[11]. The importance of recombinational repair of UV-induced DNA damage in human skin remains to be determined, but it may obviously cause a different class of genetic alterations from point mutations caused by pyrimidine dimers.

Fully in line with the predominance of pyrimidine dimers it was found that UVB radiation induced point mutations almost exclusively at di-pyrimidine sites, with mostly C to T transitions[12] (note that with random point mutations 75% would be expected to be associated with di-pyrimidic sites). Despite the frequent dimer formation at these sites, adjacent thymines did not appear to be associated with mutations; possibly by default insertion of an adenine in the synthesized DNA strand opposite a "non-instructive" (i.e., damaged) base in the template strand; a typical action of polymerase-η, see above (the "A-rule"[13]).

As mentioned earlier, UV radiation can generate reactive oxygen species (ROS) which can cause oxidative damage to the DNA bases, e.g. forming 8-oxo-G.. 8-oxo-G is a miscoding lesion causing G to T transversions[14]. Curiously enough, UVA irradiation was found to yield many T to G transversions in the *aprt* gene of CHO cells[15]. An explanation could be the mispairing of A to an 8-oxo-G that was formed in the nucleoside pool instead of the genomic DNA[16].

More recent experimental studies have clearly shown that UV radiation can also cause crude chromosomal aberrations in mammalian cells, such as detected by micro-nucleï formation, i.e., UV radiation is clastogenic[17-18]. It is, however, unknown which primary DNA lesions are responsible for these gross effects on chromosomes.

3. Oncogenes and tumor suppressor genes in skin cancer

In the cell there are various potential oncogenes and tumor suppressor genes that often play their roles in more than one signaling pathway, rather they function in signaling networks. UV radiation appears to be related to 3 types of skin cancer: the most common one, basal cell carcinoma (BCC), squamous cell carcinoma (SCC) and the most malignant one, cutaneous melanoma (CM). The question arises which genes and pathways are disrupted by UV radiation in which of these cancers.

UV mutations in P53 from BCC and SCC

The *P53* tumor suppressor gene is found to be mutated in a majority of human cancers. The p53 protein is therefore an apparently vulnerable part of the cell's signaling network: p53 plays a pivotal role in several signaling pathways related to DNA damage and expression of oncogenes[19]. Nuclear p53 expression is elevated after UV irradiation, and following a genotoxic insult p53 is involved in cell cycle arrest (late G1 and G2/M), apoptosis ('programmed cell death') and DNA repair. In SCC (about 90%) and BCC (>50%), from the US white population the *P53* gene appears to bear point mutations with the exact features of UVB-induced point mutations, i.e., associated with di-pyrimidinic sites, mostly C to T transitions and 5-10% CC to TT tandem mutations[20-21]. Evidently, the *P53* gene is also a target in UV carcinogenesis, which has been extensively confirmed in mouse experiments[22-23]. In line with this finding, it is found that the wavelength dependency of the induction of SCC closely parallels that of the induction of UV-induced DNA damage (pyrimidine dimers) in the skin, especially over the UVB and UVA2 bands[24-26].

Experiments with hairless mice show that clusters of epidermal cells with mutant p53 occur long before SCC become visible[27]; such clusters of mutant p53 have also been found in human skin[28-29]. Dysfunctional p53 is likely to affect protective responses against DNA damage and oncogenic signaling. Hence, the early occurrence of *P53* mutations may cause genomic instability and thus facilitate further carcinogenic progression. The frequency of p53-mutant cell clusters in the skin may be a direct indicator of skin cancer risk[30].

A mutation in the *P53* gene is clearly not enough to cause BCC or SCC. At the very least some oncogenic pathway has to be activated; e.g., a growth-stimulating pathway which normally starts with the activation (oligomerization) of a receptor tyrosine kinase (RTK), e.g. EGF-R, at the cell membrane, and is further mediated through proteins like RAS into the cell cytoplasm from which transcription factors are finally activated. Activating *RAS* mutations have been reported in a minority of SCC and BCC to various percentages[31-32]. These activating mutations are restricted to the codons 12, 13 and 61, and are not specific of UV radiation. The RAS-pathway may be involved in SCC and BCC, but it is not usually effected through genetic changes in the *RAS* family of genes.

BCC and the PTCH gene

Patients with Gorlin syndrome, or basal cell nevus syndrome (BCNS), suffer from multiple, familial BCC. This genetic trait was traced to mutations in the *PTCH* gene[33]. Next to frequent loss of one of the parental alleles (i.e. loss of heterozygosity, LOH) at this locus, many sporadic - non-familial - BCC showed mutations in the (remaining) *PTCH* allele[34]: 12 out of 37 tumors in SSCP screening, and 9 of these tumors showed LOH of *PTCH.* (The SSCP was apparently not sensitive enough as two tumors without variant SSCP or LOH were both found to have inactivating mutations.) Seven of 15 mutations occurred at di-pyrimidinic sites and were C to T transitions (among which 2 CC to TT tandem mutations), and could, therefore, have been caused by UVB radiation.

The *PTCH* gene is a serpentine-like receptor woven through the cell membrane. Extra-cellular 'Sonic Hedgehog' (SHH) protein couples to PTCH which triggers a pathway that ultimately activates the transcription factor GLI1, which induces epidermal hyperplasia[35], and which is expressed in almost all BCC[36]. Skin grafts of the SHH-transgenic keratinocytes onto immune-deficient mice show the specific histologic features of BCC[37]. This indicates that activation of this Sonic Hedgehog pathway is essential to the formation of BCC. This has been confirmed in transgenic mouse strains.

Ptc (murine homolog of *PTCH*) heterozygous knockout mice[38-39] develop microscopically detectable follicular neoplasms resembling human trichoblastomas and 40% develop BCC-like tumors after 9 months. Upon exposure to ionizing or UVB radiation the trichoblastomas and BCC occur earlier, and increase in size and numbers[40]. The trichoblastomas and BCC show frequent loss of the wildtype *Ptc* allele, and all the ones tested (n=12) showed expression of Gli1 (SCC, n=2, did not). Two BCC out of 5 UV-induced trichoblastoma/BCC-like tumors carried *p53* mutations (3 in total, 2 C to T and 1 C to G). These experimental data show that UV radiation can play an important role in causing or enhancing the development of BCC. Next to the induction of *p53* mutations, UV radiation could exert a more direct effect on the Sonic Hedgehog pathway by enhanced loss of the wildtype *Ptc* gene and/or possible mutation of this gene.

Melanoma and INK4a

Some familial CM are linked to the "multiple tumor suppressor" (*MTS1*) gene[41] (designated *CDKN2A* in the human genome project). It is also named *INK4a*[42] after the original finding that its product p16^{INK4a} becomes associated with cyclin dependent kinases upon transformation of human fibroblasts by SV40 virus, and acts as an inhibitor of CDK4 and CDK6. CDK4 is thus prevented from phosphorilating pRB and activating the E2F-1 transcription factor. An alternative reading frame in *INK4a* codes for the protein p14^{ARF} (p19^{ARF} in mice), which does not appear to inhibit any CDK but binds to MDM2 and thus interferes with the degradation of p53[43].

Partial or complete homozygous loss of *INK4a* is observed in a majority (about 60%) of cell lines derived from sporadic CM, and most of the remaining cell lines (e.g., 8 out of 11) bear point mutations that are typical of UV radiation, i.e., C to T transitions at dipyrimidine sites[44]. Although 60-70% of sporadic melanomas (n=62) show a lack of p16^{INK4a} expression, and all (n=5) of the metastases, the high number of homozygous losses and mutations of *INK4a* found in cell lines is not reproduced in primary CM. Homozygous deletions are found in approximately 10% and reported mutation rates range from 0 to 25 %[42]. The reason for this discrepancy is not entirely clear but it could be due to a high selection for a loss or mutation of *INK4a* in generating the cell lines. It should, however, also be noted that LOH at the *INK4a* locus in CM is quite common, and is even frequently observed in microdissected dysplastic nevi (in 75%), potential precursors of CM, as is LOH at the locus of *P53* (in 60%)[45].

Signaling pathways related to p16^{INK4a} apparently play an important role in the pathogenesis of CM. And, although the *INK4a* locus often shows LOH and less frequently mutations in primary CM, it is not clear if and to what extent solar UV radiation is responsible for these pertinent genetic changes.

RTK growth-stimulating pathway and INK4a in CM

Another family of genes that is implicated in CM are the *RAS* oncogenes, more specifically *N-RAS*. 25-70% of CM from regularly sun-exposed sites have been reported to carry activating point mutations in *N-RAS*, whereas none of the CM from irregularly exposed sites carried such mutations[46-47]. In a comparative study the percentage of *N-RAS* mutated CM from sun-exposed sites was higher in an Australian population (24%) than in a European population (12%)[48]. These mutations occur at dipyrimidine sites, the typical UV targets, but they are not dominated by C to T transitions.

As mentioned earlier, the RAS proteins function in mitogenic pathways which start by activation of RTK at the cell membrane, e.g. the receptor for epidermal growth factor, EGF-R. It is well known that oncogenic *RAS* will transform most immortal cell lines and make them tumorigenic upon transplantation into nude mice. Surprisingly, Serrano et al.[46] found that expression of oncogenic *RAS* (producing an activated H-RAS^{G12V}) in primary human or rodent cells results in a state that is phenotypically characterized as "senescence": the cells are viable and metabolically active but remain in the G1-phase of the cell cycle. This oncogenic RAS-induced arrest in G1 is accompanied by an accumulation of both p16 and p53. The link between these pathways is likely to be mediated by p14^{ARF} [49]. Inactivation of either p16 or p53 prevented this G1 arrest: the arrest did not occur in p53-/- cells, p16-/- cells, cells transfected with a dominant negative *p53* mutant (*p53*^{175H}) and cells with mutant Cdk4^{R24C} insensitive to p16. Thus, cells immortalized by dysfunctional p16 or p53 will not go into senescence upon RAS activation, but may progress to a tumorigenic state.

In a fish model (with hybrids of *Xiphophorus maculatis* and *helleri*) an RTK gene (*Xmrk*) of EGF-R family and an *Ink4a* homolog (*CdknX* or *DIFF*) appear to be important for hereditary CM[50-51]. This provides experimental evidence for the cooperation of an RTK mitogenic pathway and dysfunctional *Ink4a* in melanomagenesis. In further evidence, Chin et al.[52] demonstrated that *Ink4a*$^{-/-}$ mice in which expression of a human mutant *H-RAS*^{G12V} transgene was restricted to melanocytes, developed melanomas. Although UV irradiation did not (yet) cause CM in this mouse model, UV irradiation of hybrid *Xiphophorus* fish did cause CM[53]. UVA radiation was surprisingly effective in this model, only about 10 fold less effective than UVB radiation per unit radiant energy (J/m^2). It is, however, as yet unclear how UV radiation affected *CdknX/Ink4a* or the *Xmrk*/RTK mitogenic pathway. In the opossum *Monodelphis domestica* UVB radiation appears to induce CM[54], these tumors turned to be particularly malignant when the UV radiation is given neonatally[55]. However, UVA radiation did not induce malignant CM in this model[56], only benign melanocytic precursor lesions[57]. In a very recent study it was found that the UVB-induced CM from these opossums carried UVB-like mutations in the *CDKN2A* homolog, and that only the mutant allele was present and expressed in a metastatic cell line[58].

4. Conclusions

From the data stated above it appears that at least a combination of an activated oncogenic pathway and an inactivated tumor suppressor gene is needed in order for a skin cancer to arise: in SCC it is possibly an activated RTK/RAS pathway in combination with dysfunctional P53 tumor suppression, in BCC the Hedgehog pathway with possibly dysfunctional P53, and in CM again possibly an activated RTK/RAS in combination with inactivation of the INK4a locus. These combinations may be required, but not necessarily sufficient for the development of a tumor. Additional oncogenic events may be necessary.

Although the skin cancers appear to be related to UV radiation, the effect of UV radiation is only unambiguously clear in point mutations of *P53* in SCC and BCC. The mutations found in the other relevant genes are of a wider variety, which may (in part) be caused by solar UV radiation. Experiments are needed to clarify if and how UVB or UVA radiations can affect other relevant genes. Overall, the data presently weigh most heavily toward the carcinogenic effect of UVB radiation: the latest data on experimental induction of CM in the opossum *Monodelphis domestica* are not indicative of any important contribution of UVA radiation next to the dominant carcinogenicity of UVB radiation

Acknowledgements. The author would like to thank the Dutch Cancer Society and the European Commission for the financial support of their research groups.

References

1. Miller, S.L. and Orgel, L.E. (1974) *The origins of life on the earth*, Prentice Hall Inc., Eaglewood Cliffs (NJ).
2. Sagan, C. and Pollack, J.B. (1974) Differential transmission of sunlight on Mars: biological implications, *Icarus*, 21: 490-495.
3. Jagger, J. (1967) *Introduction to research in ultraviolet photobiology*, Prentice-Hall Inc., Eaglewood Cliffs (NJ).
4. Feynman, R.P., Leighton, R.B. and Sands, M. (1967) *The Feynman Lectures on Physics*, Vol. III, Addison-Wesley Publishing Company, Reading (MA).

5. Kielbassa, C., Roza, L. and Epe, B. (1997) Wavelength dependence of oxidative DNA damage induced by UV and visible light, *Carcinogenesis*, 18: 811-816.
6. De Laat, J.M.T. and De Gruijl FR. (1996) The role of UVA in the aetiology of non-melanoma skin cancer, *Cancer Surveys: Skin Cancer*, 26,: 173-191.
7. Kraemer KH, Lee MM. and Scotto J. Xeroderma pigmentosum; Cutaneous ocular, and neurologic abnormalities in 830 published cases, *Arch. Dermatol.*, 123: (1987) 241-250.
8. De Vries A, van Oosterom CT, Hofhuis FM, Dortant PM, Berg RJW, de Gruijl FR, Wester PW, van Kreijl CF, Capel PJ. and van Steeg H,. Increased susceptibility to ultraviolet-B and carcinogens of mice lacking the DNA excision gene XPA, *Nature*, 377: (1995) 169-173.
9. Sands AT, Abuin A, Sanchez A, Conti CJ. and Bradley A. High susceptibility to ultraviolet-induced carcinogenesis in mice lacking XPC, *Nature*, 377: (1995) 162-165.
10. Masutani C, Kusumoto R, Yamada A, Dohmae N, Yokoi M, Yuasa M, Araki A, Iwai S, Takio K. and Hanaoka F. The XPV (xeroderma pigmentosum variant) geen encodes human DNA polymerase eta, *Nature*, 399: (1999) 700-704.
11. Limoli CL, Giedzinski E, Morgan WF. and Cleaver JE. Inaugural article: polymerase eta deficiency in xeroderma pigmentosum variant uncovers an overlap between S phase checkpoint and double strand breakrepair, *Proc. Natl. Acad. Sci. USA*, 97: (2000) 7939-7946.
12. Brash DE. and Haseltine WA. UV-induced 'mutation hotspots' occur at damage 'hotspots', *Nature*, 298: (1982) 189-192.
13. Strauss B, Rabkin S, Sagher D. and Moore P. The role of DNA polymerase in base substitution mutatgenesis on non-instructional templates, *Biochimie*, 64: (1982) 829-838.
14. Shibutani S, Takeshita M. and Grollman AP. Insertion of specific bases during DNA synthesis past the oxidation-damaged base 8-oxo-G, *Nature*, 349: (1991) 431-434.
15. Drobetsky EA, Turcotte J. and Chateauneuf A. A role for ultraviolet A in solar mutagenesis, *Proc. Natl. Acad. Sci.USA*, 92: (1995) 2350-2354.
16. Poltev VI, Shulyupina NV. and Bruskov VI. The formation of mispairs by 8-oxo-guanine as a pathway of mutations induced by irradiation and oxygen radicals, *J. Mol. Recogn.*, 3: (1990) 45-47.
17. Keulers RA, de Roon AR, de Roode S. and Tates AD. The induction and analysis of micronuclei and cell killing by ultraviolet-B radiation, *Photochem. Photobiol.*, 67: (1998) 426-432.
18. Emri G, Wenczl E, van Erp P, Jans J, Roza L, Horkay I. and Schothorst A. Low doses of UVB and UVA induce chromosomal aberrations in cultured human skin cells, *J. Invest. Dermatol.*, 115: (2000) 435-440.
19. Vogelstein B, Lane D. and Levine AJ. Surfing the p53 network, *Nature*, 408: (2000) 307-310
20. Brash DE, Rudolph JA, Simon JA, Lin A, McKenna GJ, Baden HP, Halperin AJ. and Ponten J. A role for sunlight in skin cancer: UV-induced *p53* mutations in squamous cell carcinomas, *Proc. Natl. Acad. Sci. USA*, 88: (1991) 10124-10128.
21. Ziegler AD, Leffel DJ, Kunala S, Sharma WG, Simon JA, Halperin AJ, Shapiro PE, Bale AE. and Brash DE. Mutation hotspots due to sunlight in the *p53* gene of skin cancers, *Proc. Natl. Acad. Sci. USA*, 90: (1993) 4216-4220.
22. Kanjilal S, Pierceall WF, Cummings KK, Kripke ML. and Ananthaswamy HN. High frequency of *p53* mutations in ultraviolet radiation-induced skin tumors: evidence for strand bias and tumor heterogeneity, *Cancer res.*, 53: (1993) 2961-2964.
23. Dumaz N, Van Kranen HJ, de Vries A, Berg RJW, Wester PW, van Kreijl CF, Sarasin A, Daya-Grosjean L. and de Gruijl FR. The role of UVB light in skin carcinomas through the analysis of *p53* mutations in squamous cell carcinomas of hairless mice, *Carcinogenesis*, 18: (1997) 897-904.
24. De Gruijl FR. Skin cancer and solar UV radiation (Millennium review), *Eur. J. Cancer*, 35: (1999) 2003-2009.
25. De Gruijl FR and Van der Leun JC. Estimate of the wavelength dependency of ultraviolet carcinogenesis in humans and its relevance to risk assessments of a stratopheric ozone depletion. *Health Phys.* 67 (1994) 319-325.
26. Freeman SE, Hacham H, Gange RW, Maytum DJ, Sutherland JC. and Sutherland BM. Wavelength dependence of pyrimidine dimer formation in DNA of human skin irradiated in situ with ultraviolet light. *Proc. Natl. Acad. Sci. USA* 86 (1989) 5605-5609.
27. Berg RJW, van Kranen HJ, Rebel HG, de Vries A, van Vloten WA, van Kreijl CF, van der Leun JC. and de Gruijl FR. Early p53 alterations in mouse skin carcinogenesis by UVB radiation: immunohistochemical detection of mutant p53 protein in clusters of preneoplastic epidermal cells, *Proc. Natl. Acad. Sci. USA*, 93: (1997) 274-278.

28. Jonason AS, Kunala S, Price GJ, Restifo RJ, Spinelli HM, Persing JA, Leffell DJ, Tarone RE. and Brash DE. Frequent clones of *p53*-mutated keratinocytes in normal human skin, *Proc. Natl. Acad. Sci. USA*, 93: (1996) 14025-14029.
29. Ren ZP, Ponten F, Nister M. and Ponten J. Two distinct p53 immunohistochemical patterns in human squamous cell skin cancer, precursors and normal epidermis, *Int. J. Cancer*, 69: (1996) 174-179.
30. Rebel H, Mosnier LO, Berg RJ, Westerman-de Vries A, van Steeg H, van Kranen HJ. and de Gruijl FR. Early p53-positive foci as indicators of tumor risk in ultraviolet-exposed hairless mice: kinetics of induction, effects of DNA repair deficiency, and p53 heterozygozity, *Cancer Res.*, 61: (2001) 977-983.
31. Pierceall WE, Goldberg LH, Tainsky MA, Mukhopadhyay T. and Ananthaswamy HN. *Ras* gene mutation and amplification in human nonmelanoma skin cancers, *Mol. Carcinog.*, 4: (1991) 196-202.
32. Cambell C, Quinn AG. and Rees JL. Codon 12 Harvey-ras mutations are rare events in non-melanoma human skin cancer, *Br. J. Dermatol.*, 128: (1993) 111-114.
33. Hahn H, Wicking C, Zaphiropoulos PG, Gailini MR, Shanley S, Chidambaram A, Vorechovsky I, Holmberg E, Unden AB, Gilles S, Nagus K, Smyth I, Pressman C, Leffell DJ, Gerrard B, Goldstein AM, Dean M, Toftgard R, Chenerix-Trench G, Wainright B. and Bale AE. Mutations of the human homologue of *Drosophila* Patches in the Nevoid Basal Cell Carcinoma Syndrome, *Cell*, 85: (1996) 841-851.
34. Gailini MR, Stahle-Bäckdahl M, Leffell DJ, Glynn M, Zaphiropoulos PG, Pressman C, Unden AB, Dean M, Brash DE, Bale AE. and Toftgard R. The role of the human homologue of *Drosophila* Patched in sporadic basal cell carcinomas, *Nature Gen.*, 14: (1996) 78-81.
35. Fan H. and Khavari PA. Sonic hedgehog opposes epithelial cell cycle arrest, *J. Cell Biol.*, 147: (1999) 71-76.
36. Dahmane N, Lee J, Robins P, Heller P. and Ruiz i Altaba A, Activation of the transcription factor Gli1 and the Sonic Hedgehog signalling pathway in skin tumours, *Nature*, 389: (1997) 876-881.
37. Fan H, Oro AE, Scott MP and Khavari PA, Induction of basal cell carcinoma features in transgenic human skin expressing Sonic Hedgehog, *Nature Med.*, 3: (1997) 788-92.
38. Goodrich LV, Millenkovic L, Higgins KM. and Scott MP. Altered neural cell fates and medullablastomas in mouse patched mutants, *Science*, 277: (1997) 1109-1113.
39. Hahn H, Wojnowski L, Zimmer AM, Hall J, Miller G. and Zimmer A. Rhabdomyosarcomas and radiation hypersensitivity in a mouse model of Gorlin syndrome, *Nature Med.*, 4: (1998) 619-622.
40. Aszterbaum M, Epstein J, Oro A, Douglas V, LeBoit PE, Scott MP. and Epstein EH Jr. Ultraviolet and ionizing radiation enhance the growth of BCCs and trichoblastomas in patched heterozygous knockout mice, *Nature Med.*, 5: (1999) 1285-1291.
41. Kamb A, Gruis NA, Weaver-Feldhaus J, Liu Q, Harshman K, Tavtigian SV, Stockert E, Day RS III, Johnson BE. and Skolnick MH. A cell cycle regulator potentially involved in geneisis of many tumor types, *Science*, 264: (1994) 436-440.
42. Ruas, M.. and Peters G. The p16^{INK4a}/CDKN2A tumor suppressor and its relatives, *Biochim. Biophys. Acta*, 1378: (1998) F115-F177.
43. Zhang H, Xiong Y. and Yarbrough WG. ARF promotes MDM2 degradation and stabalizes p53: *ARF-INK4a* locus deletion impairs both Rb and p53 tumor suppressor pathways, *Cell*, 92: (1998) 725-734.
44. Pollock PM, Yu F, Qui L, Parsons PG. and Hayward NK. Evidence for u.v. induction of CDKN2 mutations in melanoma cell lines, *Oncogene*, 11: (1995) 663-668.
45. Lee JY, Dong SM, Shin MS, Kim SY, Lee SH, Kang SJ, Lee JD, Kim CS, Kim SH. and Yoo NJ. Genetic alterations of p16INK4a and p53 genes in sporadic dysplastic nevus, *Biochem. Biophys. Res. Comm.*, 237: (1997) 667-672.
46. Van Elsas A, Zerp SF, van der Flier S, Kruse KM, Aarnoudse C, Hayward NK, Ruiter DJ. and Schrier PI. Relevance of ultraviolet-induced N-ras oncogene point mutations in development of primary human cutaneous melanoma, *Am. J. Path.*, 143: (1996) 883-893.
47. Van 't Veer LJ, Burgering BMT, Versteeg R, Boot AJM, Ruiter DJ, Osanto S, Schrier PI. and Bos JL. N-*ras* mutations in human cutaneous melanoma from sun-exposed body sites, *Mol. Cell. Biol.*, :(1989) 3114-3116.
48. Serrano M, Lin WA, McCurrach ME, Beach D. and Lowe SW. Oncogenic *ras* provokes premature cell senescence associated with accumulation of p53 and p16^{INK4a}, *Cell*, 88: (1997) 593-602.
49. Sharpless NE. and DePinho RA. The *INK4A/ARF* locus and its two gene products, *Curr. Opin. Genes Dev.*, 9: (1999) 22-30.
50. Wittbrodt J, Adam D, Malitschek B, Mäueler W, Raulf F, Telling A, Robertson SM. and Schartl M. Novel putative receptor tyrosine kinase encoded by melanoma-inducing *Tu* locus in *Xiphophorus*, *Nature*, 341: (1989) 415-421.

51. Kazianis S, Gutbrod H, Nairn RS, McEntire BB, Della-Colletta L, Walter RB, Borowsky RL, Woodhead AD, Setlow RB, Schartl M. and Morizot DC. Localization of a *CDKN2* gene in linkage group V of Xiphophorus fishes defines it as a candidate for the *DIFF* tumor suppressor, *Genes Chromosomes Cancer*, 22: (1998) 210-220.
52. Chin L., Pomerantz J, Polsky D, Jacobson M, Cohen C, Cordon-Carco C, Horner JW II. and DePinho RA. Cooperative effects of *INK4a* and *ras* in melanoma susceptibility in vivo, *Genes Dev.*, 11: (1997) 2822-2834.
53. Setlow RB, Woodhead AD. and Grist E. Animal model for ultraviolet radiation-induced melanoma: platyfish-swortail hybrid, *Proc. Natl. Acad. Sci. USA*, 86: (1989) 8922-8926.
54. Ley RD, Applegate LA, Padilla RS. and Stuart TD. Ultraviolet radiation-induced malignant melanoma in Monodelphis domestica, *Photochem. Photiobiol.*, 50: (1989) 1-5.
55. Robinson ES, Hubbard GB, Colon G. and Vandeberg JL. Low-dose ultraviolet exposure early in the development can lead to widespread melanoma in the opossum model, *Int. J. Exp. Path.*, 79: (1998) 235-244.
56. Robinson ES, Hill RH Jr, Kripke ML. and Setlow RB. The Monodelphis melanoma model: initial report on large ultraviolet A exposures of suckling young, *Photochem. Photobiol.*, 71: (2000) 743-746.
57. Ley RD. Dose response for ultraviolet radiation A-induced focal melanocytic hyperplasia and nonmelanoma skin tumors in Monodelphis domestica, *Photochem. Photobiol.*, 73: (2001) 20-23.
58. Chan J, Robinson ES, Atencio J, Wang Z, Kazianis S, Coletta LD, Nairn RS. and McCarrey JR. Characterization of the CDKN2A and ARF genes in UV-induced melanocytic hyperplasias and melanomas of an opossum (Monodelphis domestica), *Mol. Carcinog.*, 31: (2001) 16-26.

ULTRAVIOLET RADIATION AND THE EYE

DAVID H. SLINEY
US Army Center for Health Promotion and Preventive Medicine
Aberdeen Proving Ground, MD, USA

1. Introduction

The human eye is exquisitely sensitive to light (i.e., visible radiant energy), and when dark adapted, the retina can detect a few photons of blue-green light[1]. It is therefore not at all surprising that ocular tissues are also more vulnerable to solar ultraviolet radiation (UVR) damage than the skin. For this reason, we have evolved with certain anatomical, physiological and behavioral traits that protect this critical organ from the ultraviolet (UV) damage that would otherwise be certain from the intense bath of overhead UVR when we are outdoors during daylight. For example, the UV exposure threshold dose for photokeratitis (also known as "snow blindness" or "welder's flash") - if measured by an outdoor, global UVR meter designed to respond as the action spectrum for photokeratitis - would be reached in less than 10 minutes around midday in the summer sun[2,3]. There are three critical ocular structures that could be affected by UV exposure: the cornea, the lens and the retina. Figure 1 shows a simple diagram of the human eye and points to these three critical and complex ocular tissues. The eye is about 25 mm in diameter and has an effective focal length in air of about 17 mm. Very little UV reaches the retina. The cornea transmits radiant energy only at 295 nm and above[1]. The crystalline lens absorbs almost all incident energy to wavelengths of nearly 400 nm. In youth a very small amount of UV-A reaches the retina, but the lens becomes more absorbing with age. Thus there are intra-ocular filters that effectively filter different parts of the UV spectrum and allowing only of the order of 1% or less to actually reach the retina[4].

The acute phototoxic effect of UVR on the eye, photokeratitis, has long been recognized. Less obvious are potential hazards to the eye from chronic exposure. Certain age-related changes to the cornea, conjunctiva and lens have also been thought to be related to chronic exposure to solar UVR in certain climates. Determining environmental ocular exposure can be quite difficult and this exposure to sunlight has been misjudged in many epidemiological studies[2-8]. There are many occasions where one views bright light sources such as the sun, arc lamps and welding arcs; but such viewing is normally only momentary, since the aversion response to bright light and discomfort glare limits exposure to a fraction of a second. Delayed effects are almost exclusively considered to result from environmental UVR exposure. Hence, the increased terrestrial UVR related to ozone depletion has been one cause for health concern[9].

F. Ghetti et al. (eds.), Environmental UV Radiation: Impact on Ecosystems and Human Health and Predictive Models, 259–278.

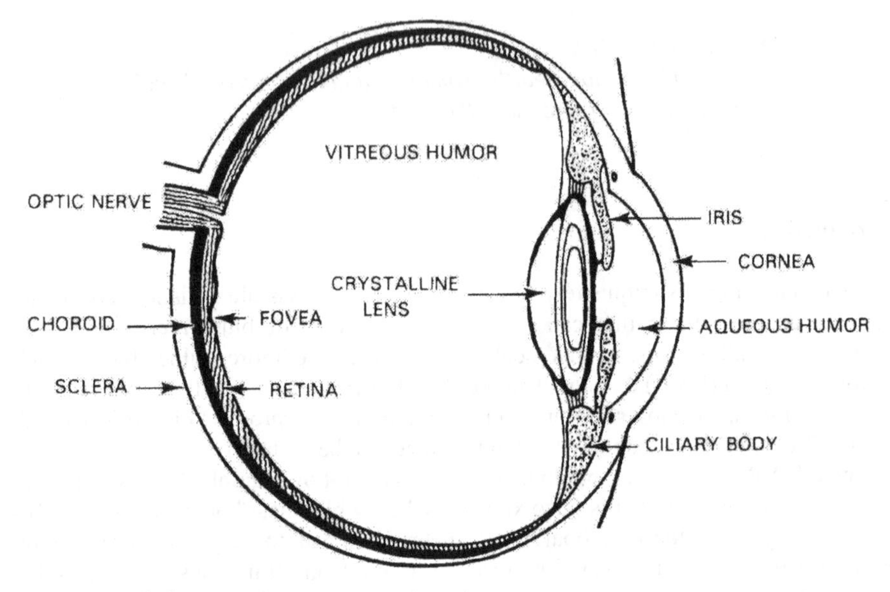

Figure 1. The human eye

2. Effects from environmental exposure

Human exposure to sunlight, particularly the UVB component, is believed to be associated with a variety of eye disorders, including damage to the cornea, lens and retina[4]. Photokeratitis (snowblindness) is clearly related to UVR exposure, whereas, cataracts (opacities of the lens) are the most frequently cited delayed consequence. Certain keratopathies (corneal degenerations) and pterygium (a fleshy growth on the conjunctiva) also are believed to result from excessive UVB exposures. Cataracts are a major cause of blindness in both developed and developing countries. However, the relative importance of different wavelengths in cataractogenesis, as well as the dose response curve, remain uncertain[4,9].

The geographical variation in the incidence of age-related ocular changes such as presbyopia and cataracts and diseases such as pterygium and droplet keratopathies have led to theories pointing to sunlight, UVR exposure and ambient temperature as potential etiological factors[5,9,10]. Some epidemiological evidence also points to an association of age-related macular degeneration (AMD) to sunlight (particularly blue-light) exposure[11]. The actual distribution of sunlight exposure of different tissues within the anterior segment of the eye is more difficult to assess than one would expect. Of greatest importance are the geometrical factors that influence the selective UVR exposure to different segments of the lens, cornea and retina. Studies show that sunlight exposure to local areas of the cornea, lens and retina varies greatly in different environments. Perhaps for this reason, the epidemiological studies of the potential role of environmental UVR in the development of ocular diseases such as cataract, pterygium, droplet keratopathies and age-related macular degeneration have produced surprisingly inconsistent findings[7-9,12]. All of these ocular diseases vary geographically; however, the lack of consistent epidemiological results is almost certainly the result of either incomplete or erroneous estimates of outdoor UVR exposure dose[2,3,5,13-15]. Geometrical factors dominate the determination of UVR exposure of the eye. The degree of lid opening limits ocular exposure to rays entering at angles near the horizon. Clouds redistribute overhead UVR to the horizon sky. Mountains, trees and building shield the eye from direct sky exposure. Most ground surfaces reflect only a small fraction of incident UVR. The result is that the highest UVR exposure appears to occur during light overcast where the horizon is visible and ground surface reflection is high. By contrast, exposure in a high mountain valley (lower ambient temperature) with green foliage results in a much lower ocular dose. Other findings of these studies show that retinal exposure to light and UVR in daylight occurs largely in the superior retina[14].

3. Epidemiological evidence

So why is it important to study the spectral and geometrical exposure of the human eye from solar UVR? It has long been suggested that the great latitudinal variation of "the world's most blinding disease," cataract[9], is strongly suggestive (as in the case of skin cancers) to result from UVR. The World Health Organization (WHO) estimates that the worldwide prevalence of cataract exceeds 50 million[9]. Indeed, if this dependence is examined along with a wide variety of laboratory studies, it suggests that environmental factors surely plays a major role in the time of onset of lenticular opacities. However, most epidemiologic evidence points to UVR in sunlight as a significant risk factor only in one type of cataract, i.e., cortical cataract[12]. The evidence for UVR as an etiological factor in at least two adverse changes in the cornea, i.e., droplet keratopathies and pterygium is much stronger[9,16]. Furthermore, the environmental studies of Sasaki[10] have provided very important insights into the geographical variations in the incidence of cataract by examining the different types of cataract characteristic of different latitudes[10]. Nuclear cataract was shown to be more common in the tropics; cortical cataracts, more common in mid-latitudes, and posterior sub-capsular cataracts were not so clearly related to latitude. Despite this latitudinal variation, some other published epidemiological studies do not appear to show a relation between UVR and cataract. Nevertheless, a wide variety of scientific evidence, from laboratory studies of the UV photochemistry of lens proteins to a number of different animal exposure studies all provide support for the hypothesis that UVR should play a far greater role in cataractogenesis. Most age-related changes in the skin (from accelerated aging to skin cancer) have been conclusively shown to result from excessive exposure to solar UVR (or "sunlight exposure"). Although no one questions that UVR exposure produces the acute effects of “sunburn” (erythema) and snowblindness (photokeratitis), some have questioned whether pterygium and droplet keratopathies are clearly related to UVR exposure[7,9]. Even more under debate are theories that suggest that UVR and light may affect retinal diseases such as age-related macular degeneration (AMD)[9]. A better resolution of these questions requires far better ocular dosimetry.

With the strong geographical variations in the incidence of nuclear cataract[10], it is surprising that most epidemiological studies show only weak or no apparent relation between UV or sunlight exposure and the incidence of nuclear cataract[12]. Perhaps it is necessary to consider the other important environmental parameter-environmental temperature. Both sunlight and ambient temperature have been cited as potential etiological factors in several age-related ocular diseases[5]. However, the combined role of these physical factors and their possible synergisms generally has not been carefully examined. These environmental co-factors, together with geometrical factors, have recently been examined to give some suggestions of the etiology of these ocular diseases[15].

4. Animal studies

Injury thresholds for acute UVR injury to the cornea, conjunctiva, lens and retina in experimental animals have been corroborated for the human eye from laboratory and accidental injury data[13-20]. Guidelines for human health and safety limits for exposure to UVR are based upon this knowledge.[2,17-25]

There are several types of hazards to the eye from intense UVR exposure, and the dosimetric concepts applicable to the anterior segment differs from quantities applicable to retinal exposure. The following effects are considered in current safety standards, which depend upon the spectrum of the UVR exposure and the temporal and geometrical characteristics of the exposure.[1,2,13,17,23-25]

5. Photokeratoconjunctivitis

Ultraviolet photokeratoconjunctivitis (more simply, “photokeratitis”, also known as "welder's flash" or "snow-blindness" - 180 nm to 400 nm) is a temporary photochemical injury of the cornea (photokeratitis) and conjunctiva (photoconjunctivitis) that appears several hours after the acute exposure; symptoms generally last for only 24 to 48 hours.[19-22] Because of the very superficial nature of the corneal injury and the rapid turnover of surface epithelial cells of the cornea (about 48-hour cycle), the effect is transient.

The symptoms of photokeratitis are severe pain and, in some cases, blepharospasm (uncontrolled blinking). The onset of symptoms, like erythema, is delayed for several hours after the UVR exposure. The signs and symptoms last for a day or two, and the haze appears in the region of the cornea (the superficial surface layer of the eye) within the palpebral fissure (i.e., within the lid opening). The condition is almost always reversible and it has generally been accepted that the condition is without sequelae. However, the laboratory studies of Ringvold[21] show damage of both the keratinocytes and endothelial cells of the cornea, and it could be- speculated that repeated, severe episodes of snowblindness may well increase the risk of delayed corneal pathologies such as pterygium and droplet keratopathies.

6. Lenticular opacities

Although not directly related to the safety assessment of the light adjustable lens (LAL), some research studies related to UVR and cataract are of interest, since exposures of all other anterior structures would also be revealed in such research, and all animal studies showed that the cornea was injured at lower exposures than the lens. Ultraviolet cataract can be produced in animals from acute exposures to UVR of wavelengths between 295 nm and 325 nm (and perhaps to 400 nm). Action spectra can only be obtained from animal studies and the cataract is generally anterior sub-capsular, produced by intense exposures delivered over a period of days.[20, 26-27]. Human cortical cataract has been linked to chronic, life-long UV-B radiation exposure.[4,5,9,12] These suggest that it is primarily UV-B radiation in sunlight, and not UV-A, that is most injurious to the lens, biochemical studies suggest that UV-A radiation may also contribute to accelerated ageing of the lens.[28]

One area of UV cataract research is of note when considering lens exposure geometry of both the crystalline lens and the LAL. Since the only direct pathway of UVR to the inferior germinative area of the lens is from the extreme temporal direction, it has been speculated that side exposure is particularly hazardous.[29-30]

7. Ophthalmoheliosis--The Coroneo Effect

For any greatly delayed health effect, such as pterygium, droplet keratopathies, cataract or retinal degeneration, it is critical to determine the actual dose distribution at critical locations. A factor of great practical importance is the actual UVR that reaches the germinative layers of any tissue structure. In the case of the lens, the germinative layer where lens fiber cell nuclei are located is of great importance. The DNA in these cells is normally well shielded by the parasol effect of the irises. However, Coroneo[28] has suggested that focusing of very peripheral rays by the temporal edge of the cornea--those which do not even reach the retina, can enter the pupil and reach the equatorial region as shown in Figure 2.

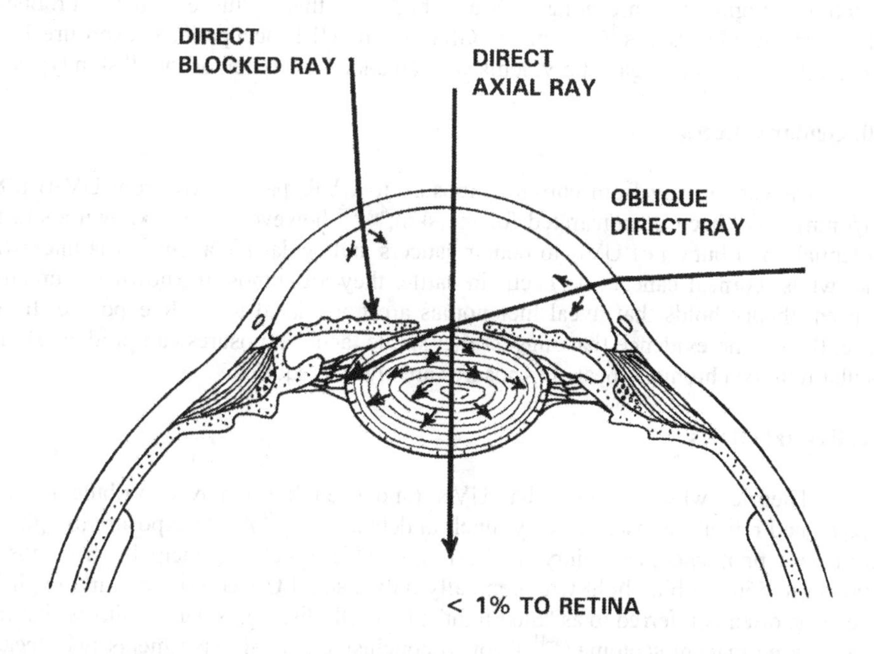

Figure 2. Coroneo Effect. Oblique rays striking the peripheral cornea can be refracted into the pupil and irradiate the sensitive germinative equatorial region of the human lens. It is this region where cell nuclei with DNA reside.

Coroneo terms this effect, which can also produce a concentration of UVR at the nasal side of the limbus and lens, "ophthalmoheliosis". He notes the more frequent onset of cataract in the nasal quadrant of the lens and the formation of pterygium (described below) in the nasal region of the cornea.[30]

8. Pterygium and droplet keratopathies

The possible role of UVR in the etiology of a number of age-related ocular diseases has been the subject of many medical and scientific papers. However, there is still a debate as to validity of these arguments. Although photokeratitis is unquestionably caused by UVR reflection from the snow,[5,31-32] pterygium and droplet keratopathies are less clearly related to UVR exposure.[9] Pterygium, a fatty growth originating in the conjunctiva that produces visual loss as it progresses over the cornea, is most common in ocean island residents (where both UVR and wind exposure is prevalent). UVR is a likely etiologic factor, [4,6,9,30] and the Coroneo Effect may also play a role.[30]

9. Erythema

Ultraviolet erythema ("sunburn" or reddening of the skin - 200 nm to 400 nm) applies to the lids of the eye. This effect appears several hours after an acute exposure and generally lasts from 8 hours to 72 hours depending upon degree exposure and spectral region. Thresholds are higher than those for producing photokeratoconjunctivitis.[1,4,33] The ACGIH and ICNIRP occupational exposure limits for UVR also protects against erythema with an added safety factor for all skin types.

10. Ocular cancers

Cancers arising from chronic exposure to UVR, particularly from UV-B (280-315 nm), have been demonstrated for the skin;[4,34,37] however little is known about the potential contribution of UVR to ocular cancers and ocular melanoma. It is interesting that whilst corneal cancers do occur in cattle, they are almost unknown in humans.[16] Current theory holds that uveal melanomas are not related to UVR exposure. In any case, there is no evidence that suggest that single, acute exposures can produce skin or ocular tumors; chronic, repeated exposures are always required.

11. Retinal effects

Theories which suggest that UVR (and even light) may contribute to some age-related retinal diseases are very much in debate.[1,4,9,16,30] Acute exposure to light can produce a photochemical injury to the retina. This effect is principally from visible 400 nm to 550 nm blue light with generally only a small UV contribution in the phakic eye. This often is referred to as "blue light" photoretinitis, e.g., solar retinitis which may lead to a permanent scotoma.[38-41] Prior to conclusive animal experiments two decades ago, solar retinitis was thought to be a thermal injury mechanism.[1] Unlike thermal retinal injury, there is no image-size dependence. The ICNIRP/ACGIH $B(\lambda)$ weighted

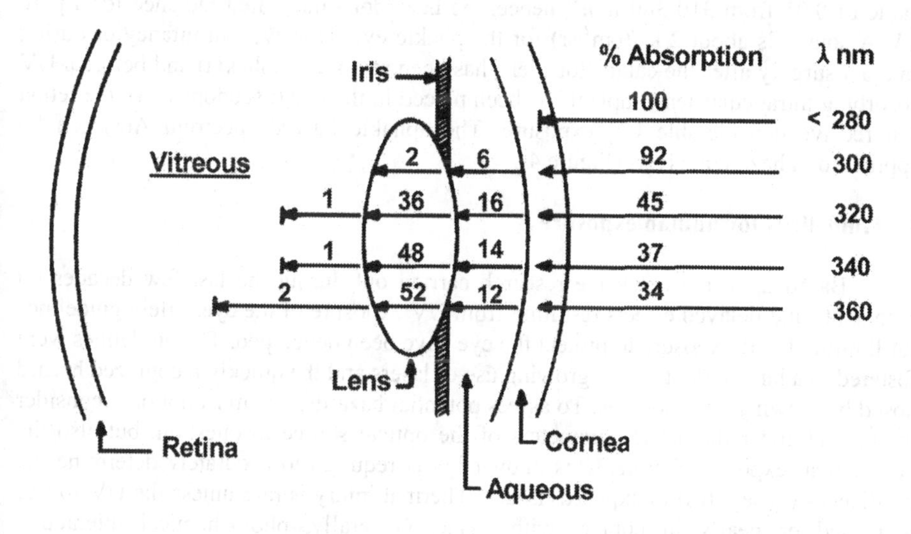

Figure 3. Spectral absorption of UV radiant energy within ocular structures.

integrated radiance limit L_B is 20 J/(cm^2·sr) averaged over a right circular cone of 0.011 radian for durations up to 10^4 seconds (2.8 hours); although a larger averaging angle is generally used for durations of the order of 1000 s or greater where a non-fixation visual task is involved. Normally the cornea, aqueous, and crystalline lens absorb 99% or more of the UVE that could enter the eye (as shown in Figure 3). The B(λ) function has a

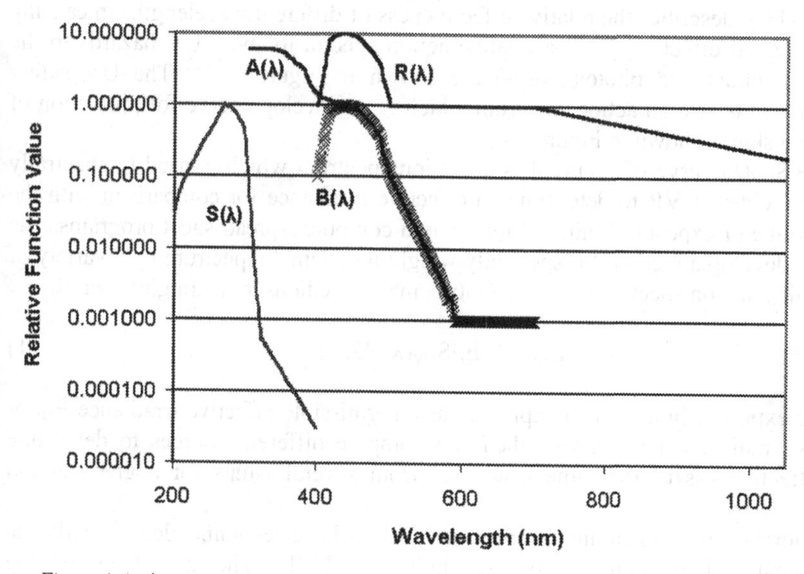

Figure 4. Action spectra.

value of 0.01 from 310-380 nm;[41] hence, the hazardous integrated radiance for a pure UV-A source is about 2 kJ/(cm^2·sr) for the phakic eye. However, in infancy or during cataract surgery after the cataractous lens has been removed (aphakia) and before a UV absorbing intraocular lens implant has been placed in the eye (pseudophakia), the retina can receive considerable UV exposure. The aphakic hazard spectrum A(λ) can be applied for a hazard analysis (Figure 4).

12. Guidelines for human exposure

Based upon the extensive research carried out during the last few decades on both acute and delayed effects resulting from UV exposure of the eye, safety guidelines for limiting UVR exposure to protect the eye have been developed. The guidelines were fostered to a large extent by the growing use of lasers and the quickly recognized hazard posed by viewing laser sources. To assess potential hazards, one must not only consider the optical and radiometric parameters of the optical source in question, but also the geometrical exposure factors. This knowledge is required to accurately determine the irradiances (dose rates) to exposed tissues. Thermal injury is rare unless the UV source is pulsed or nearly in contact with tissue. Generally, photochemical interaction mechanisms dominate in the UV spectrum where photon energies are sufficient to alter key biological molecules. A characteristic of photochemically initiated biological damage is the reciprocity of exposure dose rate and duration of exposure (the Bunsen-Roscoe Law), and acute UV effects are therefore most readily observed for lengthy exposure durations of many minutes or hours. The current guidance for UV exposure at wavelengths greater than 315 nm (UV-A) is 1 J/cm^2, and this was based upon conservative assumptions designed to protect the intact crystalline lens from both thermal and photochemical stress.

As with any photochemical injury mechanism, one must consider the action spectrum, which describes the relative effectiveness of different wavelengths in causing a photobiological effect.[1,4,42-45] The relative action spectra for both UV hazards to the eye (acute cataract and photokeratitis) are shown in Figure 4.[18-22] The UV safety function $S_{UV}(\lambda)$ is also an action spectrum which is an envelope curve for protection of both eye and skin is shown in Figure 4.

The $S_{UV}(\lambda)$ curve of Figure 4 is an action spectrum which is used to spectrally weight the incident UVR to determine an effective irradiance for comparison with the threshold value or exposure limit.[17] With modern computer spread-sheet programs, one can readily develop a method for spectrally weighting a lamp's spectrum by a variety of photochemical action spectra. The computation may be tedious, but straightforward:

$$E_{eff} = \sum E_\lambda \cdot S_{UV}(\lambda) \cdot \Delta\lambda \tag{1}$$

The exposure limit is then expressed as a permissible effective irradiance E_{eff} or an effective radiant exposure. One then can compare different sources to determine relative effectiveness of the same irradiance from several lamps for a given action spectrum.

A number of national and international groups have recommended virtually the same occupational or public exposure limits for UVR. The guidelines of the

International Commission for Non-Ionizing Radiation Protection[17,41-46] and the American Conference of Governmental Hygienists (ACGIH)[23] are by far the widest known. Both groups have recommended essentially the same limit based in large part on ocular injury data from animal studies and human accidental injury studies. The guideline to protect the skin, lens and cornea is an $S_{UV}(\lambda)$ weighted daily (8-hour) exposure H_{eff} of 3 mJ/cm^2 or 30 J/m^2 normalized at 270 nm. This corresponds to a limit of 27 J/cm at 365 nm. This limit is just below the level that produces a barely detectable increase in corneal light scatter and substantially below levels that produce clinically significant photokeratitis at 270 nm. The daily exposure limit is also about 1/3 to 1/4 of a minimal erythemal dose and less than 1/2 the exposure necessary for clinically reported keratitis. Annex A presents the ACGIH/ICNIRP human exposure limits based upon $S_{UV}(\lambda)$ for wavelengths greater than 250 nm.

A number of field survey measuring instruments have been developed which employ detectors that match the $S_{UV}(\lambda)$ action spectrum as shown in Figure 4 (See Annex Table for listed values). However, the geometry of the measurement is also of enormous importance when assessing risk of UV exposure to the eye. Outdoor safety assessments and epidemiologic studies can arrive at erroneous conclusions if measurements ignore geometrical factors and epidemiological assignments of exposure are seriously in error. A number of "reasonable" assumptions previously made by some epidemiologists regarding relative exposures have been shown to be false[6].

13. The challenge of measuring actinic UVR in sunlight

Both the *quantity* (irradiance) and *quality* (spectrum) of terrestrial ultraviolet radiation varies with the *solar zenith angle (Z)*, i.e., the angular position of the sun below the zenith (where $Z = 90^\circ$ – elevation angle above the horizon). The sun's position varies with time-of-day, the day of the year, and latitude. This variation is particularly striking in the UV-B spectral region (280-315 nm) because of the greater atmospheric attenuation along the direct atmospheric path. Stratospheric ozone absorption and molecular (Rayleigh) scattering by atmospheric N_2 and O_2 combine to attenuate the global UVR in this spectral region. However, in the troposphere, further absorption by air pollutants such as the oxides of nitrogen, sulfur-dioxide and ozone, and Mie scattering by water vapor in clouds and particulates can significantly add to the attenuation. Clouds and haze reduce the global UV (ground, horizontal) irradiance. However, haze and clouds redistribute the UVR, so that the UVR (i.e., the sky "brightness," or more correctly, radiance) of the horizon-sky can actually increase in comparison to a clear, blue-sky day. Since water vapor in clouds greatly attenuates infrared radiation, but does not significantly attenuate ground-level UVR, the warning sensation of heat to reduce the risk of sunburn can be absent on an overcast day. A light overcast, or light clouds scattered over a blue sky, do little to attenuate the UV global irradiance (unless a dense cloud lies directly over the sun), and many severe sunburns occur at the beach under such conditions. A light cloud cover may reduce the terrestrial global UVR (measured on a horizontal surface) to about half of that from a clear sky, although the horizon-sky UVR radiance does not decrease—and can even increase. Even under a heavy cloud-cover the scattered ultraviolet component of sunlight (termed the diffuse component," or "skylight") is seldom less than 10% of that under clear sky[1]. Only very heavy storm clouds can virtually eliminate terrestrial UV during summertime conditions. Although a higher global erythemal effective irradiance is measured at high altitude[46], the atmospheric scatter (diffuse component) is less, and the horizon-sky UVR does not increase by climbing mountains[5]. All of these observations point to the complex geometrical factors that challenge any outdoor UVR measurements.

Since the eye receives most of its outdoor UVR exposure from ground reflectance, the proper measurement of reflected sunlight adds a further challenge. The ground reflection of solar UV-B radiation varies over 100-fold. Green grass reflects less than 1% and most artificial surfaces and rock normally reflects less than 10%. Two exceptions are ocean surf and white, gypsum sand, which reflect about 20-25%. Fresh snow reflects about 85%, thus producing "snow blindness." Open water reflects the entire sky (diffuse plus direct component) and can be of the order of 20%, although the direct, specular reflection from the sun's image is only about 2-3% of the incident, direct UV radiation. Table 2 provides more detailed values for ground reflectance

Ocular exposure is far more affected than skin exposure by these geometrical factors. For an industrial UV source such as a welding arc, the cornea is shown to be more sensitive to UVR injury than the skin, but photokeratitis seldom accompanies summer sunburn of the skin. This seeming paradox is explained by the fact that people do not look directly overhead when the sun is very hazardous to view, whilst most people may stare at the sun when it is comfortable to observe near the horizon. Fortunately, at sunset, the filtering of UVR and blue light by the atmosphere allows us to directly view the sun. When the solar elevation angle exceeds 10 degrees above the horizon, strong squinting is observed which effectively shields the cornea and retina from most direct exposure. These factors reduce the exposure of the cornea to a maximum 5% of that falling on the exposed top of the head. However, if the ground reflectance exceeds 15%, photokeratitis may be produced after a few hours exposure. If one were to ignore the squinting factor and proper instrument field-of-view, the photokeratitis threshold would be achieved in less than 15 minutes for midday summer sunlight!

When wearing sunglasses, the pupil dilates proportionally to the darkness of the sunglasses[47]. Coroneo et al[29-30] have shown that very oblique temporal rays can be refracted into the critical nasal equatorial region of the lens and this could explain the increased incidence of opacification originating in the nasal sector of the lens in cortical cataract. The protective value of upper and lower lids, when they close down during squinting, determines the ocular UVR exposure dose in different environments. A brimmed hat or other headwear - associated with or without dark sunglasses - will modify greatly the UVR exposure dose. The geometrical factors that should be modeled by a radiometer now begin to appear as almost insurmountable. What can be done? Even geometrical positioning of the body greatly affects the solar exposure of human skin.

14. Biologically relevant measurements

With ocular UVR exposure so dependent upon geometry, the measurement challenge is not only to employ detectors with a spectral response that matches the action spectrum for the biological effect, but also provide a match of the geometry.

Several types of badges that measure total effective irradiance on the badge surface have been developed and can match the geometry of the skin, but not the eye.

15. Polysulfone film badge dosimeters

To date, the most widely used UV film-badge dosimeter has been the thermoplastic film, polysulphone. When polysulfone is exposed to UVR (particularly UVB), its UV absorption (generally calibrated at 330 nm) increases. This increase in absorbance has a useful linear-response range. Most typically, a 40-µm thick piece of polysulphone film is mounted in a cardboard or plastic mount, and these are worn by the experimental subjects on anatomic sites of interest. Sydehham constructed contact lenses of polysulfone and was able to corroborate the low UV exposure compared to that of the skin.

16. Direct clinical measurement of UVR exposure

The action spectrum and threshold for photokeratitis (snow-blindness) has been carefully studied, and was the basis of the action spectrum $S_{UV}(\lambda)$, hence, where individuals develop a threshold, just-detectable photokeratitis, the cornea itself is acting as a dosimeter. Sliney used this approach to estimate UV exposure to the crystalline lens or intra-ocular lens implant, and showed it to be an extremely small dose compared to that of the skin[49]. Erythema has long been used as a measure of individual, acute exposure of the skin to UVR, and dermatologists routinely examine benign markers of sun damage of skin on the face and on the back of the hands to judge the accelerated aging of the skin, and thereby estimate the risk for non-melanoma skin cancer. However, previous attempts to correlate the incidence of cataract with the skin-exposure estimates have been unsuccessful, presumably because of the lack of any direct relation between skin exposure and direct ocular exposure with the lids open. Another factor of interest is the impact of hats. Hats (particularly with wide brims) and other headwear greatly affect exposure to the face; however, hats which shade the eyes, may actually lead to greater lid opening, rendering the eye more vulnerable to ground reflections[50].

17. Ocular exposure dosimetry

Very pertinent data could be given by UV-sensitive contact lens dosimeters[48], but most of the pertinent data will come from studies of lid opening conditions combined with directional field measurements and different environmental conditions. Also the ground reflection will enter in these evaluations. Tables 2 and 3 from Sliney[3] provide information on the reflectance of ACGIH/ICNIRP-weighted solar UVB and the measured ACGIH/ICNIRP effective UVR from the sky with a 40° cone field-of-view. He effectively measured the UV radiance over a 40° averaging (acceptance) angle. This was later used to calculate ocular dose. Note that the relative effective UV irradiance near the horizon (i.e., within the eye's field of view) did not show big variations if the sky was visible and could increase with haze. Indeed, on an overcast day with the eyelids more open, the actual UV-B dose rate to the eye from the sky scatter can increase.

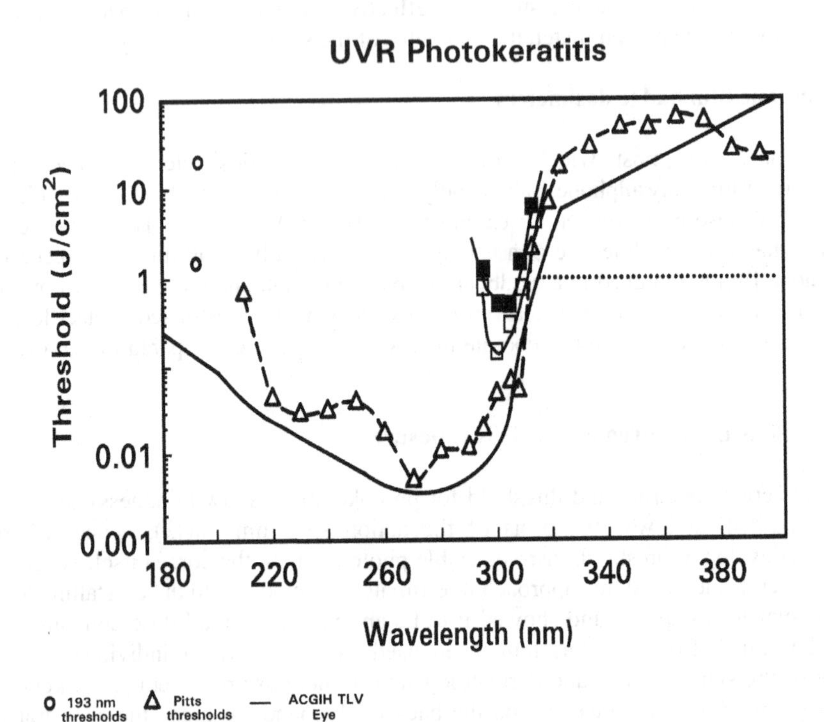

Figure 5. The solid line shows the exposure limit, expressed as the reciprocal of the spectral response curve $S_{UV}(\lambda)$ compared to the photobiological action spectra (not normalized) for erythema, and photokeratitis expressed as thresholds. The exposure limits for ultraviolet radiation were extended into the UV-A region in 1985. Open triangles are the threshold data for photokeratitis. Solid squares are the data for permanent cataract, and open squares are for reversible cataract.

The natural protection against overhead UVR afforded to the eye by the upper lid and brow ridge may be of little value for exposure to open-arc sources in the work place, since the source may frequently be within the normal horizontal line-of-sight. Hence greater eye protection is routinely required in industrial applications. Goggles and face shields are required when working around electric arc welding, which produces high levels of all UVR wavelengths. Close fitting facemasks with low transmittance to UVR, visible and infrared radiation are used for protection[1].

18. Conclusions

The design of instrumentation intended to measure the photobiological dose to ocular tissues must take into account a number of geometrical shading factors, such as the eyelids and brow ridge, as well as behavioral aspects of vision. These geometrical and imaging factors challenge the task of attempting to accurately measure the photobiologically significant exposure of the cornea, lens and retina to ultraviolet, visible and infrared radiation. The human eye is actually exposed to a very small fraction of the global ultraviolet irradiance (diffuse plus direct radiation incident upon a horizontal surface). Therefore, the acceptance angle (field-of-view) should mimic those of the human eye. However, this acceptance angle varies with sky brightness. Thus, the instrument field-of-view must be adjustable in an instrument designed to really measure the UVR exposure dose to the cornea and lens. The UV exposure to the anterior segment of the eye can be measured in a mannequin fitted with UV detectors at the ocular positions, and measurements can be made with and without UV-absorbing spectacles, or as would occur when a person wears different types of sunglasses. The exposure is greatly affected by the type of sunglass frame and partially to the UV transmittance of the sunglass lenses. In some instances, UV exposures of some specific ocular tissues can actually equal or exceed those when not wearing sunglasses.

The retina is exposed to visible light and some IR-A radiant energy within an imaged scene. Although the lid opening varies with ambient scene luminance (brightness), it is possible to mathematically predict the opening of the lids and the angular field-of-view from studies of lid opening. The light exposure to the retina is not at all uniform in outdoor daylight conditions. The central and superior regions of the retina receive much more light than the inferior retina (Figure 6). Thus, instrumentation should simulate these geometrical factors.

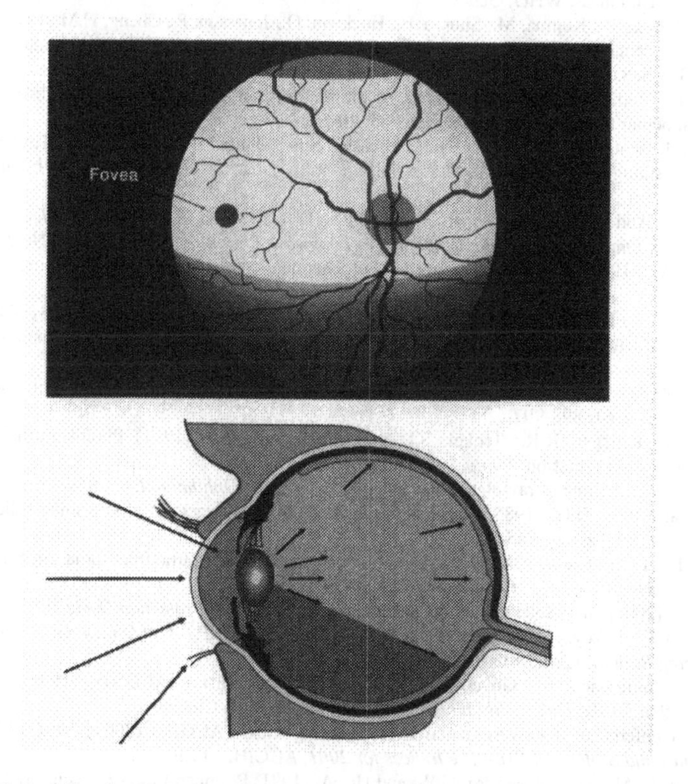

Figure 6. Retinal illumination pattern. During most outdoor conditions, much of the inferior retina is shaded, and only under conditions of bright sand or snow do both lids close to allow only the illumination of the macula and a horizontal band on either side.

References

1. Sliney, D.H. and Wolbarsht, M.L. (1980) *Safety with Lasers and Other Optical Sources*, Plenum Publishing Corp, New York.
2. Sliney, D.H. (2000) Ultraviolet radiation exposure criteria, *Radiation Protection Dosimetry*, 91: 213-222.
3. Sliney, D.H. (1995) UV Radiation Ocular Exposure Dosimetry, *J Photochem. Photobiol. B: Biol*, 31: 69-77.
4. World Health Organization (WHO) (1995) *The Effects of Solar UV Radiation on the Eye, Report of an Informal Consultation, Geneva, August – September 1993,* Publication WHO/PBL/EHG/94.1, Program for the Prevention of Blindness, WHO, Geneva.
5. Sliney, D.H. (1986) Physical factors in cataractogenesis: ambient ultraviolet radiation and temperature, *Inv. Ophthalmol Vis Sci* 27: 781-790.
6. Sliney, D.H. (1994) Epidemiological studies of sunlight and cataract: the critical factor of ultraviolet exposure geometry, *Ophthalmic Epidemiology*, 1: 107-119.
7. Klein, B.E.K., Klein, R. and Linton, K.L.P. (1992) Prevalence of age-related lens opacities in a population, the Beaver Dam Eye Study, *Ophthalmology* 44: 546-522.
8. Dolin, P.J. (1994) Ultraviolet radiation and cataract: A review of the epidemiological evidence, *Br J Ophthalmol* 78: 478-482 and reprinted in *Optometry Today* Nov/Dec 1997.
9. Taylor, H.R., West, S.K., Rosenthal, F.S., Munoz, G., Newland, H.S., Abbey, H. and Emmett, S.K. (1989) Effect of ultraviolet radiation on cataract formation, *New Engl J Med* 319: 1429-1433.
10. World Health Organization (1995) *The Effects of Solar UV Radiation on the Eye, Report of an Informal Consultation, Geneva, August – September 1993,* Publication WHO/PBL/EHG/94.1, Program for the Prevention of Blindness, WHO, Geneva.
11. Sasaki, K., Sasaki, H., Kojima, M., Shui, Y.B., Hockwin, O., Jonasson, F., Cheng, H.M., Ono, M. and Katoh (1999) N, Epidemiological studies on UV-related cataract in climatically different countries. *J Epidemiol (Japan)* 9:S33-S38.
12. Sliney, D.H. (2002) Geometrical gradients in the distribution of temperature and absorbed ultraviolet radiation in ocular tissues, *Dev. Ophthalmol.*, 35: 40-59.
13. West, S.K., Rosenthal, F.S., Bressler, N.M., Bressler, S.B., Munoz, B., Fine, S.L. and Taylor, H.R. (1989) Exposure to sunlight and other risk factors for age-related macular degeneration, *Arch Ophthalmol*,107: 875-879.
14. Parrish, J.A., Anderson, R.R., Urbach, F. and Pitts, D. (1978) *UV-A, Biological Effects of Ultraviolet Radiation with Emphasis on Human Responses to Longwave Radiation*, Plenum Press, New York.
15. Zuclich, J.A. (1989) Ultraviolet-induced photochemical damage in ocular tissues. *Health Phys* 56: 671-682.
16. Pitts, D.G. (1974) The human ultraviolet action spectrum, *Am J Optom Physiol Optics* 51: 946-960.
17. Pitts, D.G, Cullen, A.P. and Hacker, P.D. (1977) Ocular effects of ultraviolet radiation from 295 to 365 nm, *Inv Ophthalmol Vis Sci* 16: 932-939.
18. Ringvold (1983) A., Damage of the cornea epithelium caused by ultraviolet radiation, *Acta Ophthalmologica*, 61: 898-907.
19. Sliney, D.H., Krueger, R.R., Trokel, S.L. and Rappaport, K.D. (1991) Photokeratitis from 193 nm argon-fluoride laser radiation, *Photochem Photobiol*, 53: 739-744.
20. Bachem, A., Ophthalmic ultraviolet action spectra (1956) *Am J Ophthalmol*, 41: 969-975.
21. Jose, J.G. and Pitts, D.G. (1985 Wavelength dependency of cataracts in albino mice following chronic exposure, *Exp Eye Res*, 41: 545-563.
22. Sliney, D.H. (1972) The merits of an envelope action spectrum for ultraviolet radiation exposure criteria, *Amer Industr Hyg Assn J,* 33: 646-653.
23. Sliney, D.H. (2000) Ultraviolet radiation exposure criteria, *Radiation Protection Dosimetry,* 91: 213-222.
24. Duchêne, A.S., Lakey, J.R.A. and Repacholi, M.H. (1991) *IRPA Guidelines on Protection against Non-Ionizing Radiation*, McMillan, New York.
25. American Conference of Governmental Industrial Hygienists (ACGIH) (1992) *ACGIH 1992 Documentation for the Threshold Limit Values 4th Edn.*, ACGIH, Cincinnati.
26. American Conference of Governmental Industrial Hygienists (ACGIH) (2001) *2001 TLV's, Threshold Limit Values and Biological Exposure Indices for 2001*, ACGIH, Cincinnati.
27. Coroneo, M.T., Müller-Stolzenburg, N.W. and Ho, A. (1991) Peripheral light focusing by the anterior eye and the ophthalmoheliosis, *Ophthalmic Surg,* 22: 705-711.
28. Coroneo, M.T. (1993) Pterygium as an early indicator of ultraviolet insolation: an hypothesis, *Br. J. Ophthalmol*, 77: 734-739.

ANNEX 1 - Applying the Ultraviolet Radiation Exposure Limits

The ACGIH/ICNIRP EL for exposure to the eye and skin to UVR is 3 mJ/cm^2-effective, when the spectral irradiance E_λ at the eye or skin surface is mathematically weighted against the hazard sensitivity spectrum $S(\lambda)$ from 180 nm to 400 nm as follows:

$$E_{eff} = E\lambda \cdot S(\lambda) \cdot \Delta\lambda \quad (1)$$

The permissible exposure duration, t_{max}, in seconds, to the spectrally weighted UVR is calculated by:

$$t_{max} = (3\ \text{mJ/cm}^2) / E_{eff}\ (\text{W/cm}^2) \quad (2)$$

In addition to the above requirement, the ocular exposure is also limited to 1 J/cm^2 for periods up to 1000 s (16.7 min) by ACGIH and to 30,000 s by ICNIRP. ACGIH limits the total irradiance to 1 mW/cm^2 for periods greater than 1,000 s. For this requirement, the total irradiance, E_{uva}, in the UV-A spectral region is summed from 315 nm to 400 nm:

$$E_{UVA} = \sum_{315}^{400} E_\lambda \cdot \Delta\lambda \quad (3)$$

where E_λ is the spectral irradiance in W/(cm^2-nm).

One way to express the ACGIH limit for the UV-A is: if the total irradiance exceeds the 8-hour criterion of 1 mW/cm^2, the maximum exposure must also be less than:

$$t_{max} = (1\ \text{J/cm}^2) / E_{UVA}\ (\text{W/cm}^2) \quad (4)$$

Equation (4) applies to all periods up to 8 h for the ICNIRP guidelines.

Table A1.1 UVR Hazard Spectral Weighting Function.

Wavelength (nm)	UVR Hazard Function $S_{UV}(\lambda)$	Wavelength (nm)	UVR Hazard Function $S_{UV}(\lambda)$	Wavelength (nm)	UVR Hazard Function $S_{UV}(\lambda)$
180	0.012	260	0.650	335	0.00034
190	0.019	265	0.810	340	0.00028
200	0.030	270	1.000	345	0.00024
205	0.051	275	0.960	350	0.00020
210	0.075	280	0.880	355	0.00016
215	0.095	285	0.770	360	0.00013
220	0.012	290	0.640	365	0.00011
225	0.015	295	0.540	370	0.000093
230	0.019	300	0.300	375	0.000077
235	0.240	305	0.060	380	0.000064
240	0.300	310	0.014	385	0.000053
245	0.360	315	0.003	390	0.000044
250	0.430	320	0.001	395	0.000036
*254	0.500	325	0.0005	400	0.000030
255	0.520	330	0.00041		

*Low-pressure mercury germicidal lamp emission.

Table A1.2: Reflectance of ACGIH/ICNIRP-effective solar UV-B from terrain surfaces

Terrain surfaces	Diffuse reflectance effective solar UV-B (ACGIH/ICNIRP spectral weighting)
Green mountain grassland	0.8 - 1.6 %
Dry grassland	2.0 - 3.7 %
Wooden boat dock	6.4 %
Black asphalt	5 - 9 %
Concrete pavement	8 - 12 %
Atlantic beach sand (dry)	15 - 18 %
Atlantic beach sand (wet)	7%
Sea foam (surf)	25 - 30 %
Dirty snow	59%
Fresh snow	88%

Source: Sliney, D.H., 1983

Table A1.3: Measured ACGIH effective UVB from the sky with a 40° cone filed of view.

Sky conditions location, elevation	Zenith reading (μWcm-² sr^{-1})	Directly at sun (μWcm-² sr^{-1})	Opposite sun (μWcm-² sr^{-1})	Horizon sky (μWcm-² sr^{-1})
Clear sky, dry, sea level	0.1	1.4 Z=70°	0.22	0.27
Clear sky, humid, sea level	0.27	4.1 Z=50°	0.27	0.24
Ground fog, sea level	0.04	0.19 Z=75°	0.04	0.03
Hazy humid, sea level	0.014	1.4 Z=70°	0.22	0.54
Cloudy bright, 700 m	0.54	0.44 Z=45°	0.27	0.05 (tree-line)
Hazy beach	0.54	0.60 Z=75°	0.54	0.60
Hazy beach	0.38	3.5 Z=40°	0.54	0.44
Clear mountain top (2750 m)	0.54	1.6 Z=25°	0.82	0.08

Source: Sliney, D.H., 1983.

ANNEX 2 - Additional References

Anderson, J.R., A pterygium map, *Acta Ophthalmol.*, 3: 1631-1642,1954.

Baasanhu, J., Johnson, G.J., Burendei, G., and Minassian, D.C., Prevalence and causes of blindness and visual impairment in Mongolia, *Bull. WHO*, in press.

Bachem, A., Ophthalmic ultraviolet action spectra, *Am J Ophthalmol*, 41: 969-975, 1956.

Bergmanson, J.P., Pitts, D.G. and Chu, L.W., Protection from harmful UV radiation by contact lens, *J Am Optom Assoc*, 59: 178-182, 1988.

Bhatnagar, R., West, K.P., Vitale, S., Sommer, A., Joshi, S.and Venkataswamy, G., Risk of cataract and history of severe diarrheal disease in Southern India, *Arch Ophthalmol*, 109: 696-699, 1991.

Blumthaler, M., Ambach, W. and Daxecker, F., On the threshold radiant exposure for keratitis solaris, *Invest Ophthalmol Vis Sci*, 28: 1713-1716, 1987.

Bochow, T.W., West, S.K., Azar, A., Munoz, B., Sommer, A. and Taylor, H.R., Ultraviolet light exposure and risk of posterior subcapsular cataracts, *Arch Ophthalmol*, 107: 369-372, 1989.

Boettner, E.A. and Wolter, J.R., Transmission of the ocular media, *Invest Ophthalmol Vis Sci*, 1: 776-783, 1962 .

Booth, F., Heredity in one hundred patients admitted for excision of pterygia, *Aust NZ J Ophthalmol*, 13: 59-61, 1985.

Brilliant, L.B., Grasset, N.C., Pokhrel, R.P., Kolstad, A., Lepkowski, J.M., Brilliant, G.E., Hawks, W.N.. and Pararajasegaram, R., Associations among cataract prevalence, sunlight hours, and altitude in the Himalayas, *Am J Epidemiol*, 118: 250-264, 1983.

Cameron, M.E., *Pterygium throughout the world*, Springfield, IL., Charles Thomas, 1965.

Chatterjee, A., Milton, R.C. and Thyle, S., Prevalence and aetiology of cataract in Punjab, *Br J Ophthalmol*, 66: 35-42, 1982.

Cogan, D.G. and Kinsey, V.E., Action spectrum for keratitis produced by ultraviolet radiation, *Arch Ophthalmol*, 35: 670-677, 1946.

Colin, J., Bonissent, J.F. and Resnikoff, S., Epidemiology of the exfoliation syndrome, *Proc. 17th Congr. European Soc. Ophthalmol,* Helsinki: 230-231, 1985.

Collman, G.W., Shore, D.L., Shy, C.M., Checkoway, H. and Luria, A.S., Sunlight and other risk factors for cataract: an epidemiological study, *Am J Public Health*, 78: 1459-1462, 1988.

Cruickshanks, K.J., Klein, R. and Klein, B.E., Sunlight and age-related macular degeneration: the Beaver Dam Eye Study, *Arch Ophthalmol*, 111: 514-518, 1993.

Darrell, R.W. and Bachrach, C.A., Pterygium among veterans, *Arch Ophthalmol*, 70: 158-169, 1963.

Detels, R. and Dhir, S.P., Pterygium: a geographical study, *Arch Ophthalmol*, 78: 485-491, 1967.

Dhir, S.P., Detels, R. and Alexander, E.R., The role of environmental factors in cataract, pterygium and trachoma, *Am J Ophthalmol*, 64: 128-135, 1967.

Dolezal, J.M., Perkins, E.S. and Wallace, R.B., Sunlight, skin sensitivity, and senile cataract, *Am J Epidemiol*, 129: 559-568, 1989.

Doll, R., Urban and rural factors in the aetiology of cancer, *Int. J. Cancer*, 47: 803-810, 1991.

Elliott, R., The aetiology of pterygium, *Trans Ophthalmol Soc NZ*, 13: 22-41, 1961.

Eye Disease Case-Control Study Group, Risk factors for neovascular age-related macular degeneration, *Arch Ophthalmol*, 110: 1701-1708, 1992.

Fitzpatrick, T.B., Pathak, M.A., Harber, L.C., Seiji, M. and Kukita, A., *Sunlight and Man*, Tokyo, University of Tokyo Press, 1974.

Forsius, H., Exfoliation syndrome in various ethnic populations, *Acta Ophthalmol*, 68(Suppl. 184): 71-85, 1988.

Gallagher, R.P., Elwood, J.M., Rootman, J., Spinelli, J.J., Hill, G.B., Threlfall, W.J. and Birdsell, J.M., Risk factors for ocular melanoma: Western Canada Melanoma Study, *J Natl Cancer Inst*, 74: 775-778, 1985.

Garner, A., The pathology of tumours at the limbus, *Eye*, 3: 210-217, 1989.

Gray, R.H., Johnson, G.J. and Freedman, A., Climatic droplet keratopathy, *Surv Ophthalmol*, 36: 241-253, 1992.

Guex-Crosier, Y. and Herbort, C.P., Presumed corneal intraepithelial neoplasia associated with contact lens wear and intense ultraviolet light exposure, *Br J Ophthalmol*, 77: 191-192, 1993.

Ham, W.T. Jr, Ruffolo, J.J. Jr, Mueller, H.A., *et al.*, Histologic analysis of photochemical lesions produced in rhesus retina by short wavelength light, *Invest Ophthalmol*, 17: 1029-1035, 1978.

Ham, W.T. Jr., Mueller, H., Ruffolo, J., Guerry, D. and Guerry, R., Action spectrum for retinal injury from near-ultraviolet radiation in the aphakic, *Am J Ophthalmol*, 93: 299-306, 1982.

Hiller, R., Giacometti, L. and Yuen, K., Sunlight and cataract: an epidemiologic investigation,. *Am J Epidemiol*, 105: 450-459, 1977.
Hiller, R., Sperduto, R.D. and Ederer, F., Epidemiologic associations with cataract in the 1971-1972 national health and nutrition examination survey, *Am J Epidemiol*, 118: 239-249, 1983.
Hiller, R., Sperduto, R.D. and Ederer, F., Epidemiologic associations with nuclear, cortical, and posterior subcapsular cataracts, *Am J Epidemiol*, 124: 916-925, 1986.
Hollows, F. and Moran, D., Cataract - the ultraviolet risk factor, *Lancet*, 2: 1249-1250, 1981.
Holly, E.A., Aston, D.A., Char, D.H., Kristiansen, J.J. and Ahn, D.K., Uveal melanoma in relation to ultraviolet light exposure and host factors, *Cancer Res*, 50: 5773-5777, 1990.
Hyman, L.G., Lilienfeld, A.M., Ferris, F.L. and Fine, S.L.,, 1983 Senile macular degeneration: a case-control study, *Am J Epidemiol*, 118: 213-227.
International Agency for Research on Cancer, *IARC monographs on the evaluation of carcinogenic risk to humans. Vol 55: Solar and ultraviolet radiation*, Lyon, IARC, 1992.
Italian-American Cataract Study Group, Risk factors for age-related cortical, nuclear, and posterior subcapsular cataracts, *Am J Epidemiol*, 133: 541-553, 1991.
Johnson, G.J., Aetiology of spheroidal degeneration of the cornea in Labrador, *Br J Ophthalmol*, 65: 270-283, 1981.
Johnson, G.J. and Overall, M., Histology of spheroidal degeneration of the cornea in Labrador, *Br J Ophthalmol*, 62: 53-61, 1978.
Johnson, G.J., Paterson, G.D., Green, J.S. and Perkins, E.S., Ocular conditions in a labrador community, in: *Circumpolar Health 81*, Harvald B & Hansen JP (Eds), Copenhagen, Nordic Council for Arctic Medical Research, 1981.
Johnson, G.J., Minassian, D.C. and Franken, S., Alterations of the anterior lens capsule associated with climatic keratopathy, *Br J Ophthalmol* 73: 229-234, 1989
Jose, J.G., Posterior cataract induction by UVB radiation in albino mice, *Exp Eye Res*, 42: 11-20, 1986.
Jose, J.G. and Pitts, D.G., Wavelength dependency of cataracts in albino mice following chronic exposure, *Exp Eye Res*, 41: 545-563, 1985.
Karai, I. and Horiguchi, S., Pterygium in welders, *Br J Ophthalmol*, 68: 347-349, 1984.
Kinsey, V.E., Spectral transmission of the eye to ultraviolet radiation, *Arch Ophthalmol*, 39: 505-513, 1948.
Lee, G.A. and Hirst, L.W., Incidence of ocular surface epithelial dysplasia in metropolitan Brisbane: a 10-year survey, *Arch Ophthalmol*, 110: 525-527, 1992.
Lerman, S., Human ultraviolet radiation cataracts, *Ophthalmic Res*, 12: 303-314, 1980.
Leske, M.C., Chylack, L.T., Wu, S. and The Lens Opacities Case-Control Study Group, The lens opacities case-control study: risk factors for cataract, *Arch Ophthalmol*, 109: 244-251, 1991.
Lindberg, J.G., *Kliniska undersökningar över depigmenteringen av pupillarranden och genomlysbarheten av iris vid fall av åldstarr samt i normala ögon hos gamla personer*, Dissertation, Helsingfors, Finland, 1917.
Lischko, A.M., Seddon, J.M., Gragoudas, E.S., Egan, K.M. and Glynn, R.J., Evaluation of prior primary malignancy as a determinant of uveal melanoma. A case-control study, *Ophtalmology*, 96: 1716-1721, 1989.
Mack, T.M. and Floderus, B., Malignant melanoma risk by nativity, place of residence at diagnosis, and age at migration, *Cancer Causes Control*, 2: 401-411, 1991.
Mackenzie, F.D., Hirst, L.W., Battistutta, D. and Green, A., Risk analysis in the development of pterygia, *Ophthalmol*, 99: 1056-1061, 1992.
Mainster, M.A., Ham, W.T. and Delori, F.C., Potential Retinal Hazards, Instrument and environmental light sources, *Ophthalmology*, 90: 927-931, 1983.
Mao, W. and Hu, T., An epidemiologic survey of senile cataract in China, *Chinese Med J*, 95: 813-818, 1982.
Milham, S. Jr, *Occupational Mortality in Washington State 1950-1979* (DHSS (NIOSH) Publ. No. 83-116), Cincinnati, OH, National Institute for Occupational Safety and Health, 1983.
Minassian, D.C., Mehra, V. and Johnson, G.J., Mortality and cataract: findings from a population-based longitudinal study, *Bull WHO*, 70: 219-223, 1992.
Mohan, M., Sperduto, R.D., Angra, S.K., Milton, R.C., Mathur, R.L., Underwood, B.A., Jaffery, N., Pandya, C.B., Chhabra, V.K., Vajpayee, R.B., Kalra, V.K., Sharma, Y.R. and The Indian-US Case-Control Study Group, India-US case-control study of age-related cataracts, *Arch Ophthalmol*, 107: 670-676, 1989.
Moran, D.J. and Hollows, F.C., Pterygium and ultraviolet radiation: a positive correlation, *Br J Ophthalmol*, 68: 343-346, 1984.
Nachtwey, D.B. and Rundel, R.D., A photobiological evaluation of tanning booths, *Science*, 211: 405-407, 1981.

Naumann, G.O.H. and Apple, D., *Pathology of the Eye*, New York, Springer-Verlag, 1986.
Noell, W.K., Walker, V.S. and Kang, B.S., Retinal damage by light in rats, *Invest Ophthalmol Vis Sci*, 5: 450-473, 1966.
Norn, M.S., Spheroid degeneration, pinguecula, and pterygium among Arabs in the Red Sea Territory, Jordan, *ACTA Ophthalmol*, 60: 949-954, 1982.
Office of Population Censuses and Surveys, *Occupational Mortality: the Registrar General's Decennia Supplement for Great Britain 1979-80, 1982-83* (Series DS No. 6), London, Her Majesty's Stationary Office, 1986.
Oldenburg, J.B., Gritz, D.C. and McDonnell, P.J., Topical ultraviolet light-absorbing chromophore protects against experimental photokeratitis. *Arch Ophthalmol*, 108: 1142-1144, 1990.
Ostendfeld-Åkerblom, A., Pseudoexfoliation in Eskimos (Inuit) in Greenland, *Acta Ophthalmol*, 66: 467-468, 1988.
Østerlind, A., Trends in incidence of ocular malignant melanoma in Denmark 1943-1982, *Int. J. Cancer*, 40: 161-164, 1987.
Østerlind, A., Olsen, J.H., Lynge, E. and Ewertz, M., Second cancer following cutaneous melanoma and cancers of the brain, thyroid, connective tissue, bone, and eye in Denmark, 1943-80, *Natl. Cancer Inst. Monogr*, 68: 361-388, 1985.
Parkin, D.M., Muir, C.S., Whelan, S.L., Gao, Y.-T., Ferlay, J. and Powell, J. (Eds), *Cancer Incidence in Five Continents, Vol. VI* (IARC Scientific Publications No. 120), Lyon, International Agency for Research on Cancer, 1992.
Pitts, D.G., The human ultraviolet action spectrum, *Am J Optom Arch Am Acad Optom*, 51: 946-960, 1974.
Pitts, D.G., The ocular effects of ultraviolet radiation, *Am J Optom Phys Optics*, 55: 19-35, 1978.
Pitts, D.G., Cullen, A.P. and Hacker, P.D., Ocular effects of ultraviolet radiation from 295 to 365nm, *Invest Ophthalmol Vis Sci*, 16: 932-939, 1977.
Ringvold, A. and Davangar, M., Changes in the rabbit corneal stroma caused by UV irradiation, *Acta Ophthalmol*, 63: 601-606, 1985.
Rosen, E.S., Filtration of non-ionizing radiation by the ocular media, in: *Hazards of light: myths and realities of eye and skin*, Cronley-Dillon J, Rosen ES, Marshall J (Eds), Oxford, Pergamon Press, pp 145-152, 1986.
Saftlas, A.F., Blair, A., Cantor, K.P., Hanrahan, L. and Anderson, H.A., Cancer and other causes of death among Wisconsin farmers, *Am J.Ind.Med.*, 11: 119-129, 1987.
Schwartz, S.M. and Weiss, N.S., Place of birth and incidence of ocular melanoma in the United States, *Int.J.Cancer*, 41: 174-177, 1988.
Seddon, J.M., Gragoudas, E.S., Glynn, R.J., Egan, K.M., Albert, D.M. and Blitzer, P.H., Host factors, UV radiation, and risk of uveal melanoma: a case-control study, *Arch Ophthalmol*, 108: 1274-1280, 1990.
Shibata, T., Katoh, N., Hatano, T. and Sasaki, K., Population based case-control study of cortical cataract in the Noto area, Japan, *Ophthalmic Res*,in press, 1993
Siemiatycki, J., *Risk factors for cancer in workplace*, Boca Raton, Fl, CRC Press, 1991.
Sliney, D.H., Estimating the solar ultraviolet radiation exposure to an intraocular lens implant, *J Cataract Refract Surg*, 13: 296-301, 1987.
Sliney, D.H., Physical factors in cataractogenesis: ambient ultraviolet radiation and temperature, *Invest Ophthalmol Vis Sci*, 27: 781-790, 1986.
Sliney, D.H., Eye protective techniques for bright light, *Ophthalmology*, 90: 937-944, 1983.
Taylor, H.R., Pseudoexfoliation, an environmental disease?, *Trans Ophthalmol Soc UK*, 99: 302-307, 1979.
Taylor, H.R., Aetiology of climatic droplet keratopathy and pterygium, *Br J Ophthalmol*, 64: 154-163, 1980a.
Taylor, H.R., The environment and the lens, *Br J Ophthalmol*, 64: 303-310, 1980b.
Taylor, H.R., West, S.K., Rosenthal, F.S., Munoz, B., Newland, H.S., Abbey, H. and Emmett, E.A., Effect of ultraviolet radiation on cataract formation, *New Engl J Med*, 319: 1429-1433, 1988.
Taylor, H.R., West, S.K., Rosenthal, F.S., Munoz, B., Newland, H.S. and Emmett, E.A., Corneal changes associated with chronic UV irradiation, *Arch Ophthalmol*, 107: 1481-1484, 1989.
Taylor, H.R., West, S., Munoz, B., Rosenthal, F.S., Bressler, S.B. and Bressler, N.M., The long-term effects of visible light on the eye, *Arch Ophthalmol*, 110: 99-104, 1992.
Tucker, M.A., Shields, J.A., Hartge, P., Augsburger, J., Hoover, R.N. and Fraumeni, J., Sunlight exposure as risk factor for intraocular malignant melanoma, *New Engl J Med*, 313: 789-792, 1985.
Turner, B.J., Siatkowski, R.M., Augsburger, J.J., Shields, J.A., Lustbader, E. and Mastrangelo, M.J., Other cancers in uveal melanoma patients and their families, *Am.J.Ophtalmol*, 107: 601-608, 1989.
Vitale, S., West, S., Munoz, B., Schein, O.D., Maguire, M., Bressler, N. and Taylor, H.R., Watermen Study II: mortality and baseline prevalence of nuclear opacity, *Invest Ophthalmol Vi Sci*, 33: 1097, 1992.

Waring, G.O., Roth, A.M. and Ekins, M.B.,, 1984 Clinical and pathological description of 17 cases of corneal intraepithelial neoplasia, *Am J Ophthalmol*, 97: 547-559.

West, S.K., Rosenthal, F.S., Bressler, N.M., Bressler, S.B., Munoz, B., Fine, S.L. and Taylor, H.R., Exposure to sunlight and other risk factors for age-related macular degeneration, *Arch Ophthalmol*, 107: 875-879, 1989.

Wittenberg, S., Solar radiation and the eye: a review of knowledge relevant to eye care, *Am J Optom Physiol Optics*, 63: 676-689, 1986.

Wong, L., Ho, S.C., Coggon, D., Cruddas, A.M., Hwang, C.H., Ho, C.P., Robertshaw, A.M. and MacDonald, D.M., Sunlight exposure, antioxidant status, and cataract in Hong Kong fishermen, *J Epidemiol Comm Health*, 47: 46-49, 1993.

Yannuzzi, L., Fisher, Y., Slakter, J. and Krueger, A., Solar retinipathy. A photobiologic and geophysical analysis, *Retina*, 9: 28-43, 1989.

Young, R.W., *Age-related cataract*, New York, Oxford University Press, 1991.

Zaunuddin, D. and Sasaki, K., Risk factor analysis in a cataract epidemiology survey in West Samatara, Indonesia, *Dev Ophthalmol*, 21: 78-86, 1991.

Zigman, S., Yulo, T. and Schultz, J., Cataract induction in mice exposed to near UV light, *Ophthalmol Res*, 6: 259-270, 1974.

Zigman, S., Graff, J., Yulo, T. and Vaughen, T., The response of mouse ocular tissue to continuous near-UV light exposure, *Invest Ophthalmol*, 14: 710-713, 1975.

STUDENT ABSTRACTS

The photochemistry of a common UV-B sunscreen absorber

BICE S. MARTINCIGH

School of Pure and Applied Chemistry, University of Natal, Durban, South Africa

The photochemistry of 2-ethylhexyl-*para*-methoxy-cinnamate (2-EHMC) is of fundamental interest as this is currently the most commonly used ultraviolet-B absorber in sunscreen and cosmetic formulations. The commercial product is available as *trans*-EHMC and is known to isomerise to the *cis*-form in sunlight, and thereby lose some of its absorbing efficiency. This phenomenon was confirmed by a spectrophotometric study. The isomers were separated and quantified by using high performance liquid chromatography (HPLC). This showed the attainment of a photostationary state between the *trans*- and *cis*-isomer. Irradiation of a concentrated solution of *trans*-EHMC with wavelengths greater than 300 nm also caused the absorber to dimerise[1] with itself via a [2 + 2] cycloaddition reaction thereby further decreasing its absorbing power. Six isomers of the dimer were separated and isolated by a gradient mode of elution in HPLC. Fourier transform infrared and proton nuclear magnetic resonance spectroscopy were used to characterise the isolated dimers. It has also been speculated that 2-EHMC can photobind to constituents of DNA in a manner similar to its self-dimerisation reaction. Any such sunscreen-DNA photoadduct could be potentially mutagenic and carcinogenic. The formation of such adducts has been investigated and the identity of that formed on irradiation of thymidine-5'-monophosphate in the presence of 2-EHMC was confirmed by capillary zone electrophoresis-mass spectrometry analysis.

Reference

1. Broadbent JK, Martincigh BS, Raynor MW, Salter LF, Moulder R, Sjöberg P, Markides KE (1996) Capillary supercritical fluid chromatography combined with atmospheric pressure chemical ionisation mass spectrometry for the investigation of photoproduct formation in the sunscreen absorber 2-ethylhexyl-para-methoxycinnamate, *Journal of Chromatography A* 732: 101-110.

F. Ghetti et al. (eds.), Environmental UV Radiation: Impact on Ecosystems and Human Health and Predictive Models, 279–288.

Oxidative stress caused by tropospheric ozone on some biomolecules

MARIA GIUBELAN, AURELIA MEGHEA, MARIA GIURGINCA, ALIN MIHALY
University "Politehnica" of Bucharest, National Consultancy Center for Environmental Protection, Bucharest, Romania

Living cells acting as "cell factories" are directly involved into fundamental processes of metabolism. Sometimes in such biochemical reactions the active species of oxygen are also involved. As a result of oxidative reactions accelerated by stress and pollution, a number of short life radicals and oxygenated compounds are formed which affect the metabolism of the main constituents of cells, protein and aminoacids, giving rise to premature aging and diseases by damaging the functioning of vital organs. The aim of this paper is to evidence the involvement of ozone during oxidative process of some aromatic aminoacids, tryptophane (Trp), tyrosine (Tyr) and hystidine (His).

The UV-VIS spectroscopy and chemiluminescence experiments were used in this study. The main kinetic characteristics were determined to interpret the mechanism of the processes. The ozone was generated in ozonation plant. The tryptophane (see Figure) changes through by ozonation for example; the 271 and 287 nm bands change hypsochromously by 20 nm and presents a new band at 314 nm.

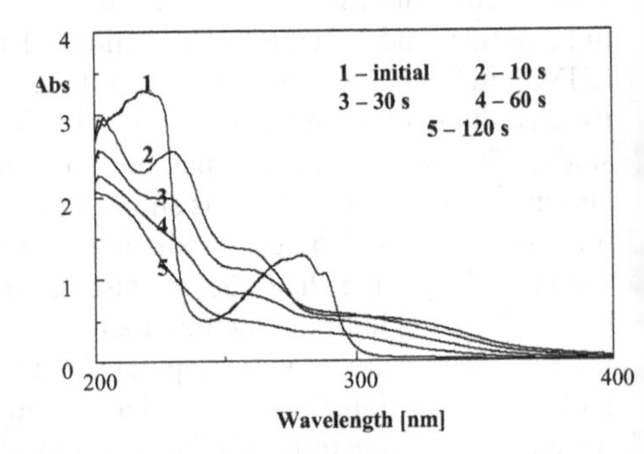

Kinetic characteristics of ozonation process is presented in the following table for the debit (5 L/h) of ozone. The resulting products by ozonation were tested by chemiluminescence technique. The conclusion is a decrease in the antioxidant effect of the tryptophane. This characteristic is presented also by other aminoacids as a results of the prooxidant effect of the newly formed ketonic compounds.

AA	Reaction order	Rate constant	Comments
His	1	0.0893	-
Trp	2	0.0027	-
Tyr	1.2 1.3	0.0059 0.0054	first 30 s after 30 s

The oxidative stress of ozone on some biomolecules results in structural changes in the initial molecules. The kinetic of ozone reaction with the tested biomolecules has a reaction order 1 or 2, showing the complexity of the process. The effect of ozone on the tested biomolecules is change of the antioxidant character into prooxidant.

UV-B-induced oxidative stress and damage in the cyanobacterium *Anabaena* sp.

YU-YING HE, DONAT-P. HÄDER
Friedrich-Alexander-Universität, Institut für Botanik und Pharmazeutische Biologie, Erlangen, Germany

The increased UV-B irradiance on Earth due to the depletion of stratospheric ozone[1] is detrimental to all forms of life including cyanobacteria, the most ancient photosynthetic organisms. Under UV-B stress, the leakage of electrons from the photosynthetic electron transport chain to oxygen[2], the photosensitization of photosynthetic pigments such as chlorophylls, phycobiliproteins and quinones[3] enhances the formation of reactive oxygen species (ROS) and thus exert oxidative stress and oxidative damage to biomolecules such as lipids, proteins and DNA. Here we report the UV-B-induced oxidative stress and oxidative damage including lipid peroxidation and DNA strand breaks in the cyanobacterium *Anabaena* sp. Protective effects of ascorbic acid were also investigated.

UV-B-induced production of ROS was detected *in vivo* by using the ROS-sensitive probe 2′,7′-dichlorodihydrofluorescein diacetate (DCFH-DA). Photooxidative damages by UV-B radiation were determined. Thiobarbituric acid reactive substances (TBARS) and fluorometric analysis of DNA unwinding (FADU) methods were adapted to measure lipid peroxidation and DNA strand breaks in *Anabaena* sp. under UV-B stress. Our results suggested that moderate UV-B radiation results in an increase of ROS production, enhanced lipid peroxidation and DNA strand breaks and a decreased survival. The addition of ascorbic acid, an antioxidant which can scavenge ROS either directly or via the ascorbate-glutathione cycle, counteracted the UV-B effect with respect to the ROS enhancement and protected the test organisms from oxidative stress and the resultant lipid peroxidation as well as DNA strand breaks, which allowed better survival under UV-B stress.

Acknowledgements. The work outlined here was financially supported by an Alexander-von-Humboldt Research Fellowship to Y.-Y. He and the European Union (DG XII, Environment programme, ENV4-CT97-0580) to D.-P. Häder.

References

1. Kerr JB, McElroy CT (1993) Evidence for large upward trends of ultraviolet-B radiation linked to ozone depletion, *Science* 262: 1032-1034.
2. Jordan BR (1996) The effects of ultraviolet-B radiation on plants: a molecular perspective, *Adv. Bot. Res.* 22, 97-162.
3. Franklin LA, Forster RM (1997) The changing irradiance environment: consequences for marine macrophyte physiology, productivity and ecology, *Eur. J. Phycol.* 32: 207-232.

A polychromatic action spectrum of MAA synthesis in the dinoflagellate *Gyrodinium dorsum*

MANFRED KLISCH, DONAT-P. HÄDER
Friedrich-Alexander-Universität, Institut für Botanik und Pharmazeutische Biologie, Erlangen, Germany

Ultraviolet radiation, especially in the UV-B (280-315 nm) wavelength range poses significant stress to aquatic life. One important tolerance strategy in phytoplankton is the synthesis of mycosporine-like amino acids (MAA)[1]. So far an action spectrum for the synthesis of MAA has only been available for a certain cyanobacterium[2]. A polychromatic action spectrum has been determined in this study because polychromatic radiation is a closer approximation to natural conditions than monochromatic radiation[3].

Gyrodinium dorsum was grown in F/2 medium[4]. A solar lamp (Dr. Hönle, Martinsried, Germany) was used to provide PAR and UV radiation. The samples were covered by filters (Schott & Gen., Germany) of different cut-off wavelengths from 665 nm to 225 nm, a UV-band pass filter (UG 11) or placed in darkness, respectively. Samples were taken at 24 h-intervals for 72 h and extracted in 100 % methanol. Spectrophotometric scans of the extracts from 250 nm to 750 nm were performed, and a polychromatic action spectrum of MAA synthesis was determined.

The polychromatic action spectrum revealed maximal sensitivity of MAA synthesis to radiation around 310 nm. Wavelengths >340 nm are only slightly effective while short wavelength UV-B radiation inhibits MAA production. Thus increased UV-B might even counteract this protective mechanism. On the other hand organisms located at greater depths may produce MAAs in response to the radiation penetrating to their level in the water column and thus become more resistant to surface level UV-B radiation.

Acknowledgements The work outlined here was financially supported by the European Union (DG XII, Environmental Programme, ENV4-CT97-0580).

References

1. Sinha RP, Klisch M, Gröniger A, Häder D-P (1998) Ultraviolet-absorbing/screening substances in cyanobacteria, phytoplankton and macroalgae, *J. Photochem. Photobiol. B: Biol.* 83-94.
2. Portwich A, Garcia-Pichel F (2000) A novel prokaryotic UVB photoreceptor in the cyanobacterium *Chlorogeopsis* PCC 6912, *Photochem. Photobiol.* 493-498.
3. Rundel RD (1983) Action spectra and estimation of biologically effective UV radiation, *Physiol. Plant.* 360-366.
4. Guillard RRL, Ryther JH (1962) Studies of marine planktonic diatoms I. *Cyclotella nana* Hustedt and *Detonula convervacea* (Cleve) Gran., *Can. J. Microbiol.* 229-239.

Mycosporine-like amino acids (MAAs) protect against UV-B-induced damage in *Gyrodinium dorsum*

P.R. RICHTER, M. KLISCH, R. P. SINHA, D.-P. HÄDER
Friedrich-Alexander-Universität, Institut für Botanik und Pharmazeutische Biologie, Erlangen, Germany

Gyrodinium dorsum is a marine motile dinoflagellate which is adapted to high light conditions. In response to UV radiation the cells produce mycosporine-like amino acids (MAAs), which show a strong absorbance in the UV range[1-2]. Experimental evidence for their protective role against UV is still scarce. In the present study the effect of an increased MAA content against excessive UV-B irradiation was tested.

The cells were grown in artificial seawater in a Kniese apparatus. In order to stimulate the synthesis of MAAs in the cells the organisms were irradiated with artificial solar radiation with 320 nm (UV-A + PAR) or 395 nm (PAR only) cut-off filters, respectively, for a period of up to 48 h. Two cultures (with high MAA and low MAA content, respectively) were exposed to high UV-B irradiation from a transilluminator or solar radiation. The motility and velocity of the cells were analyzed at regular time intervals.

The cells, which received UV-A during the induction synthesized large amounts of MAAs, while the cells which were exposed to PAR only showed a marginal increase in their MAA content. Five MAAs, shinorine (334 nm), porphyra-334 (334 nm), palythine (320 nm) and two unidentified MAAs having λ_{max} at 310 and 331 nm, respectively, were detected by HPLC analyses. Cells, which were pretreated with UV-A and PAR were considerably more resistant against high artificial UV-B radiation, as can be seen from the kinetics of motility and velocity. While the cells with low MAAs content showed a complete loss of motility within 3 h, the cells with high MAAs content survived at least 2 times longer. Also the decrease in velocity was much faster in the induced cells. A protective effect of MAAs was also detected against solar radiation. Induced and not-induced cells (same treatment as in the previous experiment, MAA synthesis was similar) were exposed to solar radiation. While cells which were pretreated with UV and PAR were not affected by solar radiation, the not-induced (PAR only) cells lost their motility completely within about 150 min. The experiments were repeated several times with consistently the same results.

The data from these experiments are in good agreement with a detected screening effect of MAAs against inhibition of photosynthesis in the dinoflagellate *G. sanguineum*[4]. The results indicate that MAAs function as UV absorbing/screening compounds in *Gyrodinium dorsum* and help the cells to survive in a high UV radiation regime.

References

1. Klisch M, Häder D-P (2000) Mycosporine-like amino acids in the marine dinoflagellate *Gyrodinium dorsum*: induction by ultraviolet irradiation, *J Photochem Photobiol B.: Biol* 55: 178-182.
2. Sinha RP, Klisch M, Gröniger A, Häder D-P (1998) Ultraviolet-absorbing/screening substances in cyanobacteria, phytoplankton and macroalgae, *J Photochem Photobiol B: Biol* 47: 83-94.
3. Klisch M, Sinha RP, Richter PR, Häder D-P (2001) Mycosporine-like amino acids (MAAs) protect against UV-B-induced damage in *Gyrodinium dorsum* Kofoid, *J Plant Physiol* 158: 1449-1454.
4. Neale PJ, Banaszak AT, Jarriel CR (1998) Ultraviolet sunscreens in *Gymnodinium sanguineum* (Dinophyceae): mycosporine-like amino acids protect against the inhibition of photosynthesis, *J Phycol* 34: 928-938.

Screening differences in photosynthesis and fluorescence characteristics of plant leaves by UV-A and blue light fluorescence imaging

FATBARDHA BABANI[1], MARTIN KNAPP[2], HARTMUT K. LICHTENTHALER[2]
[1]*Biological Research Institute, Academy of Sciences, Tirana, Albania*
[2]*Botanical Institute II, University of Karlsruhe, Karlsruhe, Germany*

Green plants exited by UV-radiation emit blue (F440) and green fluorescence (F520) as well as red (F690) and far-red (740) chlorophyll (Chl) fluorescence. The blue-green fluorescence is primarily emitted by ferulic acid covalently bound to cell wall carbohydrates[1]. The red and far-red fluorescence is emitted by chlorophyll a in the chloroplasts of the green mesophyll cells. High resolution multi-colour fluorescence imaging techniques, developed over the past seven years, offer the new possibility to study the distribution and gradients of fluorescence signatures over the whole leaf area[2,3,4]. We have described the differences in the UV-A induced fluorescence imaging of sun and shade leaves in the fluorescence bands blue, green, red and far-red including the fluorescence ratio images blue/green, blue/red, blue/far-red and red/far-red. The blue light induced images of the Chl fluorescence decrease ratio (R_{Fd} = (Fm-Fs)/Fs) are used to show the differences and gradients in photosynthetic activity.

Fluorescence images of sun and shade beech leaves showed that the blue (F440) and green (F520) fluorescence emitted from the upper leaf side in both leaf types were almost the same and much lower than that of the lower leaf side whereas the lower leaf side of sun leaves showed considerably lower fluorescence yield than in shade leaves. The lower intensity of the red and far-red Chl fluorescence of sun leaves as compare to shade leaves is due to the fact that they possess in their epidermis cells many soluble flavonols and other phenolic substances, which absorb the UV-A used for fluorescence excitation but do not fluoresce themselves[1]. Thus, less UV-A reaches the mesophyll chloroplasts for Chl fluorescence excitation. Also thicker epidermis cell walls of sun leaves, a considerable wax layer of sun leaves and an enhanced scattering can influence the amount of UV-radiation entering the leaf. A second cause for the lower Chl fluorescence yield is that in sun leaves with their higher Chl content the red Chl fluorescence is reabsorbed to a higher degree than in shade leaves. Shade leaves are characterized by significantly lower values of the fluorescence ratios blue/red (F440/F690) and blue/far-red (F440/F740) at each leaf side than sun leaves. The Chl fluorescence ratios red/far-red are lower in the upper side of both leaf types as compared to the lower side. Thus, the fluorescence ratio F690/F740, as indicator of the *in vivo* Chl content of leaves[2], showed that Chl content is highest in the upper side of sun leaves and is higher in the upper than in the lower side of both leaf types. A higher Chl fluorescence yield at Fp (1 s after onset of illumination) and at Fs (reached after 5 min of illumination) of shade leaves as compared to sun leaves was observed. The Chl fluorescence ratios, R_{Fd} and Fp/Fs, which are indicators of the photosynthetic quantum conversion and photosynthetic activity of leaves[3], showed the expected higher values in sun than shade leaves of beech.

The much lower Chl fluorescence yield of sun leaves by UV-A excitation demonstrates that they are better protected against UV-radiation than shade leaves. The photosynthetic apparatus of sun leaves has a higher capacity for photosynthetic quantum conversion than shade leaves as seen in the R_{Fd}-images.

References

1. Lichtenthaler HK, Schweiger J (1998) J Plant Physiol 152: 272-282.
2. Buschmann C, Lichtenthaler HK (1998) J Plant Physiol 152: 297-314.
3. Lichtenthaler HK, Babani F (2000) Plant Physiology Biochemistry 38: 889-895.
4. Lichtenthaler H K, Babani F, Langsdorf G, Buschmann C (2000) Photosynthetica 38: 521-529.

Repair processes in plant following UV-irradiation photoreactivation of DNA damages

NADIA GONCHAROVA
International Sakharov Environmental University, Minsk, Belarus

Photoreactivation is a substantial increase of survival of UV-irradiated cells under the effect of visible light. Object of this study are the shoots of barley (*Hordeum vulgare* L.). The seeds shot in darkness in thermostat at 24-26 °C. Westinghouse FS-40 and Philips TL 40/12 lamps were used for UV irradiation. Doses were measured with an UVX-radiometer dosimeter (UV Ltd., USA). To assess the duration of cellular cycle and proliferative activity after UV irradiation and in normal conditions, the shoots with 0.5-1 cm long radicles were incubated with ^{3}H-methionine (0.2 MBq/ml) for 30 min. They were washed, put into the solution of unmarked thymidine and incubated in darkness. One and half hour after UV irradiation and every 2 hours during the following 25 hours the radicles were fixed with an ethanol/acetic acid solution (3:1) and stained with Falgen. The fixed preparations from crushed radicles were covered with a special photoemulsion for 7 days. The number of mitoses was calculated in developed preparations. The duration of separate phases and of whole cellular cycle was calculated by the curves of marked mitoses. The principal radicles were fixed and stained with acetocarmine, and, after maceration, the mitotic index (MI; %) was determined from the mitosis number in temporary preparations. The shoots were irradiated with short-wave UV radiation at 3.5 and 5.4 x 10^3 J/m^2 doses. The principal radicles were fixed every 2-3 hours in a 20 hours period. The chromosome aberrations were analyzed by means of the anaphase method in temporary preparations stained with acetocarmine and crushed. Immediately after UV irradiation the shoots were irradiated with photoreactivating light of different spectral composition (300-390 nm, 330-460 nm, 380-1500 nm) and fixed 7-20 hours after irradiation. DNA was isolated from marked plants. UV-induced damages of DNA sites were determined by their sensitivity to the action of UV endonuclease. It was found that the photoreactivating light is able to induce chromosome aberrations by itself and to inhibit cell proliferation. Mainly the aberrations of chromatid were observed in UV-irradiated cells. Single fragments constituted 85% of chromosome aberrations. The maximum photoreactivating effect was observed in cells irradiated in the S phase; the minimum in cells irradiated in the G_1 phase. In cells irradiated in phases G_1 and S, the photoreactivation increased the yield of chromosome aberrations by 21% and 55%, respectively. The photoreactivating light and UV irradiation did not change the mitotic activity of the cells. This suggests that the decrease of the number of chromosome aberrations after photoreactivation is connected to DNA photoreactivation. The observed photoreactivation process might indicate that a small part of UV-induced thymine dimers constitutes the primary damage of chromosome aberration. We tested this hypothesis determining the number of sites vulnerable to UV-endonuclease action. There was a significant degradation of DNA irradiated at doses of 2.2-17.0 x 10^2 J/m^2, which allowed to determine the number of sites vulnerable to UV-endonuclease. A linear dependence was found between the number of sites and the UV dose.

The comparison between the curves of the UV-dose dependence of site number and of chromosome aberrations indicates a correlation between induction of DNA damage and induction of chromosome aberrations.

Effects of ultraviolet-b radiation and drought stress on *Ocimum basilicum* leaves

A. SZILÁGYI[1], J.F. BORNMAN[1], È HIDEG[2]
[1] *Lund University, Lund, Sweden;* [2] *Biological Researcher Center, Szeged, Hungary*

A reduction in stratospheric ozone results in a selective increase in the amount of ultraviolet-B (UV-B, 280-315 nm) radiation reaching the earth's surface due to the absorbing properties of ozone. Plants have evolved a range of mechanisms to protect themselves against UV-B induced damage. One of them is an alteration of the secondary metabolic pathways, including the enhancement of phenolic compounds such as flavonoids. Certain of these chemical compounds have been shown to have antioxidant properties.

Different light combinations have been tested in a nutritionally important plant species, *Ocimum basilicum*, to maximise growth conditions for antioxidant-flavonoid production. Plants were grown under visible light either alone or with an additional low level of UV-B radiation. Three-week-old plants were exposed to oxidative stress by applying (a) 13 kJ m^{-2} day^{-1} biologically effective (BE) UV-B for 8 days, (b) drought for 4 days and (c) drought and 13 kJ m^{-2} day^{-1} UV-B_{BE} also for 4 days.

The photosynthetic activity of both control and UV-acclimated plants was monitored during the stress experiments by measuring chlorophyll fluorescence parameters of the leaves (F_0, F_m, F_v/F_m, yield, non-photochemical quenching). The effect of UV-B radiation on the accumulation of total flavonoids was spectrophotometrically measured during the stress treatments. Chemical assays were applied to detect the formation of different reactive oxygen species (ROS). ROS production was also estimated by measuring the EPR spectra. Results from plants exposed to UV-B are shown in Figure 1. Acclimation to low levels of UV-B during growth enhanced the ability of the plants to withstand a second, more severe stress by UV-B radiation. This was reflected in smaller loss of photosynthetic yield and lower levels of lipid peroxidation products and ROS in UV-acclimated plants than in non-acclimated controls.

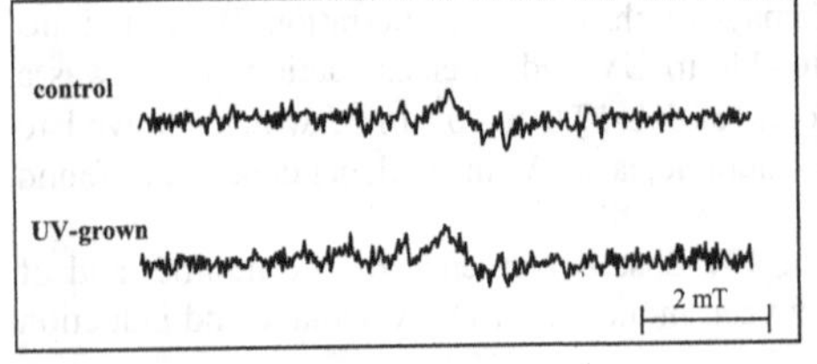

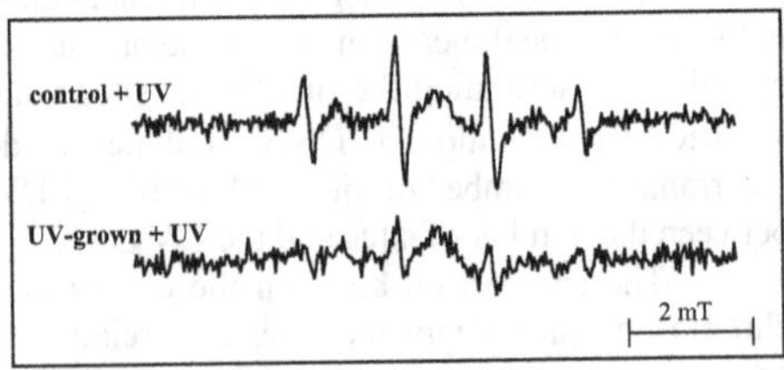

Figure 1. The EPR spectra show the amount of detected hydroxyl radicals in untreated (left panel) and treated samples (right panel). The applied UV stress resulted in a higher amount of hydroxyl radical production in the control basil leaves than in the UV-grown leaves.

Can UV-B alter the carbon gas (CO_2, CH_4) dynamics of peatland ecosystems?

RIIKKA NIEMI[1], PERTTI J. MARTIKAINEN[2], JOUKO SILVOLA[3], ANU WULFF[1], TOINI HOLOPAINEN[1]

[1]*Department of Ecology and Environmental Science, University of Kuopio, Kuopio, Finland*

[2]*Department of Environmental Sciences, University of Kuopio, Kuopio, Finland*

[3]*Department of Biology, University of Joensuu, Joensuu, Finland*

Northern peatlands have major atmospheric importance. Since the last glacial period, they have accumulated 455 Pg carbon in the peat deposits. Besides sinks for CO_2, peatlands are one of the most important sources of CH_4 [1]. The importance of plants that possess gas conducting aerenchymatous tissue for CH_4 emission is well documented[2]. Peatlands cover large areas in Siberia, Scandinavia and the northern North America. Although great increases in UV-B are being predicted for this region, there is no published data on the UV effects on peatland carbon dynamics so far.

To evaluate whether UV could alter the peatland carbon gas dynamics, we exposed peatland microcosms to modulated supplemental (30%) UV radiation at an open field facility in Kuopio, Central Finland (62°13′N, 27°35′E) in two 3-month-long experiments. The microcosms were intact peat monoliths (depth 40 cm, ∅ 11 cm) with a continuous *Sphagnum* moss matrix and some aerenchymatous plants (mainly *Eriophorum vaginatum*) and dwarf shrubs.

At the end of the first experiment, CH_4 emission was 30% lower from the UV-B treated monoliths than from the monoliths receiving only ambient radiation[3]. Gross photosynthesis and net CO_2 exchange were significantly lower in the UV-B treatment than in the UV-A control. The changes were most probably associated with the UV-B induced morphological changes in *E. vaginatum* i.e. reduced leaf cross section (-25%) and percentage of the aerenchyma (-14%). In the second experiment, we observed no clear UV effects on the carbon gas dynamics. The disparity between the experiments may be explained by different UV doses: during the first experiment, the weather was mostly warm and sunny, while during the second experiment it was cool and cloudy. There is a great need for future research in this complex area.

References

1. Bartlett KB, Harriss RC (1993) Review and assessment of methane emissions from wetlands, *Chemosphere* 26: 261–320.
2. Schimel JP (1995) Plant transport and methane production as controls on methane flux from arctic wet meadow tundra, *Biogeochemistry* 28: 183–200.
3. Niemi R, Martikainen PJ, Silvola J, Wulff A, Turtola S, Holopainen T (2002) Elevated UV-B radiation alters fluxes of methane and carbon dioxide in peatland microcosms, *Global Change Biology*, 8: 361-371

ECOTOX - a biomonitoring system for UV-effects and toxic substances

C. STREB, P. RICHTER, D.-P. HÄDER
Friedrich-Alexander-Universität, Institut für Botanik und Pharmazeutische Biologie, Erlangen, Germany

In view of increasing water pollution and UV radiation, the importance of biomonitoring is growing. These bioassays determine the responses of living organisms to stressors. The biological test system ECOTOX can be used in ecotoxicology as a fully automatically early warning system to analyze the water quality, but also in toxicology to study direct influences of chemical substances or radiation.

Euglena gracilis is a unicellular photoautotrophic freshwater flagellate. In its natural environment the organism reaches the optimal position for its photosynthetic apparatus in the water column by photo- and gravitaxis. At low irradiances the organism swims towards the light, at high irradiances it shows negative phototaxis, until this effect is balanced by negative gravitaxis. In darkness, like in the cuvette of the ECOTOX-system, cells show a precise negative gravitaxis[1].

The movement behaviour such as swimming velocity, direction of movement, precision of gravitactic orientation, percentage of motile cells and the cell shape can be determined by the ECOTOX-system. The optical apparatus includes the observation cuvette, an infrared diode, a miniaturized microscope, and a CCD camera coupled to a framegrabber card in the computer. With the real-time image analysis software the picture can be optimized, a threshold to distinguish between background and objects can be adjusted. Cell culture and sample water are pumped in automatically, the organisms are tracked, and significant parameters are displayed. The results are stored in ASCII files and can be examined in a spreadsheet program, where dose-effect relationships can easily be calculated. *Euglena gracilis* was found to be highly sensitive to external influences in the used parameters and therefore is an ideal bioassay. Toxic effects of pollution in aqueous habitats, waste water and freshwater with heavy metal ions or organic solvents and effects of UV radiation can be determined[2].

References

1. Tahedl H, Häder D-P (1999) Fast examination of water quality using the automatic biotest ECOTOX based on the movement behavior of a freshwater flagellate, *Wat Res* 33: 426-432.
2. Tahedl H, Häder D-P (2000) The use of image analysis in ecotoxicology, in D-P Häder (ed.) *Image analysis methods and applications*, CRC Press, Boca Raton, pp 447-458.

9 781402 036965